QD
469
C74
1979

10 21

Coulson's
Valence

ROY McWEENY

THIRD EDITION

D0145690

OXFORD UNIVERSITY PRESS

1979

Oxford University Press, Walton Street, Oxford OX2 6DP

OXFORD   LONDON   GLASGOW

NEW YORK   TORONTO   MELBOURNE   WELLINGTON

KUALA LUMPUR   SINGAPORE   JAKARTA   HONG KONG   TOKYO

DELHI   BOMBAY   CALCUTTA   MADRAS   KARACHI

IBADAN   NAIROBI   DAR ES SALAAM   CAPE TOWN

© Roy McWeeny, 1979

All rights reserved. No part of this publication may be reproduced, stored in a retrieval system, or transmitted, in any form or by any means, electronic, mechanical, photocopying, recording, or otherwise, without the prior permission of Oxford University Press

First Edition 1952
Second Edition 1961
Third Edition 1979

British Library Cataloguing in Publication Data
Coulson, Charles Alfred
Coulson's valence.—3rd ed.
1. Valence (Theoretical chemistry).
I. Title   II. McWeeny, Roy
541′.224   QD469   78-40323

ISBN 0-19-855144-4
ISBN 0-19-855145-2   Pbk.

Printed in Great Britain
by Thomson Litho Ltd., East Kilbride, Scotland

Preface

The greater part of *Valence*, in its earlier editions, was written before 1950. It contained the first broad and authoritative account of molecular orbital ideas in the quantum theory of valence and it was written for chemistry students of modest mathematical attainment. It was for many years unrivalled, was translated into six languages, and won the affection of students and their teachers the world over. Its impact on chemistry was enormous; but by 1975 it was a great book grown old. Even ten years after the 1961 edition the need for a complete revision and rewriting of almost the whole book was already urgent. Its author had no doubts about how much would be required; sadly, he did not live to take on the task.

In this Third Edition I have tried to reconstruct the book broadly as, I believe, its original author would have wished. My personal links with Charles Coulson lasted from the days I spent as a research student in Oxford, during 1946–48, until a few days before his death in 1974, a period which took theoretical chemistry from the era of calculating machines with handles to the age of the computer. During that time he never lost faith in what he called 'primitive patterns of understanding' and had he lived to write a Third Edition there is no doubt that this faith would have remained evident. In trying to update the book, in the face of the fearsome mathematical developments of the last twenty-five years, I have done my best to capture the spirit of the original, to emulate its high standards of clarity and style, and to avoid overburdening the text with equations whenever a pictorial argument would suffice.

The plan of the book follows quite closely that of previous editions. Chapter 1 has been little changed, except for the addition of two short sections to provide a more modern perspective on valence theory. Chapters 2 and 3 have been substantially rewritten to give fuller coverage of the basic ideas of quantum mechanics, thus keeping the book self-contained; only one section (§3.10) is of a more difficult nature and this may be passed over on first reading with no great loss. In subsequent chapters some material has almost disappeared whilst other sections have expanded considerably to take account of recent developments. In particular, theories of chemical reactivity now receive a whole chapter and the treatment of energy band theory has been extended in view of the growing importance of solid-state chemistry. A final chapter has been added for readers who may want 'to obtain some appreciation of current methods of performing electronic structure calculations. The three appendices are also new and problems

D. HIDEN RAMSEY LIBRARY
U.N.C. AT ASHEVILLE
ASHEVILLE. N. C. 28814

have been set at the end of chapters. SI units have been employed throughout.

The literature of valence theory is now enormous and my debt to others will be apparent; but this is a book for chemistry students and their teachers, not for experts in molecular quantum mechanics, and I have therefore been sparing in references to original work. The ones I have given merely provide access points to some of the main lines of development and indicate the origin of the main ideas and concepts. I have not attempted to provide a survey of theories and calculations.

I hope *Coulson's Valence* will be a worthy successor to the two earlier editions of *Valence*. In a real sense it is the result of a collaboration, though unhappily I alone must bear full responsibility for the errors and obscurities which surely remain. To repeat the words of an earlier Preface 'I should be grateful to be told of places where I could do better'.

I am grateful to Mrs. Eileen Coulson for encouraging me to undertake this revision and for kindly putting at my disposal annotated personal copies of previous editions, along with an English text of the section on *Compounds of the rare gases* prepared for the German edition; the penetrating comments in a familiar scribble gave me constant reassurance. I am grateful, too, to Dr. P. W. Atkins for painstakingly reading the whole manuscript and sending a long list of critical remarks which he modestly described as 'puny and piffling' but which were nonetheless valuable; to those authors and publishers who have allowed me to reproduce material; Oxford University Press for editorial help and for gently keeping up the pressure when my spirits were flagging; and to Mrs. S. P. Rogers for turning my handwritten copy into immaculate typescript.

Contents

1
Theories of valence

1.1. Essentials of any theory of valence

This book is concerned with the theory of valence. It is therefore useful first to recall some of the chief phenomena to be explained. We are not concerned primarily with the details of molecular structure, interesting as these may be, but rather with the main principles of molecule formation, for in so far as the details are significant, they must follow from the basic principles.

In the first place, we must show why molecules form at all. Why, for example, do two hydrogen atoms come together to form a permanent and stable compound, molecular H_2, while two helium atoms do not form a corresponding He_2 (except of so transitory a character that it cannot be called a chemical species)? Secondly, why do atoms form compounds in definite proportions? From the earliest days of Gay-Lussac and Berzelius the law of multiple proportions has insisted on the fundamental character of the combining ratio for different atoms. But the picture of an atom which resulted from the work of subsequent investigators such as Frankland and Kekulé was rather like that of a tiny sphere on the surface of which a certain number of 'knobs' (i.e. valences) determined the number of other atoms that could be directly attached to it. This picture is inconsistent with known chemical behaviour; it is true that H (with one knob) would combine with another H to form H_2 but not with *two* more H's to form H_3—this being the 'saturation' property of chemical valence—but how could we account for the existence of *both* CO and CO_2? A satisfactory theory must explain not only the numerical ratios of the numbers of atoms combining to form a molecule but also the *variability* of such ratios (i.e. the existence of *multiple* valences) and the relationship of such behaviour to the position of an atom in the periodic table.

Thirdly, a complete theory of valence should be able to account for the stereochemistry of molecules. The shapes and dimensions of molecules can now be inferred with increasing precision by a wide variety of spectroscopic and other physical techniques, and such knowledge places new demands on a theory of valence. Why, for example, do the H—C—H angles in methane (CH_4) all have the tetrahedral value of $109° 28'$, while in chloroform ($CHCl_3$) the corresponding Cl—C—Cl angles are increased to about $110\frac{1}{2}°$? Why is CO_2 linear, while H_2O is V-shaped?

A good theory of valence should also give a *unified* explanation of these three main aspects of molecular structure; it should show *if* an atom has the power to attach another atom, or group, to itself, *how many* such atoms it may attach, and *in what geometrical arrangement*.

Finally, new physical techniques give a very intimate knowledge of the nature and properties of bonds, of their shapes and individual features, far beyond that implied by drawing a single line to represent a single bond and two lines for a double bond. These intimate details should also be accounted for by the theory.

1.2. Electronic character of valence

In a profound sense the description of bonds in a molecule is simply the description of the distribution of electrons around the nuclei. The early attempts to develop an electronic theory of valence were based on geometrical models in which the electrons were regarded as static point charges placed at the corners of cubes or tetrahedra with the nuclei at the centres; they gave an interpretation of the octet rule, in which each atom achieved a configuration with eight outermost electrons located at the corners of a cube, and they suggested a connection with the periodic table introduced by Mendeleev in 1869. But such models violate a fundamental theorem in electrostatics which states that no static distribution of charges can be in stable equilibrium—the charges must be *moving*. Even the earliest ideas, going back to Berzelius (1819), of the bonding in ionic compounds suffered from this defect. The attractions among ions with dissimilar charges suggested the electrostatic origin of bonding, but some new repulsive force was then necessary to prevent the collapse of the whole structure; the equilibrium in ionic crystals could not be accounted for by electrostatics alone. In molecules such as H_2, with covalent bonds, electrostatic forces of the Coulomb type were even less able to account for the attraction between the individual electrically neutral atoms.

The first attempt at a dynamic model was due to Bohr (1913) who assumed that the electron in the hydrogen atom moved in an *orbit* around the positively charged nucleus. By applying the classical (Newtonian) laws of motion, coupled with a postulated 'quantum condition', he obtained a discrete set of permitted orbits, each with a characteristic energy. Transitions from one energy level to another then accounted almost perfectly for the emission and absorption of corresponding energy quanta of radiation, as observed in spectroscopy. But the theory broke down completely for systems with more than one electron, or with more than one nucleus (e.g. in H_2^+).

Bohr's theory failed because it applied the laws of classical physics, established by observing the motion of masses in the laboratory, to electrons and nuclei—particles so small that their motion, as classically envisaged (e.g. in a precisely defined orbit), could never actually be observed in any conceivable experiment. We cannot expect that the same laws necessarily apply at the electronic level, or even that the classical concepts retain their validity. It is now known that at this level we must employ the laws of quantum mechanics—of which one particular formulation, wave mechanics, was introduced by Schrödinger in 1926. The laws of quantum mechanics reduce to those of classical mechanics for systems of large mass† (e.g. everyday objects in the laboratory), but they also apply, with a validity and accuracy which is now beyond question, to the electrons in atoms and molecules. On the basis of quantum mechanics it is now possible in principle, and to a considerable extent in practice, to build up a coherent and complete theory of molecular structure and properties.

1.3. Importance of the energy

Broadly speaking, two atoms form a molecule because there is a lowering of the total energy when they come together. For most purposes the energy of a molecule may be regarded as the sum of an electronic energy (comprising the kinetic energy of the electrons and their potential energy, both mutual and in the field of the nuclei, calculated as if the nuclei were fixed in their equilibrium positions), together with the mutual potential energy of the nuclei and the energies of vibration, rotation, and translation of the molecule as a whole (associated almost entirely with the nuclei, with their comparatively large masses). The energy due to vibration, rotation, and translation of the molecule, although significant in spectroscopy, is normally a very small fraction of the total, and the total energy is thus almost synonymous with the electronic energy, supplemented by the Coulombic repulsion energy of the nuclei. To avoid constant explicit reference to the latter, it is convenient to include the nuclear repulsion energy as an extra term in the electronic energy; references to the total electronic energy will usually be understood in this sense, i.e. as the total energy *in the absence of nuclear motion*. This 'electronic energy' then plays the dominant role in determining molecular structures. For example, the fact that hydrogen forms diatomic aggregates such as H_2 rather than triatomic aggregates such as H_3 follows from the fact that the electronic energy of H_3 is greater than the sum of that of H_2 and H. Similarly the HOH angle in water is about $104\frac{1}{2}°$ and the two OH bond lengths are each 0·096 nm because it is for these values of the internal co-ordinates of the molecule that the energy is lowest. A satisfactory theory of valence therefore must be able to show how the electronic energy depends on the positions of the nuclei, and thus to predict not only the equilibrium configuration but also how the energy changes when the molecule is distorted. This means we can obtain the restoring force in any deformation and have available all the information required for calculating the normal modes of vibration. Consequently, the theory of valence has

† It should be noted that the forces associated directly with the mass (i.e. *gravitational* forces) are entirely negligible for electrons. To separate the two atoms in H_2 to infinity against only their gravitational attraction, each being of mass $1·7 \times 10^{-27}$ kg, would require about $2·5 \times 10^{-54}$ J, but the energy actually required, as determined thermochemically, is about $6·7 \times 10^{-19}$ J—larger by a factor of more than 10^{35}! The observed energies are thus entirely *electrical* in origin.

immediate implications for infrared and Raman spectroscopy (in which characteristic frequencies depend on vibrational force constants) and in the discussion of rotational fine structure (determined by moments of inertia, and hence by the positions of the nuclei) and of a multitude of other geometry-dependent effects.

In discussing the formation and dissociation of molecules we are concerned with energy *differences*, usually small differences between very large quantities. Consequently, nuclear motion may sometimes need to be taken into account. The difference between the total electronic energy of the molecule (remembering that (p. 3) this conventionally includes the repulsion energy between all pairs of nuclei) and that of its component atoms is the *electronic binding energy* of the molecule. This is not the same as the experimentally observed dissociation energy (i.e. the energy required to break up the molecule into its constituent atoms) when allowance has been made for nuclear motion which, in practice, is always present. The three terms to consider are as follows: (i) the zero-point energy of vibration, which for large polyatomic molecules may reach a total value comparable with the energy required to break any one bond, though for diatomic molecules the ratio is usually of the order $\frac{1}{10}$ to $\frac{1}{20}$; (ii) the translational energy, which is equal to $\frac{3}{2}kT$ both for the original molecule and for each separate fragment; (iii) the energy of rotation of the molecule as a whole, which, except very near $0\,\mathrm{K}$, is $\frac{3}{2}kT$ for a non-linear system and kT for a linear one. For a diatomic molecule the dissociation energy may be called the bond energy, since it represents the energy required to break the bond. Because of the effects (i)–(iii) this bond energy depends slightly on the temperature, though of course the electronic binding energy does not. For a polyatomic molecule we can still speak of the dissociation energy of any one bond, for this represents the energy needed to break the molecule at this bond into two fragments. However, on account of the fact that after the break-up there is often a considerable electronic rearrangement in the two parts, sometimes resulting in the recovery of quite a large amount of energy, the total energy of dissociation of a molecule is not usually the same as the sum of the separate dissociation energies of each bond. A simple example of this distinction between the total binding energy and the sum of the dissociation energies of the separate bonds is found in the water molecule H_2O. The total binding energy is about $9\cdot49\,\mathrm{eV}$, so that, if we wished, we could speak of an average bond energy of $4\cdot75\,\mathrm{eV}$. However, the energy needed to break either of the O—H bonds separately is no less than $5\cdot18\,\mathrm{eV}$. The explanation of the apparent discrepancy is that after we have broken the first O—H bond we only require $4\cdot31\,\mathrm{eV}$ to break the second, and the sum of $5\cdot18$ and $4\cdot31$ is the same as the original $9\cdot49$.

There is yet one further distinction to be mentioned. Bond energies are usually inferred from heats of reaction, adjusted so as to refer to a constant pressure of 1 atm and a temperature of $25\,^\circ\mathrm{C}$. These heats of reaction represent changes in the enthalpy $H = U + PV$, rather than the internal energy† U, and if, as usually happens in a dissociation, there is a change of volume, we ought to correct for the presence of the term PV.

Many of these corrections to the original electronic binding energy of the molecule would be unnecessary if all values were extrapolated to zero temperature, but such corrections are usually difficult to make with complete certainty and when made they do not appear to cause any great relative changes in bond energies. For this reason tables of bond energies given in the literature are for standard conditions of temperature and pressure, and are not therefore strictly bond energies at all! In very accurate numerical work care is needed to distinguish between energy and enthalpy, but it will not be necessary to do so in this book where we shall be concerned almost entirely with electronic binding energies.

It may serve to emphasize the relative importance of some of these energy terms if we show their values for the particular case of H_2, taken in part from Herzberg and Monfils (1960). The temperature-dependent terms are estimated for $T = 291$ K.

Some energy values for H_2	
Total electronic energy of H_2	$3098\cdot3\,\mathrm{kJ\,mol}^{-1}$
Electronic energy of two H atoms	$2642\cdot6\,\mathrm{kJ\,mol}^{-1}$
Electronic binding energy	$458\cdot1\,\mathrm{kJ\,mol}^{-1}$
Zero-point vibrational energy	$25\cdot9\,\mathrm{kJ\,mol}^{-1}$
Rotational energy of H_2	$2\cdot5\,\mathrm{kJ\,mol}^{-1}$
Translational energy of H_2	$3\cdot8\,\mathrm{kJ\,mol}^{-1}$
Correction for PV term	$2\cdot1\,\mathrm{kJ\,mol}^{-1}$
Bond energy of H_2	$435\cdot1\,\mathrm{kJ\,mol}^{-1}$

Two conclusions follow immediately from these figures. In the first place the corrections for rotation and translation and the PV term are small, but that for the zero-point vibration is more significant. In the second place the electronic binding energy is only a small proportion (here about one-seventh) of the total electronic energy. If we had chosen heavier atoms the fraction would have been even smaller. Thus for Li_2 the ratio is 1 in 14 and for methane it is no more than 1 in 38. The binding energy is thus the difference of two much larger quantities, and if we want to calculate it with a reasonable accuracy we must be able to compute these other quantities (electronic energy of the molecule and total electronic energy of the separate atoms) with even more precision. As we shall see, this imposes very severe restrictions on *ab initio* calculations of binding

† The thermodynamic internal energy U refers to a bulk sample (e.g. 1 mol) whereas we have been discussing the energy E of a single atom or molecule. For 1 mol of a dilute gas (intermolecular forces negligible) $U = LE$ where L is the Avogadro constant ($6\cdot022 \times 10^{23}\,\mathrm{mol}^{-1}$). The bond energies etc. quoted above are more usually quoted in $\mathrm{kJ\,mol}^{-1}$ (e.g. 916 kJ is the binding energy of $6\cdot022 \times 10^{23}$ H_2O molecules, while that for a single molecule is $9\cdot49$ eV). Tables of units and conversion factors appear at the end of this book.

energies, though fortunately it does not affect the main basis of the theory of valence.

There is, one interesting corollary which follows from the energy values given above. Since the binding energy is a small difference between larger quantities, we must expect it to be very sensitive to the atoms being bonded. There is, in fact, a specificity about bond energies which is quite absent from most other types of force. This was recognized as long ago as the time of Berzelius, though he could not relate it, as we can do, to subtle differences in electronic behaviour. A few examples will serve to illustrate this specificity.

(*a*) The strongest known single bond in a diatomic molecule is the HF bond whose energy ($563.4\,\text{kJ mol}^{-1}$) greatly exceeds that of either HH or FF (432.0 and $154.9\,\text{kJ mol}^{-1}$ respectively). Evidently hydrogen and fluorine are mutually adapted to the formation of a strong bond much better than are either of them to bond formation with another atom of the same kind. An explanation of this will be found in Chapter 5.

(*b*) Phosphorus and nitrogen show single-bond energies (P—P and N—N) which are not very different (~ 200 and $167\,\text{kJ mol}^{-1}$, respectively) and the structure of the atom is much the same in both cases, yet phosphorus forms a stable tetrahedral molecule P_4 and nitrogen does not.

(*c*) The 'inert gas' atoms such as Ne, Xe, Kr,... were long thought to be incapable of forming chemical bonds, but in recent years many stable molecules, such as XeF_2, have been synthesized with bond energies as large as $200\,\text{kJ mol}^{-1}$.

The bonding thus depends on a specific pairing of atoms or on a specific geometry of a compound, rather than on intrinsic properties of the individual atoms. To be wholly successful, a theory must be able to account for all such peculiarities.

1.4. Energy diagrams

Energy relationships in a molecule are often summarized pictorially in an energy diagram. Two such diagrams are shown in Fig. 1.1. The first curve shows the potential energy V of a particle bound to the origin ($x = 0$) by an elastic force proportional to the displacement $F = -kx$. A positive force component means one directed along the positive x axis, and the minus sign thus indicates that the force is towards the origin. The work done by an equal and opposite force in pulling the particle away from the origin to a distance x is thus $\int_0^x kx\,\mathrm{d}x = \tfrac{1}{2}kx^2$; this is the *potential energy* as a function of x, and $V = V(x) = \tfrac{1}{2}kx^2$ gives the parabola shown. According to classical mechanics, which turns out to be a good first approximation in dealing with relatively massive particles such as nuclei, the vibrational motion corresponding to a total energy E may be discussed as follows. We draw a horizontal line in the diagram at height E and invoke the result $E = T + V$ (conservation of energy)

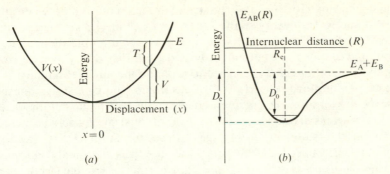

FIG. 1.1. Energy diagrams: (*a*) particle oscillating about the origin ($x = 0$) with energy E; (*b*) potential energy curve for a diatomic molecule AB showing E_{AB} as a function of internuclear distance R.

in which $T = \frac{1}{2}mv^2$ is the *kinetic energy*. For any value of x (i.e. position of the particle) the two segments of a vertical line (as shown in Fig. 1.1 (*a*)) then indicate the simultaneous values of the kinetic and potential energies. Evidently when the particle passes through the origin $V = 0$ and its energy is wholly kinetic, but as its displacement increases T diminishes to zero, and at the extremities of its motion the particle comes momentarily to rest—its energy then being entirely potential energy. The particle cannot *pass* the points where $T = 0$, because to do so its kinetic energy would have to become negative—which is impossible for any real value of the velocity. The points at which $T = 0$ therefore indicate the extremities of the motion, according to classical physics, and hence the *amplitude* of the vibration. If we give the system more energy, i.e. increase the value of E, the amplitude of the motion increases, but the particle is still confined to a classically 'allowed' region between the points at which the line at height E is intercepted by the potential energy curve. The force acting on the particle is $F = -\mathrm{d}V/\mathrm{d}x$ and therefore vanishes where the potential energy curve has a stationary point; when the stationary point is a *minimum*, as at the origin in Fig. 1.1 (*a*), a particle at rest at that point would be in *stable equilibrium*, any small displacement resulting in a force directed towards restoring the equilibrium. Energy diagrams retain their usefulness in quantum mechanics, with a modified interpretation which we discuss in Chapter 2.

The second diagram, Fig. 1.1 (*b*), shows a potential energy curve for a typical diatomic molecule AB; the quantity plotted is $E_{AB}(R)$, the total electronic energy discussed in §1.4, which plays the part of a potential energy when we are thinking of the motion of the nuclei. When R is very large $E_{AB} \rightarrow E_A + E_B$, the sum of the electronic energies of the separate atoms A and B. As the atoms approach, E_{AB} reaches a minimum at some value $R = R_e$, the equilibrium internuclear distance in the stable molecule AB. At distances less than R_e, the energy curve begins to rise, any bonding effect produced by the electrons ultimately being outweighed by the strong mutual repulsion of the positively charged nuclei. The energy relationships discussed in §1.3 are clarified in the figure. Around the minimum the curve is roughly parabolic and the vibrational energy of the nuclei is indicated, as in Fig. 1.1 (*a*), by a horizontal line; the amount of energy required to separate the nuclei to infinity, in the absence of vibration, is clearly D_e (the electronic binding energy), but in the presence of vibration the energy needed is only D_0 (the actual dissociation energy). Whereas in classical physics the vibrational energy may be increased continuously by supplying energy, the observed vibrational energy takes only certain specific values $E_1, E_2, \ldots$; such discrete values are usually indicated by drawing *energy levels* in the diagram. The existence of discrete levels, referred to as 'quantization', is one of the characteristic features of quantum mechanics. The true dissociation energy is defined, as indicated, for a molecule in its *lowest* vibrational level (E_1), although the term is often loosely applied to D_e itself.

One of the major problems of valence theory is to predict the exact form of the potential energy curve for a given molecule. Empirically, such curves may often be fitted by an expression due to Morse (1929). If we define energies relative to that at infinite separation, putting $E(R) = E_{AB}(R) - (E_A + E_B)$, the Morse curve is given by

$$E(R) = D_e[\exp\{-2a(R - R_e)\} - 2\exp\{-a(R - R_e)\}]$$

where a is an adjustable constant (a 'parameter'). The minimum value is clearly $-D_e$, at $R = R_e$, and near the minimum

$$E(R) = -D_e + D_e a^2 (R - R_e)^2 + \ldots$$

which is a parabola with a minimum at $E = -D_e$.

For a polyatomic molecule the corresponding problem is to define a potential energy *surface*, for then E will depend on *several* independent variables, namely a number of bond lengths and angles. Such surfaces are of great importance in the theory of chemical reactions.

1.5. Quantum chemistry

The theory of valence is one aspect of a much bigger subject, for the laws of quantum mechanics allow us, in principle, to calculate not only the energy of an electronic system—and hence to predict the geometries and structures of molecules—but also all other properties of such systems. We should be able to calculate excited states and the energies of spectroscopic transitions, the intensities of spectral lines and how they depend on the interaction between the molecule and the radiation field, the optical activity of molecules, the interactions between different molecules, the results of collisions between molecules, and the nature of the chemical reactions which may take place. The study of all such properties and processes, using the methods of quantum mechanics, has come to be known collectively as 'quantum chemistry'. Before turning specifically to the study of chemical bonds, it is useful to have some general appreciation of the status of quantum chemistry and of the great advances which have taken place in recent years.

Quantum theory has had a two-fold impact on chemistry. The first is *conceptual*, for there is now hardly an area of chemistry left in which we do not use language and ideas that come from quantum mechanics, ideas which must in large part be formulated in later chapters. The second has to do with the possibility, which with present-day computers is fast becoming a reality, of actually calculating from first principles many of the things we wish to know about molecules. This book is concerned mainly with the physical concepts and their use in establishing what have been called 'primitive patterns of understanding', but such an approach is in no sense inferior to one which requires powerful computers to solve the equations of quantum mechanics. The two approaches are complementary. If our concepts and 'patterns of understanding' are sound they will be supported by rigorous calculation; if not they must be rejected. Conversely, the physical idea often suggests the most appropriate mathematical attack or the most satisfactory approximation. The situation has been described (Coulson 1960) in the following way:

'Chemistry is an experimental subject, whose results can be built into a pattern around quite elementary concepts. The role of quantum chemistry is to understand these concepts and to show what are the essential features of chemical behaviour. To say that the electronic computer shows that $D(\text{H—F}) \gg D(\text{F—F})$ is not an explanation at all, but merely a confirmation of experiment. Any acceptable "explanation" must be in terms of repulsions between non-bonding electrons, dispersion forces between atomic "cores", and the like.'

We shall be concerned mainly with explanations of that kind, but it is nevertheless important to realize that quantitative calculations *can* nowadays be made, sometimes with a precision higher than can be obtained in the laboratory. This is certainly true for the hydrogen molecule (Kolos and Wolniewicz 1968) and for a considerable number of small molecules and ions, many of which are of astrophysical importance (see, for example, Hammersley and Richards 1974). In such cases, the energy levels, and the fine structures of spectral lines, have been calculated even with the inclusion of the extremely small corrections required by relativity theory. Such work, apart from its intrinsic value, is also important in so far as it offers a stringent experimental test of the equations and methods of quantum mechanics; so far, there is no evidence to suggest that there is anything significantly wrong with the equations. As Dirac asserted in 1929:

'The underlying physical laws necessary for the mathematical theory of a large part of physics *and the whole of chemistry* are thus completely known'.

The difficulties we encounter in applying the laws to complicated molecules are of a practical nature, and it is these difficulties alone which compel us in most cases to seek an understanding of chemical valence and of molecular properties at a more qualitative level.

How can such an understanding be achieved? What do we mean by 'understanding' the bonds in a molecule, or even by the 'size' and 'shape' of a molecule? Let us return for a moment to the general picture adopted in §1.3. We can never see a molecule, so all our information is inferred from a wide variety of experiments. The development of powerful physical techniques such as X-ray and electron diffraction, optical, ultraviolet (UV), and infrared (IR) spectroscopy, nuclear magnetic resonance (NMR) and electron spin resonance (ESR) spectroscopy, nuclear quadrupole resonance (NQR), and photoelectron and Auger spectroscopy, to mention just a few, has, however, given an immense amount of information about the positions and motions of electrons and nuclei in a molecule.

The picture which emerges is the one assumed in §1.3, which we now develop a little further. The nuclei vibrate around certain well-defined equilibrium positions (which determine the equilibrium geometry of the molecule) but their vibrational amplitudes seldom exceed about 0·01 nm, whereas the separations of neighbouring nuclei (i.e. the bond lengths) are in the range 0·1–0·3 nm. In speaking of the structure of a molecule we can therefore think of the nuclei as occupying their mean, or equilibrium, positions; for many purposes we can think of them as fixed. The electrons, however, are best thought of as a 'charge cloud' whose density varies from place to place. The exact meaning of this interpretation will become clear later, but for the present we can picture the charge cloud as what we should see if we could take a time-exposure photograph of the rapidly moving electrons; the image would then be blurred and would be most dense in those regions where the electrons spent most of their time or, in other words, were most likely to be found. The charge cloud is characterized by its density P, whose value at any point represents the number of electrons per unit volume at that point.

The form of the electron charge cloud in a molecule may be obtained experimentally, for X-rays are scattered by electrons and from measurements of the scattered intensity at different angles it is possible to infer the charge density of the scatterer. The most familiar way of exhibiting the results is by means of a contour map. The contour maps produced by X-ray crystallographers provide an experimental realization of the charge density P. A contour map for the anthracene molecule is shown in Fig. 1.2. Each contour joins points in the molecular plane at which the total charge density has the same value. There are obvious peaks at the positions of the 14 carbon nuclei from which the internuclear distances (i.e. bond lengths) can be inferred. It will be noticed that on this diagram the positions of the hydrogen atoms are far from precise. This is because a hydrogen atom provides only one electron compared with six from a carbon atom, so the intensity of X-rays scattered from that part of the molecule is small, leading to considerable uncertainty in position.

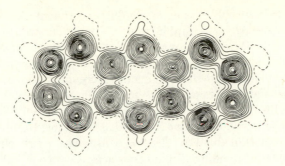

FIG. 1.2. Electron density contours for anthracene determined by X-ray methods. (From *Endeavour* **25**, 129 (1969), courtesy of G. E. Bacon).

The charge density P can also be *calculated* from quantum mechanics, and generally speaking there is good agreement between observed and computed densities. The importance of the charge-density concept can hardly be exaggerated because, as can be proved theoretically, a molecule does behave for many purposes exactly as if its electrons really were 'smeared out' into a continuous distribution of charge of density P (in electrons per unit volume). For example, the forces which hold the nuclei together in the molecule, against their mutual repulsions, may be correctly calculated using classical electrostatics provided the charge density is known. Chemical binding thus results from the attractive forces between the positively charged nuclei and the negatively charged electron cloud which surrounds them. As Fig. 1.2 shows the electron density is strongly concentrated around each nucleus, as it would be in a free atom, but there is still sufficient density in the bond regions (indicated by the col between each pair of peaks) to provide the attraction necessary for bonding.

Electron density must clearly play a central role in our discussions of chemical valence, but it also determines an enormous range of other electronic properties of molecules, many of which (like the scattering of X-rays) provide a basis for experimental investigations of molecular structure. We shall end this chapter by giving a few more examples to show the enormous conceptual value of the charge cloud as a means of interpreting experimental results.

(i) If one atom in a molecule attracts electrons towards it more strongly than another it will take a larger share of the electron density; a 'polarization' will occur and the molecule will acquire a dipole moment. The dipole moment can be calculated as if the system were simply a set of positive point charges embedded in a continuous distribution of electrons of density P. Changes in measured dipole moments, arising for example from substitution, may thus be related to polarization of the charge cloud. Higher moments (e.g. quadrupole moments) can be measured and calculated similarly.

(ii) In nuclear quadrupole resonance (NQR) experiments, a *nucleus* with a non-zero electric quadrupole moment interacts with the electron distribution. (When the positive charge is not distributed spherically the field produced may be the same as that of a *quadrupole*, e.g. two dipoles with their positive ends together.) The interaction energy can be measured and gives information on the non-uniformity of the electric field arising from the electron cloud at the position of the nucleus. This field can be calculated from a knowledge of P; thus, NQR experiments yield information about the electron density against which we can test our theories of bonding. The nucleus acts as a 'probe' with which we can investigate the charge cloud.

(iii) The ionization of a molecule occurs when an electron is given sufficient energy (e.g. by absorption of radiation) to overcome the forces which bind it to the molecule; any excess energy appears as kinetic energy of the emergent electron (Fig. 1.3) and can be measured. An electron tightly bound to a nucleus (i.e. in an 'inner shell') requires a lot of energy to get it out, but it is found that the precise amount depends on the environment of the atom from which it comes. In photoelectron spectroscopy the variable ionization energies give rise to corresponding peaks in the measured kinetic energy of the emergent electrons. Figure 1.4 shows peaks corresponding to the removal of an inner-shell electron from the four different carbon atoms in the molecule shown, all arising from absorption of the same amount (one photon) of radiation energy. Why is there not just one peak characteristic of a carbon inner-shell electron? Again, we think of the electron distribution; the fact that fluorine has a very strong electron attracting power means that electron density will be pulled away from the first carbon; an electron of its inner shell will therefore experience a lowering of energy, i.e. it will become more tightly bound than in a free carbon atom. (The potential energy of charges q_1 and q_2, at a distance r, is $q_1q_2/4\pi\varepsilon_0 r$.) The carbon to which oxygen atoms are attached is similarly affected, but not so much, and the third and fourth carbons are even less affected. Consequently, we expect four slightly different ionization potentials and four kinetic energy peaks. Again the nuclei act as probes, telling us how the charge cloud is polarized by different substituent atoms. Figure 1.5 shows another example, this time involving nitrogen in place of carbon. The fact that the peaks are so different in size gives further information about the molecule, and is the subject of Problem 1.8 at the end of this chapter.

Such examples illustrate clearly the 'primitive patterns of understanding' with which we shall be concerned throughout the remaining chapters. Our 'understanding' of valence and of electronic properties must be in terms of

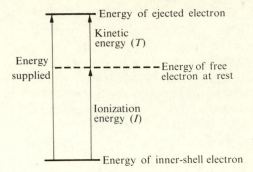

FIG. 1.3. Energy diagram for photoelectron spectroscopy. Knowledge of the quantum of energy supplied and measurement of the kinetic energy of the emergent electron allow us to infer the ionization energy.

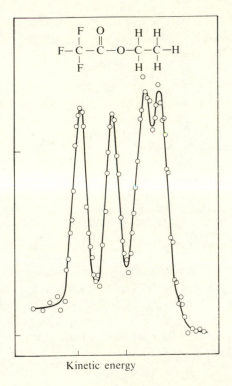

FIG. 1.4. Carbon 1s photoelectron spectrum of ethyl trifluoroacetate. The rate of ejection of 1s electrons is plotted against their kinetic energy; there is one peak for each carbon, depending on its electronic environment. (From J. M. Hollander and W. L. Jolly, *Acct. of chem. Res.* **3**, 193 (1970), with permission.)

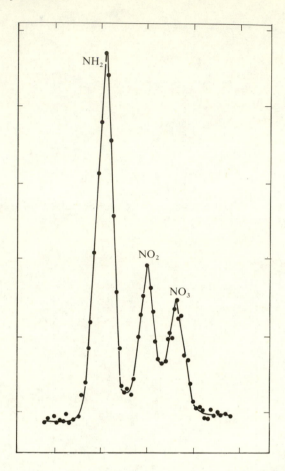

Fɪɢ. 1.5. Nitrogen 1s photoelectron spectrum of *trans*-dinitrobis(ethylenediamine)cobalt (III) nitrate [Co(NH$_2$·CH$_2$·CH$_2$·NH$_2$)$_2$(NO$_2$)$_2$]NO$_3$. (From J. M. Hollander and W. L. Jolly, *Acct. of chem. Res.* **3**, 193 (1970), with permission.)

simple physical models which can readily be visualized, and not in terms of numbers gushing from a computer! Before, however, we can formulate our ideas more precisely we must start at the beginning, with the concepts of quantum mechanics and with some of the simplest solvable problems; only then shall we be able to approach the more formidable problems associated with molecular bonding.

Problems

1.1. Sketch the form of the potential energy function $V = V(r)$ for two hard spheres, each of radius R, attracting each other with a force inversely proportional to the square of the distance r between their centres.

1.2. The spheres in Problem 1.1 are released from rest at a distance $r = r_0$. Discuss what happens subsequently, using an energy diagram. What would happen if the spheres were projected towards each other from a very large distance?

1.3. Suppose the spheres in the two preceding problems were compressible and repelled each other strongly when the distance between the centres was less than $2R$. How would the form of the potential energy curve be modified? Can you suggest a mathematical form for the repulsive term in the potential energy?

1.4. The potential energy function for two helium atoms is fairly well represented as

$$V = \frac{A}{r^{13}} - \frac{B}{r^6}$$

with $A/a_0^{13} = 0\cdot2135$ eV and $B/a_0^6 = 0\cdot0193$ eV, a_0 being the Bohr radius ($a_0 = 0\cdot0529$ nm). Plot V against r/a_0 (i.e. the distance measured in units of a_0). Show that the atoms might be expected to form a molecule and determine the expected equilibrium bond length (R_e) and dissociation energy (D_e).

1.5. Show how R_e and D_e in Problem 1.4 may be determined without drawing the curve. (Hint: Differentiate to find the minimum.)

1.6. Suppose the thermal energy (vibrational energy of the nuclei) of the system in Problem 1.4 is of the order of kT, namely $2\cdot6 \times 10^{-2}$ eV at 300 K. Is the molecule He_2 likely to be stable against dissociation into two helium atoms? Draw a line on the energy diagram (Problem 1.4) to represent a vibrational state for which the dissociation energy would be one-half of the electronic binding energy. What would be the corresponding amplitude of vibration according to classical mechanics (which is approximately valid for heavy particles)?

1.7. Two helium atoms are in head-on collision. How much kinetic energy must they have if their centres are to approach to within a distance of (a) $0\cdot9R_e$ and (b) $0\cdot8R_e$? Would you describe a helium atom as 'hard' or 'soft'?

1.8. Can you explain (i) the relative positions and (ii) the relative sizes of the three peaks in Fig. 1.5? (Hint: You may assume that the inner-shell electrons of all the nitrogen atoms, whether in NH_2, NO_2, or NO_3, are equally likely to be ejected by the incident radiation.)

2

Wavefunctions: Atomic orbitals

2.1. Wavefunctions

Before we can discuss the behaviour of electrons in a molecule, we must understand their behaviour in isolated atoms. In particular we must learn how quantum mechanics describes the motion of a single electron in a central field of attraction provided, for example, by a nucleus—as in the hydrogen atom. To do so it is necessary to review briefly the essential ideas of quantum mechanics, or rather wave mechanics, the particular formulation of quantum mechanics due to Schrödinger, which is used throughout this book. By discussing first the motion of one electron in a central field we shall meet in their simplest context all the ideas needed for discussion of the electronic structures both of atoms and of molecules.

In Bohr's early work it was supposed that the electron in a hydrogen-like system (H, He$^+$, Li^{2+}, ...) moved around the nucleus, just as a planet moves around the sun, in some definite orbit which could be calculated using the equations of classical mechanics (Newton's laws). It was Heisenberg (1927) who first showed clearly the logical inconsistency of applying this picture to a particle as small as an electron. Briefly, there is no way of measuring simultaneously the exact velocity and position of an electron; the more closely we attempt to measure its position, the more we shall disturb its motion and the less accurately, therefore, shall we be able to define its corresponding velocity. Heisenberg's 'uncertainty principle', which gives this idea mathematical form, therefore denies the possibility of describing the motion in terms of a classical orbit in which position and momentum are exactly and simultaneously specified at all points on the path. The classical description must be abandoned and Newton's equations replaced by new equations—those of quantum mechanics—which admit the idea of uncertainty and are thus of an essentially *statistical* nature.

Wave mechanics, in particular, has a direct experimental basis and arose from Schrödinger's attempts to reconcile the apparent coexistence of wave and particle properties in a moving electron. We consider three examples of 'non-classical' behaviour, all concerned with the wave–particle duality.

(i) *The photoelectric effect.* Light of frequency v falls on a clean metal surface *in vacuo*. If v is greater than a certain 'threshhold value' v_0, electrons are ejected and their kinetic energy is found to be

$$\tfrac{1}{2}mv^2 = h(v - v_0) = hv - W \tag{2.1}$$

where h is *Planck's constant* with the dimensions of energy × time;

$$h = 6 \cdot 625 \times 10^{-34} \,\text{J s}.$$

The energy of an ejected electron depends only on the *frequency*, though the *number* of electrons ejected is proportional to the intensity of the radiation.

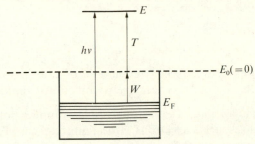

FIG. 2.1. Energy diagram for the photoelectric effect.

The interpretation of the experiment in terms of an energy diagram (§1.4) is shown in Fig. 2.1, where the meaning of the various quantities is as follows: E_0, energy of an electron *at rest* 'at infinity' (i.e. outside the metal), taken as zero of potential energy; E, energy of ejected electron (kinetic energy $\tfrac{1}{2}mv^2$); E_F, energy of the highest-energy electrons inside the metal (the so-called Fermi energy); W, minimum energy needed to get an electron out of the metal (the 'work function'). If the intensity of radiation is extremely low electrons can be ejected one at a time; the energy transferred from the radiation field to the electron is then

$$E - E_F = \tfrac{1}{2}mv^2 + W = hv.$$

If the intensity is increased nothing is changed except that the *number* of energy units absorbed, each of magnitude hv, is increased, each one ejecting an electron. Here we neglect the possibility that some energy may be absorbed in lattice vibrations, heating the metal. The important thing is that energy is absorbed from the radiation field in *quanta*, the size of a quantum of energy being hv. A quantum of field energy is called a *photon*, and each photon may be visualized as a strongly localized 'wave packet' or 'pulse' of radiation whose properties may in some ways resemble those of a particle.

(ii) *The Compton effect.* A photon, with energy $E = hv$, does in fact possess a definite momentum p. If a photon collides with an electron, both particles are scattered and the momentum of the photon can be inferred. This is the Compton effect. According to relativity theory mass and energy are inter-related, and for a free particle $E = (m^2c^4 + p^2c^2)^{1/2}$ where m is the 'rest mass'. For a particle at rest ($p = 0$) this gives the Einstein mass–energy relation

$$E = mc^2 \tag{2.2}$$

while for a photon (assuming *zero* rest mass, $m = 0$) $E = (p^2 c^2)^{1/2}$ and thus

$$p = E/c. \tag{2.3}$$

However, from the Planck relation $E = h\nu$, and the photon momentum is thus $p = h\nu/c$. Since a wave travels with (phase) velocity $c = \nu\lambda$, where λ is the wavelength, this implies

$$\lambda = h/p. \tag{2.4}$$

This relationship between the wavelength of a beam of light and the momentum of its photons was accurately verified by Compton.

(iii) *Electron diffraction.* The relationship (2.4) applies not only to photons but also to particles such as electrons with a finite rest mass. Experiment shows that electrons coming through a hole are *diffracted* (Fig. 2.2). In practice slits and holes are not convenient; a crystal lattice is used as a 'diffraction grating', but the principle is the same. The distribution of intensity on a target (photographic plate) is just what would be expected for a beam of radiation with wavelength given by (2.4). Since, for a material particle† $p = mv$, this means

$$\lambda = \frac{h}{mv}. \tag{2.5}$$

For an electron, the particle velocity v is easily worked out from the voltage used to accelerate it; if the potential drop is V the work done on the electron is eV and this is converted into kinetic energy $\frac{1}{2}mv^2$. Equation (2.5) was first proposed by de Broglie and subsequently verified experimentally by Davisson and Germer (1927) and by Thomson (1928).

It is worth remarking, in passing, that the experimental techniques associated with the photoelectric effect and with electron diffraction are now firmly established in chemistry. Electron diffraction is used in determining molecular dimensions (see for example, Bauer 1970). The photoelectric effect is the basis of photoelectron spectroscopy (see for example Baker 1970). Here the metal is replaced by a single molecule and W (the work function) is replaced by I (an ionization potential, i.e. the energy required to remove an electron from the energy level it occupied in the molecule); Fig. 2.1 should be compared with Fig. 1.3. Even the Compton effect is now being used in studying the electron distribution in molecules. Recent developments in theory and techniques are reviewed in Williams (1977).

† Assumed moving at 'non-relativistic' velocity (i.e. velocity small compared with that of light).

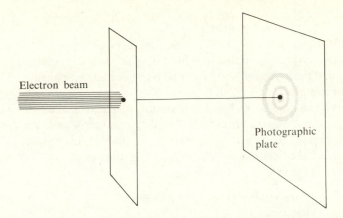

Fig. 2.2. Schematic illustration of electron diffraction.

How can the wave and particle aspects of the behaviour of an electron be reconciled? On the one hand, the degree of blackening at some point of a photographic plate due to a diffracted beam of electrons is proportional to the number of electrons arriving there: on the other hand, it can be correctly calculated by the methods of physical optics, treating the beam as an incident wave described by a wavefunction ψ, with wavelength given by (2.5), and assuming the degree of blackening is determined by the *intensity* ψ^2. It follows that the number of electrons found at a particular point is proportional to the square of a wavefunction evaluated at that point. The wavefunction for particles moving in a beam, each with momentum p, has wavelength $\lambda = h/p$ and must satisfy some kind of wave equation—namely Schrödinger's equation, considered in Chapter 3. What is important here is that no attempt is made to define a precise path, or orbit, for a particular electron, and that the result of the experiment is described *statistically*. When the beam is diffracted, essentially the same experiment (i.e. shooting an electron through the hole) is performed on each of a very large number of electrons; the *fraction* of the total to arrive at a given point is then, by definition, the *probability* that a single electron in the beam, taken at random, would arrive at that point. A resumé of the definition and properties of probabilities is given in Appendix 1. Evidently the wavefunction ψ is needed even to describe the behaviour of *one* particle; ψ is a function of position, $\psi = \psi(x, y, z)$, and $|\psi(x, y, z)|^2$ determines the probability of finding the particle at the point with co-ordinates x, y, z. In the beam experiment each electron makes only one spot on the photographic plate—more likely at a point where ψ is large, less likely where it is small; as more and more electrons arrive the whole diffraction pattern is built up, the number found at any point being proportional to $|\psi|^2$ at that point. We shall write

$$P(x, y, z) = |\psi(x, y, z)|^2 \tag{2.6}$$

and call $P = P(x, y, z)$ a *probability function*. It should be noted that in (2.6) we have used the square modulus of ψ instead of ψ^2; this simply ensures that P is always a real positive number (as it must be by definition) even if ψ should turn out to be complex.†

Strictly speaking, the probability of finding an electron at a *point* (i.e. in zero volume) is zero, and the function P is more properly called a probability *density* function, i.e. a probability per unit volume. An element of volume, using Cartesian co-ordinates, is $d\tau = dx\,dy\,dz$ (the volume of a box with sides dx, dy, dz) and thus

$$\left(\begin{array}{l}\text{probability of particle in volume}\\\text{element } d\tau \text{ at point } x, y, z\end{array}\right) = P(x, y, z)\,d\tau \tag{2.7}$$

where P is determined by (2.6). Now the probability of finding the particle *somewhere* in space must be unity, and hence (see Appendix 1)

$$\int P(x, y, z)\,d\tau = 1 \tag{2.8}$$

where the integration extends over the whole of space. This physically necessary property puts an important restriction on ψ, namely

$$\int |\psi(x, y, z)|^2\,d\tau = 1 \tag{2.9}$$

and a wavefunction satisfying this condition is said to be *normalized*. If an otherwise satisfactory solution of Schrödinger's equation does not satisfy (2.9), it can be made to do so on multiplying by a 'normalizing factor' N and requiring that

$$N^2 \int |\psi(x, y, z)|^2\,d\tau = 1 \tag{2.10}$$

provided the integral is finite; we say that wavefunctions must be *quadratically integrable*. In fact, the integral over *any* region must be finite in order to preclude the possibility (which would be nonsense) of a particle being found there with infinite probability.

Let us now summarize the procedure to be adopted if we wish to discuss the behaviour of a particle moving under the influence of given forces (e.g. an electron under the influence of one or more nuclei).

   (i) We write down the wave equation in the form appropriate to the given conditions.

---

† If ψ is a complex quantity we can write $\psi = f + ig$ where $i = \sqrt{(-1)}$ and f and g are real. On introducing the complex conjugate $\psi^* = f - ig$, the square modulus of ψ is $|\psi|^2 = \psi\psi^* = f^2 + g^2$.

(ii) We solve this equation to find a wavefunction $\psi(x, y, z)$.

(iii) The probability density is then

$$P = |\psi(x, y, z)|^2.$$

We shall not go into the details of this technique now because that will be the subject matter of Chapter 3. For the present we shall be satisfied with a qualitative account of the procedure and its results when applied to the study of atomic structure. The pictorial character of these results can easily be appreciated before we give the theoretical justification for them. However, before going any further we must point out one very important factor that has so far been omitted. The wave equation has an infinite number of solutions, many of which do not correspond to any physical or chemical reality; we call such solutions unacceptable. Acceptable wavefunctions must satisfy certain conditions. We have already remarked that the integral $\int |\psi|^2 \, d\tau$ must be finite; we must add that ψ must be everywhere finite, single valued, and smooth (that is, both the function and its slope must be continuous). These conditions seem very simple and obvious, but their consequences are very far reaching. For an electron bound to an atom or molecule they imply that ψ dies away smoothly to zero at large distances from the nucleus, and such solutions can in fact be found only for certain special 'allowed' values of the energy E. The corresponding 'quantized energy levels', successfully predicted by Bohr for the hydrogen atom but only with the assumption of certain 'quantum conditions', thus emerge very naturally in quantum mechanics from the general physical requirement that the wavefunction be mathematically 'well behaved'. The corresponding states of motion are called *stationary states*, not because the electron is not moving in such a state, but because both the energy E and the function P ($= \psi\psi^*$) are *independent of time*. Other states are possible in which the energy is not conserved and in which the state, characterized by ψ and P, evolves as time passes; such states are important in discussing scattering and collisions or the interaction of particles with oscillating fields (which may cause *transitions* from one stationary state to another), but they will be mentioned in this book only briefly and they play no direct role in the theory of valence. When we speak of 'allowed' energy levels and wavefunctions we shall invariably be referring to stationary states.

2.2. The charge-cloud interpretation of $|\psi|^2$

In quantum mechanics a moving particle is represented by a wavefunction ψ such that $|\psi|^2 \, d\tau$ is the probability that it is found in the volume $d\tau$. There is, however, an alternative and more pictorial (though less strictly accurate) interpretation of ψ which we have anticipated in §1.5. Let us deal with the case in which the moving particle is an electron. Then we imagine that this electron is spread out in the form of a cloud—we have referred to it as a charge cloud— the density of this cloud at any point being proportional to $|\psi|^2$. In places where

$|\psi|^2$ is largest the charge cloud is densest and most of the negative charge is to be found.

The essential difference between this interpretation and our earlier one is that instead of speaking of the *probability* density (such that $P\,d\tau$ is the chance of finding the electron in $d\tau$) we picture the electron as being 'smeared out' with a density P (electrons/unit volume). Equation (2.9) then indicates, correctly, that the total charge (obtained by integrating the density over all space) is just one electron. Now if a single electron is a particle, it cannot possibly be distributed over regions of the size of an atom or molecule, which are of the order of 10^{-10} m in each direction. The charge-cloud picture must therefore not be taken as literally true. Only the statistical, or probability, interpretation is really valid. A link between the two viewpoints may be found in the following way. Let us suppose that at a particular moment we were able by some means to determine exactly where the electron was, and that we represented this position by a minute dot in space. Let us now repeat the observation a very large number of times (perhaps a million times) and each time put a similar dot in the appropriate place. When we have finished, the distribution of dots will look exactly like a cloud; the densest parts of the cloud will be those where there are most dots, and these are precisely the places where our individual observations are most likely to discover the electron. Therefore the density of the charge cloud is a direct measure of the probability function; the electron density 'observed' in the X-ray diffraction experiment which led to Fig. 1.2 is really the function $P(x, y, z)$ which gives the probability per unit volume of finding an electron at the point x, y, z.

Even though it may be a fiction, the idea of a charge cloud is a very convenient fiction with a certain formal validity. Its great conceptual value, as indicated in §1.5, is that it can be visualized pictorially, unlike a many-electron wavefunction, and that many observable quantities (see pp. 69–71) may be computed by classical electrostatics *exactly as if the electron really were smeared out with density P*. Moreover, the charge density function may still be defined even for a *many*-electron system and is found to have exactly similar properties.

In the next section we shall show such contour maps for some stationary states of the single electron in a hydrogen atom, and use a similar pictorial method to display the wavefunction ψ itself.

2.3. The hydrogen atom ground state: Atomic units

The ideas discussed so far are very nicely illustrated by the example of the ground state of the hydrogen atom. We have only one electron to deal with, and this moves round the nucleus which we may take to be fixed at the origin of co-ordinates. In such a case there are many allowed wavefunctions and corresponding energies; the lowest of these is called the ground state.

According to the Bohr theory the electron moves round the nucleus in a circle whose radius is called the radius of the first orbit, or more simply 'the Bohr radius', and which is written a_0. It is found that this radius is related to the mass m and charge e of the electron by the formula

$$a_0 = \kappa_0 \hbar^2 / me^2 \tag{2.11a}$$

where we introduce two convenient abbreviations: $\hbar$ (crossed h) stands for $h/2\pi$ and is sometimes called the 'rationalized Planck constant', while $\kappa_0 = 4\pi\varepsilon_0$ (ε_0 being the permittivity of free space). The allowed energy turns out to be

$$E = -\frac{me^4}{2\kappa_0^2 \hbar^2} = -\frac{e^2}{2\kappa_0 a_0} = -13\cdot60 \text{ eV}. \tag{2.11b}$$

The minus sign here means that the energy is *lower* than the zero of energy, which conventionally corresponds to infinite separation of the proton and electron. The whole of the motion would take place in one plane. Although this theory has been discarded, the quantities defined in (2.11) are still of fundamental importance.

In quantum mechanics, the electron is described instead by a suitable wavefunction. In Chapter 3 we find

$$\psi = \left(\frac{1}{\pi a_0^3}\right)^{1/2} \exp\left(-\frac{r}{a_0}\right) \tag{2.12}$$

where r is the distance from the origin. From this it follows that the probability density is

$$P = \psi^2 = \frac{1}{\pi a_0^3} \exp\left(-\frac{2r}{a_0}\right). \tag{2.13}$$

Clearly the motion is no longer restricted to a plane, for P is spherically symmetrical around the origin and the electron is thus equally likely to be found at all points with the same value of r (i.e. on a sphere of radius r). Further, the wavefunction (2.12) is already normalized since, noting that $\int_0^\infty r^2 e^{-cr} dr = 2/c^3$ and taking for $d\tau$ a spherical shell,

$$\int \psi^2 \, d\tau = \int_0^\infty \psi^2 4\pi r^2 \, dr = \int_0^\infty 4a_0^{-3} r^2 \exp\left(-\frac{2r}{a_0}\right) dr = 1.$$

There are several ways in which we might exhibit graphically the nature of (2.12) or (2.13).

(a) We could plot a graph showing how ψ (or P) varies with r.
(b) We could draw contours of constant value of ψ (or P). In this case the contours are concentric spheres, and at all points on any one sphere ψ (or P) has the same value.
(c) We could draw the charge cloud (or, rather, a section of it by some plane through the origin), values of P being indicated by density of shading.

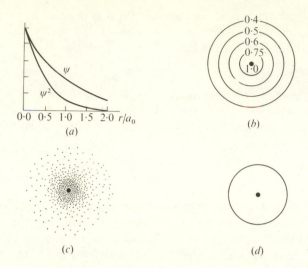

Fɪɢ. 2.3. Representations of the wavefunction for the hydrogen atom ground state: (a) plots of ψ and ψ^2 against r (scaled to unity at $r = 0$); (b) contours of ψ (similarly scaled) in a plane containing the nucleus; (c) charge cloud $P\ (=\psi^2)$; (d) boundary surface.

(d) We could draw a 'boundary surface'. This is that particular one of the contours referred to in (b) such that the total charge outside the contour is some definite small percentage (e.g. 10 per cent) of the total electronic charge.

All these possibilities are illustrated in Fig. 2.3. Of these diagrams (a) and (b) are very precise, and require an exact knowledge of ψ; (c) is able to give us a very good general impression of the charge distribution even if it does not represent the density of the cloud with complete precision; (d) is much the simplest, but for many purposes it provides a surprisingly adequate pictorial representation and we shall make frequent use of it.

There is yet another method of representing the charge density which is often used for atoms. When P is spherically symmetrical we may plot instead what is called the *radial* density $4\pi r^2 P(r)$. Since $4\pi r^2\,\mathrm{d}r$ is the volume lying between the two spheres r, $r + \mathrm{d}r$, it follows that $4\pi r^2 P(r)\,\mathrm{d}r$ is the total probability that the electron is *anywhere* within the spherical shell, thickness $\mathrm{d}r$, at distance r from the origin. Fig. 2.4 (a) shows the graph of this radial density. It is interesting, by way of a final comparison with the Bohr theory, that the maximum radial density in the new theory occurs when $r = a_0$, so that in this case the *most probable* distance of the electron from the nucleus coincides with the single precise value in the corresponding Bohr orbit.

On the right-hand side of Fig. 2.4 (b) there is shown the variation of $F(r)$, where $F(r)$ is that fraction of the total charge cloud which lies outside a sphere of radius r. Evidently $F(0) = 1$ and $F(\infty) = 0$. It will be noticed that $F(r)$ falls

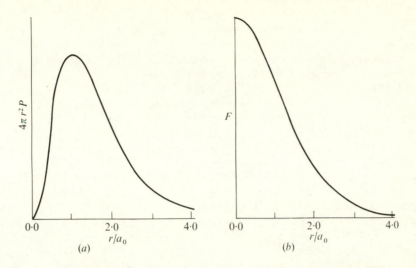

FIG. 2.4. (*a*) Radial density for the hydrogen atom ground state. (*b*) Corresponding fraction of charge cloud outside a sphere of radius *r*.

off rapidly with *r*. If, for example in Fig. 2.3 (*d*), we chose the boundary surface for which 10 per cent of the charge lay outside, its radius would be $2 \cdot 6a_0$ ($\sim 0 \cdot 14$ nm). The analytical expression for $F(r)$ is

$$F(r) = \int_r^\infty \psi^2 4\pi r^2 \, \mathrm{d}r = \left(1 + \frac{2r}{a_0} + \frac{2r^2}{a_0^2} \right) \exp \left(-\frac{2r}{a_0} \right) \qquad (2.14)$$

and it is therefore easy to draw alternative boundaries outside which there is a greater or smaller fraction of the charge.

Atomic units. It is often convenient to adopt the Bohr radius a_0, more briefly the 'bohr', as an 'atomic unit' of length, for in equations such as (2.12), (2.13), and (2.14) the distance *r* always appears in conjunction with a_0 (e.g. in r/a_0 or $(r/a_0)^2$) and r/a_0 simply measures the distance *r* 'in units of a_0'. Strictly speaking, a_0 is not a *unit* at all, for it depends on $e, m, \hbar, \kappa_0$, whose values are determined by *measurement* in terms of the primary units (coulomb, kilogram, etc.) which are fixed purely by convention, but the term 'atomic unit' is widely used and its meaning is quite clear. Similarly, energies may be measured in terms of the group of fundamental constants $e^2/\kappa_0 a_0$ in (2.11*b*), which we denote by E_h (sometimes called the 'hartree'), and E/E_h measures an energy *E* in units of E_h. This is a completely consistent system in which every type of physical quantity has an 'atomic unit'; units of charge, mass, action and permittivity are $e, m, \hbar, \kappa_0$, respectively, and all other units are expressible as appropriate combinations of these four. For example, a dipole moment has

TABLE 2.1

Some commonly used atomic units

Quantity and dimensions	Atomic unit	SI equivalent
Length (L)	a_0 (bohr)	$5 \cdot 292 \times 10^{-11}\,\mathrm{m}$
Mass (M)	m (electron mass)	$9 \cdot 110 \times 10^{-31}\,\mathrm{kg}$
Charge (Q)	e (proton charge)	$1.602 \times 10^{-19}\,\mathrm{C}$
Energy ($\mathrm{ML^2T^{-2}} = \mathrm{W}$)	E_h (hartree)	$4 \cdot 359 \times 10^{-18}\,\mathrm{J}$
Electric permittivity ($\mathrm{Q^2W^{-1}L^{-1}}$)	$e^2/a_0 E_\mathrm{h}\,(= 4\pi\varepsilon_0)$	$4\pi \times 8 \cdot 854 \times 10^{-12}\,\mathrm{F\,m^{-1}}$
Action (WT)	$\hbar$	$1 \cdot 055 \times 10^{-34}\,\mathrm{J\,s}$
Time (T)	$\hbar/E_\mathrm{h}$	$2 \cdot 419 \times 10^{-17}\,\mathrm{s}$
Electric potential ($\mathrm{WQ^{-1}}$)	E_h/e	$2 \cdot 721 \times 10^1\,\mathrm{V}$
Electric field strength ($\mathrm{WQ^{-1}L^{-1}}$)	E_h/ea_0	$5 \cdot 142 \times 10^{11}\,\mathrm{V\,m^{-1}}$
Magnetic flux density ($\mathrm{MT^{-1}Q^{-1}}$)	$\hbar/ea_0^2$	$2 \cdot 351 \times 10^5\,\mathrm{T}$

dimensions of charge × distance, and its atomic unit is thus ea_0 or, from (2.11a), $\kappa_0 \hbar^2/me$. Usually, it is convenient to adopt the set e, m, a_0, E_h (instead of $e, m, \hbar, \kappa_0$) and the corresponding expressions for the atomic units of some commonly occurring quantities are shown in Table 2.1.

To summarize, the atomic units of charge, mass, length, and energy are taken to be

$$e, \qquad m, \qquad a_0 = \frac{\kappa_0 \hbar^2}{me^2}, \qquad E_\mathrm{h} = \frac{me^4}{\kappa_0^2 \hbar^2}.$$

The charge on the electron is $-e$ (-1 'atomic unit of charge'), its mass is m (1 'atomic unit of mass'), and in the ground state of the hydrogen atom its energy is $-\frac{1}{2}E_\mathrm{h}$ ($-\frac{1}{2}$ 'atomic unit of energy' or $-\frac{1}{2}$ 'hartree'). The interpretation of the energy unit E_h is simple: the electron in its ground state in the hydrogen atom has an energy $\frac{1}{2}E_\mathrm{h}$ below the level (zero) corresponding to removal of the electron to infinity; thus $\frac{1}{2}E_\mathrm{h}$ is the ionization potential of the hydrogen atom. According to (2.11),

$$E_\mathrm{h} = 27 \cdot 20\,\mathrm{eV}$$

and is twice the hydrogen ionization potential as calculated with neglect of nuclear motion.

When using atomic units throughout a piece of work, the quantities $e, m, a_0, \hbar$, etc. are frequently given their unit numerical values and consequently disappear from all equations; *the results are then understood to be expressed in appropriate atomic units*. For instance, the probability density (2.13) would appear as

$$P = \left(\frac{1}{\pi}\right)e^{-2r}.$$

It is then important to interpret the quantities correctly. Thus, r is the numerical measure of the distance from the nucleus and really stands for the *number* r/a_0, which is called a 'dimensionless variable'. P also appears as a pure number; it is the probability density expressed as a multiple of the atomic unit of density, namely a_0^{-3}. If we wish to express P and r in absolute units we must therefore write

$$P = a_0^{-3}\left(\frac{1}{\pi}\right)\exp\left(-\frac{2r}{a_0}\right)$$

the factor a_0^{-3} putting in the correct physical dimensions (L^{-3}). We thus retrieve the original form (2.13). The use of dimensionless variables (i.e. the *numbers* which express quantities as multiples of stated units, instead of the *quantities* themselves) is convenient in numerical work, but care must be taken to be consistent throughout, *all* quantities being referred to the chosen system of units.

2.4. Atomic orbitals

The wavefunction (2.12) may be said to describe the *state* of the electron. We have given up the classical idea of motion in a precise orbit, but such a wavefunction provides the best picture we can get. We call it an *atomic orbital* or simply an AO.

The AO shown in Fig. 2.3 is only the first, that of lowest energy, out of an infinite family of solutions of the Schrödinger equation; it refers to the ground state, which is chemically the most important, but if sufficient energy is put into the atom (e.g. by radiation) it can make a transition or 'jump' into one of the 'excited states' whose AO corresponds to a higher energy. We must now begin a systematic study of the first few types of AO in this infinite set, not only for the hydrogen atom with atomic number $Z = 1$ but also for arbitrary atomic number, since we shall want to deal with all atoms in the periodic table. It is absolutely essential to have a clear mental picture of the more common AO's, and to recognize and remember their symmetry types and their names.

The AO's for an electron moving around a nucleus of charge Ze are usually expressed in polar co-ordinates r, θ, ϕ (indicated in Fig. 2.5). This form is useful in classifying the solutions, though later we shall find them easier to visualize in terms of the Cartesian co-ordinates x, y, z. For future reference we note that the co-ordinate systems are related by the equations $x = r\sin\theta\cos\phi$, $y = r\sin\theta\sin\phi$, $z = r\cos\theta$. The allowed functions are all of the form

$$\psi_{nlm}(r, \theta, \phi) = R_{nl}(r)Y_{lm}(\theta, \phi) \tag{2.15}$$

in which different solutions are distinguished by the labels n, l, m which are called *quantum numbers* and take only integral values. The factor R_{nl} is a function only of distance of the electron from the nucleus and depends on the two quantum numbers n and l, while the factor Y_{lm} supplies an angle

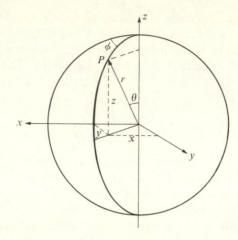

FIG. 2.5. Spherical polar co-ordinates showing their relationship to the Cartesian system.

dependence and depends on the quantum numbers l, m. While n takes values 1, 2, 3, ..., those of l and m are restricted as follows:

$$l \leqslant n-1, \qquad m = l, \quad (l-1), \quad (l-2), \ldots, -l \tag{2.16}$$

Although there are three quantum numbers, the energy depends only on one, n. It is given by

$$E_{nlm} = -\frac{1}{2}\frac{Z^2}{n^2} E_{\text{h}} = -\frac{1}{2}\frac{Z^2}{n^2}\text{hartree} \tag{2.17}$$

and reduces to the result already given for hydrogen when $Z = 1$ and $n = 1$. The number n is called the *principal* quantum number. Since, for $n > 1$ the quantum numbers l and m may take various values without changing the energy, the allowed states must fall into sets in which all members of a set have the same energy; such states are said to be *degenerate* and the number of independent states in the set is the *degree of degeneracy*.

The pattern of energy levels is displayed in Fig. 2.6. Here the levels have been given their conventional names and it should be noted that for each l value a level is $(2l+1)$-fold degenerate corresponding to the $2l+1$ possible choices of $m\,(=l, l-1, \ldots, -l)$; thus the level labelled 2p indicates the common energy of three distinct states corresponding to $m = 1, 0, -1$. The name consists of the *principal quantum number n*, which indicates the energy, followed by a letter which indicates the angular form of the wavefunction; the names s, p, d, f, ... (corresponding to $l = 0, 1, 2, 3, \ldots$) arose historically in classifying spectral series (sharp, principal, diffuse, fundamental, ...).

Before sketching the shapes of the AO's, a word must be said about the significance of the quantum number m—sometimes called the 'magnetic'

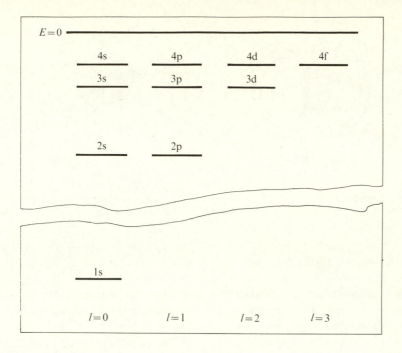

FIG. 2.6. Classification of energy levels for the hydrogen atom (for convenience, the·scale is broken between 1s and 2s levels, otherwise the 1s level would be *twice* as far below the line $E = 0$).

quantum number (again from spectroscopy, states with different m values being differently affected by a magnetic field). The three states of 2p type would be denoted, according to (2.13), by $\psi_{2,1,1}$, $\psi_{2,1,0}$, $\psi_{2,1,-1}$. They all satisfy the Schrödinger equation equally well, with the same energy value E_{2p}, but other equally acceptable possibilities also exist. If we take a 'linear combination'

$$\psi = a\psi_{2,1,1} + b\psi_{2,1,0} + c\psi_{2,1,-1} \tag{2.18}$$

(where a, b, c are numerical coefficients), this still satisfies the Schrödinger equation and is still a 2p-type function with energy E_{2p}. There is thus a certain arbitrariness in how we choose the different members of a degenerate set of states, and this can be removed only by agreeing on some convention. In fact, the most common convention in valence theory (though not in atomic spectroscopy) is to combine in pairs, taking the sum and the difference, the functions with equal and opposite m values (e.g. $\psi_{2,1,1}$ and $\psi_{2,1,-1}$). The AO's so defined are very easy to visualize and are related in a simple way to a set of Cartesian co-ordinate axes.

First we give a detailed example, taking the function $\psi_{2,1,0}$ which is of 2p type. We suppose the nucleus has charge Ze and write out the explicit result

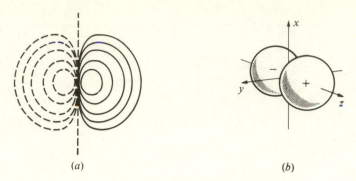

(a)   (b)

FIG. 2.7. Representations of a 2p AO: (a) contour map of ψ values in the xy plane, negative values indicated by broken lines; (b) simplified three-dimensional representation of the AO.

corresponding to (2.15), using Cartesian co-ordinates and atomic units. The radial factor, namely $R_{21}(r)$, takes the very simple form $r \exp(-Zr/2)$ where, in Cartesian co-ordinates, the radial distance is $r = (x^2 + y^2 + z^2)^{1/2}$; the angular factor is simply $\cos\theta$ (i.e. z/r from Fig. 2.5). Thus, the wavefunction has the form $z \exp(-Zr/2)$. The exponential factor is spherically symmetrical and falls away smoothly as we go away from the nucleus in any direction. However, the factor z introduces an angle dependence and also 'cuts the function in two', for at all points on the xy plane $z = 0$ and the wavefunction must vanish. When $\psi = 0$, we say the function has a *node* and the xy plane is a *nodal surface*. Moreover ψ is positive on one side of the nodal surface (z positive) but negative on the other (z negative); reflecting a point across the xy plane (i.e. reversing the sign of z but leaving x and y unchanged) simply changes the sign of ψ. One half of the function is thus identical with the other except for a sign reversal. The function is indicated in the contour map in Fig. 2.7 (a). This orbital clearly 'points along the z axis' and is therefore usually called a $2p_z$ AO. The figure only shows, of course, the values of ψ in a two-dimensional section, the xz plane, whereas the pattern of ψ values is essentially three dimensional. Evidently the contours of equal ψ values are the same for *any* plane containing the z axis, for if we fix attention on one point (given z) and move it around the z axis in a circle (i.e. change x and y, keeping the axial distance $(x^2 + y^2)^{1/2}$ constant) $z \exp\{-\frac{1}{2}Z(x^2 + y^2 + z^2)^{1/2}\}$ remains constant. When the function is unchanged in a rotation of this kind we say it has *axial symmetry*. Now let us give a simplified picture of ψ using the idea of a (conventional) boundary surface, as in Fig. 2.3 (d), outside which ψ has only small values. We could use a contour corresponding to some particular ψ value, but, more crudely, we could simply indicate the regions (e.g. by enclosing them in spheres) in which ψ is large. In this way we arrive at the representation in Fig. 2.7 (b) which indicates, in perspective, the real three-dimensional form of the AO; for qualitative purposes this type of representation of an AO is entirely adequate

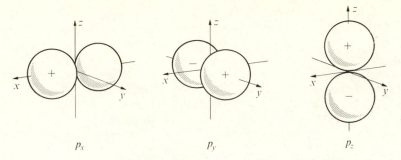

Fig. 2.8. Family of three p-type AO's.

and we shall use it constantly. It should be noted that the two 'lobes' of the AO are labelled by signs showing where ψ takes positive and negative values; ψ^2 is of course positive everywhere, but when we use the AO's in valence theory we shall find it essential to indicate the signs of the different regions of ψ.

We are now ready to display all the AO's we shall need, grouping them according to principal quantum number n. For $n = 1$ there is only one AO ($l = m = 0$); this is the 1s AO

$$\psi_{1s} = N_{1s} e^{-Zr} \tag{2.19}$$

where N_{1s} is a normalizing factor (tabulated later). ψ_{1s} has been depicted already in Fig. 2.3 (d).

For $n = 2$ there is one AO with $l = m = 0$, the 2s AO, but now, from (2.16), $l = 1$ is allowed and with it the possibilities $m = 0$ (giving the 2p AO discussed in detail above) and $m = \pm 1$. The appropriate combinations of the functions with $m = \pm 1$ (sum and difference) lead to two AO's which we designate $2p_x$ and $2p_y$ because they are exactly like $2p_z$ except that they point along the x and y axes, respectively, instead of the z axis (Fig. 2.8). The full set of AO's with principal quantum number $n = 2$ is shown in Table 2.2. We need only remark that the 2s AO has a *radial* node in the form of a sphere of radius $r = 2/Z$ (for which the factor $2 - Zr$ vanishes), while the three 2p AO's, each with an angular node, are exactly similar—like dumbells—except for orientation in

TABLE 2.2

Hydrogen-like 2s and 2p AO's in atomic units. (To convert to absolute units replace r by r/a_0, x by x/a_0, etc.)

2s orbital ($n = 2, l = 0$)	$\psi_{2s} = N_{2s}(2 - Zr)\exp(-Zr/2)$
2p orbitals ($n = 2, l = 1$)	$\psi_{2p_x} = N_{2p}x\exp(-Zr/2)$
	$\psi_{2p_y} = N_{2p}y\exp(-Zr/2)$
	$\psi_{2p_z} = N_{2p}z\exp(-Zr/2)$

TABLE 2.3

Hydrogen-like 3s and 3p AO's in atomic units. (To convert to absolute units replace r by r/a_0, x by x/a_0, etc.)

3s orbital ($n = 3, l = 0$)	$\psi_{3s} = N_{3s}(27 - 18Zr + 2Z^2r^2)\exp(-Zr/3)$
3p orbitals ($n = 3, l = 1$)	$\psi_{3p_x} = N_{3p}x(6 - Zr)\exp(-Zr/3)$
	$\psi_{3p_y} = N_{3p}y(6 - Zr)\exp(-Zr/3)$
	$\psi_{3p_z} = N_{3p}z(6 - Zr)\exp(-Zr/3)$

space. It is not surprising that the 2p AO's are degenerate, all with the same energy E_{2p} in Fig. 2.6; this degeneracy is due to *symmetry* (the spherical symmetry of the central field). However, the degeneracy between the 2s and 2p AO's, which are totally dissimilar in shape, is more remarkable; it is an 'accidental' degeneracy, arising from a peculiar property of the coulombic form of the potential, which could not have been predicted from simple symmetry considerations.

For $n = 3$ we again obtain s- and p-type AO's for $l = 0$ and $l = 1$ respectively. The 3s and 3p AO's have the forms given in Table 2.3 and clearly have the same general shapes as the 2s and 2p AO's. The main difference is that each possesses one extra radial node, for ψ_{3s} contains the quadratic factor $27 - 18Zr + 2Z^2r^2$ and vanishes for *two* values of r, while each 3p AO contains a factor $6 - Zr$ and therefore has a radial node when $r = 6/Z$. The exact

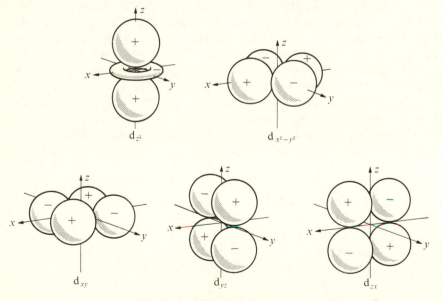

FIG. 2.9. Family of five d-type AO's.

TABLE 2.4

Hydrogen-like 3d AO's in atomic units. (To convert to absolute units replace r by r/a_0, x by x/a_0, etc.)

3d orbitals ($n = 3, l = 2$)	
$\psi_{3d_{z^2}}$	$= N_{3d}\tfrac{1}{2}(3z^2 - r^2)\exp(-Zr/3)$
$\psi_{3d_{x^2-y^2}}$	$= N_{3d}\tfrac{1}{2}\sqrt{3}\,(x^2 - y^2)\exp -Zr/3)$
$\psi_{3d_{xy}}$	$= N_{3d}\sqrt{3}\,xy\exp(-Zr/3)$
$\psi_{3d_{yz}}$	$= N_{3d}\sqrt{3}\,yz\exp(-Zr/3)$
$\psi_{3d_{zx}}$	$= N_{3d}\sqrt{3}\,zx\exp(-Zr/3)$

positions of these nodes are for most purposes unimportant. They are relatively near the nucleus and in their *outer* parts the 3s and 3p AO's are not very different in shape from 2s and 2p; the radial nodes are often ignored and the AO's are represented pictorially exactly as in Figs. 2.3 (*d*) and 2.8 by spheres and dumbells. Thus, Fig. 2.8 serves to indicate *any* set of p-type AO's.

There is, however, a new possibility with principal quantum number $n = 3$: for $l = 2$ and $m = 2, 1, 0, -1, -2$, gives a set of five degenerate 3d functions. These AO's have totally different shapes from those discussed so far; they are given in Table 2.4, again in the forms obtained by combining the $m = \pm 2$ functions and the $m = \pm 1$ functions, and are sketched in Fig. 2.9.

The naming of the AO's indicates the angular factors in the wavefunction. The AO's are all alike, apart from orientation, except the first: Why does this appear to be different and why are there not *six* exactly similar d functions

$$3d_{xy}, \qquad 3d_{yz}, \qquad 3d_{zx}, \qquad 3d_{x^2-y^2}, \qquad 3d_{y^2-z^2}, \qquad 3d_{z^2-x^2}$$

the last two again having four lobes, just like $3d_{x^2-y^2}$ except for orientation? The answer is simply that we do not *need* six; there are only five *independent* solutions of the Schrödinger equation with $n = 3$, $l = 2$, and which five we adopt is a matter of convention. To see that the three functions $3d_{x^2-y^2}$, $3d_{y^2-z^2}$, and $3d_{z^2-x^2}$ are not independent we need only add them together, obtaining a result of the form

$$(x^2 - y^2)f(r) + (y^2 - z^2)f(r) + (z^2 - x^2)f(r) = 0. \qquad (2.20)$$

In other words, any one can be expressed as -1 times the sum of the other two. The fact that we adopt $3d_{x^2-y^2}$ and $3d_{z^2}$ instead of, say, $3d_{x^2-y^2}$ and $3d_{y^2-z^2}$, is again a matter of convention; the first choice gives functions with a useful property (orthogonality) which we discuss later. In spite of its apparent difference in shape, the $3d_{z^2}$ AO can be represented as a combination or 'superposition' of two of the four-lobe functions, for

$$(z^2 - x^2)f(r) - (y^2 - z^2)f(r) = (2z^2 - x^2 - y^2)f(r)$$
$$= (3z^2 - r^2)f(r)$$

which, with a suitable normalizing factor, is the $3d_{z^2}$ AO given in Table 2.4. The five 3d AO's are thus expressible in terms of six functions which are identical except for orientation in space, and the equality of their corresponding energies is seen to be a symmetry degeneracy. The degeneracy of the 3s, 3p, and 3d functions (Fig. 2.6) is again remarkable, depending purely on the coulomb form of the central field.

We could continue, with $n = 4$, and should find 4s, 4p, and 4d AO's, but l would then be able to take values up to $l = 3$ and in that case we should again get a new type of AO—the 4f orbitals, of which there would be 7 (i.e. $2l + 1$). The s-, p-, and d-type functions are similar to those of lower principal quantum numbers; the f-type AO's are rather more complicated. We shall not make much reference to f-type orbitals and therefore do not show their forms.

TABLE 2.5

Normalizing factors for hydrogen-like AO's in atomic units. (For absolute units multiply by $a_0^{-3/2}$.)

1s	2s	2p	3s	3p	3d
$\dfrac{Z^{3/2}}{\sqrt{\pi}}$	$\dfrac{Z^{3/2}}{4\sqrt{(2\pi)}}$	$\dfrac{Z^{5/2}}{4\sqrt{(2\pi)}}$	$\dfrac{2Z^{3/2}}{81\sqrt{(3\pi)}}$	$\dfrac{2Z^{5/2}}{81\sqrt{\pi}}$	$\dfrac{Z^{7/2}}{81\sqrt{(6\pi)}}$

In conclusion we show in Table 2.5 the normalizing factors for the AO's we have listed. We also add a few general comments. In the first place we note that as n increases and the energy rises, the orbitals become more complicated, being cut into more and more regions (with alternating signs) by nodal surfaces. This is a general property of wavefunctions; the energy rises as the number of nodes increases. For hydrogen, in particular, we note that all the AO's of given n (and hence energy) have the same number of nodal surfaces, namely $n - 1$. Secondly, every orbital can be written as a function of Zx, Zy, Zz and is therefore simply reduced in scale as we increase Z; if we double Z, a given ψ will achieve the same value when the co-ordinates are divided by 2, i.e. at a point twice as close to the nucleus; in the same way, the 1s AO for Li^{2+} is like that for H, but shrunk three times closer to the nucleus. Finally, for large values of r, every AO falls to zero at a rate dominated by the exponential factor $\exp(-Zr/n)$; therefore if n is doubled r must double in order to maintain the value of this factor, i.e. ψ spreads out roughly twice as far. These two competing factors—an AO shrinking as Z increases, but expanding as n increases—will be important when we consider the variation of the sizes of atoms and ions along the rows and down the columns of the periodic table.

2.5. Electronic structure of atoms

We now turn to the quantum-mechanical description of many-electron atoms in general. First, however, we summarize the ideas developed so far, which refer to the hydrogen atom ($Z = 1$) or to hydrogen-like ions ($Z > 1$).

(a) The electron is described by a wavefunction ψ and associated energy E, which together must compromise an acceptable solution of Schrödinger's equation. The wavefunction may be called an atomic orbital (AO) and $|\psi(x, y, z)|^2$ is the probability per unit volume of finding the electron at point x, y, z.

(b) The AO's (i.e. different acceptable solutions) are labelled by quantum numbers. Of these, n ($= 1, 2, 3, \ldots$) puts the AO's in ascending order of energy and size, l ($= 0, 1, 2, \ldots$) tells us the geometrical form of the AO (s, p, d, $\ldots$ type) and the degree of degeneracy ($2l + 1$), and m, or some other equivalent label, distinguishes among the $2l + 1$ degenerate AO's of given type.

(c) The energy levels (Fig. 2.6) relate to experimentally observable quantities; for example, $E_{1s} = -I_{1s}$ where I_{1s} is the ionization potential for removing the electron from the 1s AO. Similarly, an electron in a state with energy E_1 may make a transition to a state with higher energy E_2, provided the energy needed

$$|E_1 - E_2| = h\nu$$

is supplied by a photon of radiation of frequency ν, i.e. by absorption of light, or may *fall* from energy level E_2 to energy level E_1 with *emission* of a photon. Such transitions provide the basis for interpretation of spectral series (for H, He$^+$, etc.) as indicated in Fig. 2.10. The observed frequencies and the fact that 'selection rules' govern the permitted jumps are completely accounted for by theory.

We shall find that similar ideas extend to much more complicated systems; basically because each electron moves in a sense 'independently' of the others and can therefore be described by its 'own' orbital.

In principle, the method of dealing with an atom containing several electrons differs little from that described already. We write down the appropriate wave equation and solve it. Now it is easy to write down the equation (see Chapter 3) but quite impossible to solve it in the same sort of exact way that was possible for the hydrogen atom with only one electron. The fundamental difficulty is, of course, that each electron repels every other electron with an inverse-square-law force, so that the motion of any one electron is dependent on the motion of all the others. It is this non-separability of the individual electronic motions which introduces the real complexity into the problem. At the same time it suggests a way out of the difficulty. Let us

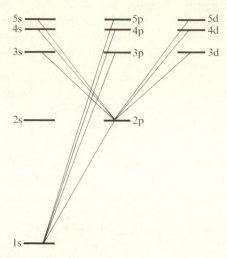

FIG. 2.10. Origin of spectral series for a hydrogen-like system. The lines between the levels indicate typical series of 'allowed' transitions.

suppose that when dealing with one of the electrons we disregard its *instantaneous* interactions with the other electrons and imagine it to move in an effective electric field which is obtained as a suitable *average* over all positions of all the other electrons. In such a case each electron would be described by a wave equation which involved only its own co-ordinates and not the co-ordinates of any other electron; the effect of all the other electrons would be taken into account in choosing the potential energy function for the first electron. Such a wave equation is vastly easier to deal with than the more strictly accurate one and can always be solved by numerical methods. Since the average potential due to the other electrons may be considered to have spherical symmetry (this is not strictly true, in general, but it is a useful approximation), the s, p, d, ... classification still remains and the wavefunctions resemble quite closely the functions already described for hydrogen.

Of course this does not solve our problem, for according to this scheme we can only write down the wave equation for any one electron if we already know the wavefunctions for all the others; and the number of possible wavefunctions for each electron is infinite! Progress becomes possible, however, when we remember that normally we are dealing with the *ground state* of the atom, in which the electrons are in the *lowest-energy* AO's subject to any other conditions which may apply. The most important such condition was known even before the days of quantum mechanics; not more than two electrons can be assigned to a state with given values of n, l, m, i.e. to a given AO. We shall see later that this is a form of *Pauli's exclusion principle* and that two electrons in the same orbital must then differ in a *fourth* quantum number referring to electron *spin*—they must have 'opposite spins'. For the present we may ignore spin since it has no appreciable effect on the orbital description of the electrons beyond limiting the number of electrons in a given state (orbital) to be 0, 1, or 2; in these three cases we say the orbital is 'unoccupied', 'singly occupied', or 'doubly occupied', respectively. In the ground state, therefore, we first assign the electrons to the 1s, 2s, 2p, ... AO's in ascending energy order, each being 'filled' when it holds two electrons, until all are assigned; a statement of the numbers of electrons in the AO's of various types is then called the *electron configuration* of the ground state.

We now approach the problem in stages, as first suggested by Hartree (1928) and Fock (1930). Suppose there are N electrons in the atom and all have been assigned to AO's to give the electron configuration. Then let us guess plausible wavefunctions for each of these electrons; this process is nothing like so difficult as it may at first appear, for after a little experience quite good wavefunctions may be guessed without much trouble. Now choose one of the electrons and find the average field provided by all the others. This average field is simply the field that would be provided if each of these electrons was a charge cloud as in §2.2. If necessary we average this field, using all orientations of the total charge cloud, to get the angle-independent field characteristic of a spherical charge distribution. This process allows us to write down, and then to solve, the wave equation for our chosen electron. We obtain what may be called a first-improved wavefunction for this electron. The same procedure may next be used to calculate the average field for a second electron, and enables us to obtain a first-improved wavefunction for this electron also. The process is continued until we have a complete bunch of first-improved orbitals. Then, starting with the new orbitals, we again improve them, one by one, and calculate second-improved AO's. This technique is continued until successive iteration makes no appreciable difference to the orbitals. We may then say that the resultant AO's are 'self-consistent'; this means that if we choose any one electron its charge cloud is precisely that which comes from solving the wave equation in which the potential field is due to the charge on the nucleus and the sum of the charge clouds of all the other electrons. This is true whichever electron we choose. For this reason it is called the method of the self-consistent field or, for brevity, the SCF method; the orbitals resulting from such a calculation are SCF AO's.

Naturally enough, there are certain errors in the above procedure. These arise chiefly from 'smoothing out' the distribution of all electrons except the one being considered. However, it may be shown that these errors are not serious and, if we are prepared to spend sufficient time, the wavefunctions obtained in this way may be used as a very good basis for more accurate calculations in which no such 'smoothing out' is invoked. It turns out that the charge density is usually altered by at most a few per cent in most regions. The simpler procedure is thus quite adequate for most purposes. The total electronic energy calculated using the self-consistent AO's is usually in error by only about 1 per cent—an extremely gratifying result.

In view of its later applicability to molecules, it may be desirable at this stage to summarize the principles which we use for describing the electronic structure of an atom. It will be recognized at once that these principles are natural extensions of those listed on p. 35 for atomic hydrogen. They are as follows.

(a) Each electron is represented by a wavefunction ψ, called an atomic orbital; this orbital is found by solution of the appropriate Schrödinger wave equation, obtained by the Hartree procedure or some similar technique, and is such that ψ^2 measures the density of the charge cloud for this electron.

(b) Each AO is designated in terms of a set of quantum numbers. First there is the principal quantum number n, which chiefly determines the energy and also the size of the orbital, so that electrons in orbitals with the same value of n are said to be in the same 'shell'. Then there is the geometrical shape, given by the quantum number l, or the label s, p, d, ..., and finally an appropriate suffix as in (p_x, p_y, p_z) to show which of the degenerate orbitals of this symmetry we are actually using.

(c) Each AO has a characteristic energy which is found from the wave equation. This energy measures very approximately the work required to remove this particular electron, i.e. to ionize the atom in a particular way. Each type of electron has its own ionization potential. The usual order of energies is (see Fig. 2.11)

$$1s < 2s < 2p < 3s < 3p < 3d \sim 4s \ldots$$

in which we now see that the 'accidental' degeneracies of AO's in the same shell but of different symmetry (Fig. 2.6) have been 'resolved' or 'broken' now that the potential field for each electron is no longer of simple inverse distance form. The total energy of an atom is now the sum of the energies of all the AO's which are occupied by electrons, corrected to include electron interactions. Without this correction, the interaction would, in fact, have been counted twice over, because the Coulomb repulsion $e^2/\kappa_0 r_{ij}$ between electrons i and j has been included in the wave equations—and hence in the energy values—of both electrons separately.

(d) When assigning electrons to the various allowed AO's we must take account of the Pauli exclusion principle. This states that in no electron configuration may an orbital contain more than two electrons, and then only if their spins are opposed; in this case we say the electrons have 'paired' spins. In the ground state the orbitals of *lowest* energy are filled,

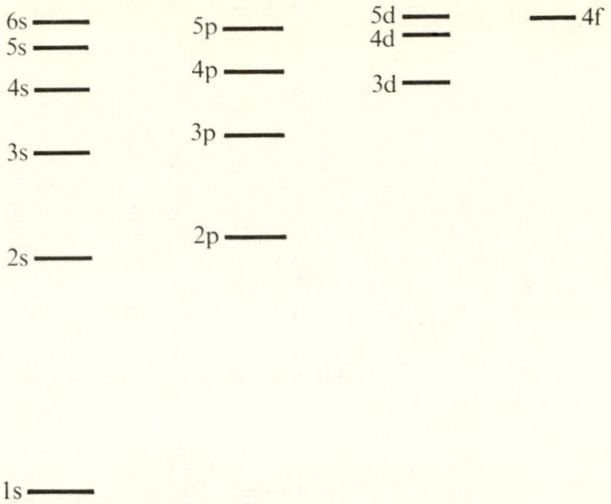

6s ——	5p ——	5d ⚌	—— 4f
5s ——		4d ⚌	
4s ——	4p ——	3d ——	
3s ——	3p ——		
2s ——	2p ——		
1s ———			

FIG. 2.11. Orbital-energy diagram for many-electron atoms. The levels are shifted, relative to those in Fig. 2.6, by electron interaction. The diagram is not to scale and is intended to show only the usual *order* of the levels.

in ascending energy order. In case of ambiguity (two assignments giving the same total orbital energy) further rules are required, as we shall see presently.

(*e*) An important class of excited states may be described by promoting an electron from a given AO to one of higher energy (i.e. by changing the electron configuration), all other AO's being left unchanged. If the AO's occupied before and after the 'jump' have energies E_1 and E_2, the corresponding spectroscopic absorption frequency is given approximately by $|E_1 - E_2| = h\nu$. Again, selection rules operate to allow certain transitions and forbid others.

The principles listed above embody what is usually known as the *aufbau* (or building-up) approach to the description of atomic structures. In this approach, which we now illustrate for a number of atoms, we start from a set of allowed orbitals (corresponding to the levels indicated in Fig. 2.11) and then feed the electrons one at a time into these orbitals, beginning with 1s which is the lowest, and satisfying the exclusion principle by allowing only up to two electrons in each one. The resultant electron configuration is indicated by listing the types of orbital occupied, with the number of electrons in each type as a superscript. Thus hydrogen is represented in its ground state by (1s), helium by $(1s)^2$, lithium by $(1s)^2(2s)$, nitrogen by $(1s)^2(2s)^2(2p)^3$, etc. The method is very simply portrayed in a diagram if we keep in mind Fig. 2.11 but replace each degenerate level by a *set* of levels, one for each orbital of the given energy. Alternatively, the set of levels is sometimes shown as a set of 'cells' or 'boxes' (as in earlier editions of this book). If an orbital is singly occupied, we

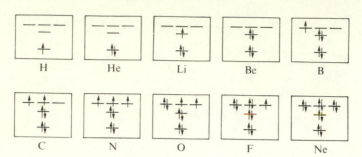

FIG. 2.12. A 'levels-and-arrows' diagram to show the ground-state electron configurations of the first 10 atoms in the periodic table.

put an arrow on the level; if it is doubly occupied, we put two arrows on the level, one pointing up and the other down to indicate the spins. Diagrams of this kind for the first 10 atoms of the periodic table are shown in Fig. 2.12. In drawing these diagrams we have had to use Hund's rules to decide, in the cases of nitrogen and oxygen and all other atoms containing incomplete groups of electrons, just which of the equivalent orbitals (here the $2p_x$, $2p_y$, $2p_z$ orbitals) are filled. These rules[†] are that for degenerate (i.e. energetically equivalent) orbitals (i) electrons tend to avoid being in the same orbital so far as is possible, and (ii) two electrons, each singly occupying a given pair of degenerate orbitals (e.g. $2p_x$, $2p_y$), tend to have their spins parallel in the state of lowest energy. The rules show us at once that in nitrogen the 1s and 2s orbitals are doubly filled and each of the $2p_x$, $2p_y$, $2p_z$ orbitals is singly occupied, the three electrons concerned having parallel spins, and that oxygen, with one more p electron, has one p orbital (say $2p_z$) doubly filled, the other two being singly occupied, with parallel spins, as shown in Fig. 2.12. The $(2p_z)^2$ group, which as we shall see later makes no contribution to the divalent character of the oxygen atom, is sometimes called a 'lone pair' of electrons. A similar name is used for the nitrogen $(2s)^2$, or for any other similar group of two mutually paired electrons in the outer electronic shell. In giving electron configurations, the inner (i.e. with principal quantum number less than that of the outermost shell) completed or 'closed' shells are sometimes omitted, and then the two configurations above would be written s^2p^3 and s^2p^4.

The *aufbau* approach gives an immediate interpretation of the periodic system of the elements. Thus the group of two electrons $(1s)^2$ completes what we might call the inner or K shell, a further group of eight electrons $(2s)^2(2p)^6$ completes the L shell, eight more complete the next subgroup of the M shell, and so on. The details should be familiar, and we do not need to repeat them.

[†] See for example Herzberg (1944), Chap. III. (We need only the rules which tell us how the AO's are occupied and not the more subtle considerations which relate to the interpretation of atomic spectra.)

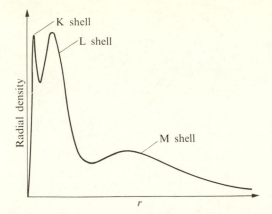

Fig. 2.13. Charge distrbution in Ar according to Hartree. The area under the curve is equal to the total number of electrons, here 18.

(See for example Puddephatt (1973)). We can see why the groups we have mentioned are referred to as shells if we use the calculated AO's to draw the electronic density functions. Fig. 2.13 shows the particular case of Ar (Hartree & Hartree 1938) and the way in which the complete charge distribution is built up by addition of the contributions from $(1s)^2$, $(2s)^2(2p)^6$, and $(3s)^2(3p)^6$. Quite evidently the K, L, and M shells are here almost distinct† and the outer part of the atom is dominated completely by the outer shell, here the M shell. It is clear that in so far as the size of an atom has any meaning, it is determined entirely by this outer shell. Indeed, calculations such as these are the basis of any theoretical estimates of the atomic radius of an atom or ion. This radius varies according to the number of electrons present and the nuclear charge. As the nuclear charge is increased each individual shell moves in towards the origin so that the rare-gas atoms, which properly come at the end of each row of the periodic table, have the smallest atomic radii and the alkali atoms which start each new row have the largest. This is well illustrated in Fig. 2.14 which shows the radial density function for the ground states of a series of ions plotted on the same scale. These ions all have a rare-gas structure (completed shells) and are relatively small. It will be noticed, however, that the heavier atoms tend to be larger than the lighter ones.

2.6. Approximate SCF orbitals.
Screening constants

Full SCF calculations are quite complicated and lengthy, though nowadays they are performed automatically by electronic computers and have therefore

† This distinction is not at all obvious if $P(r)$ is plotted instead of $4\pi r^2 P(r)$; the r^2 factor 'magnifies' the density fluctuations.

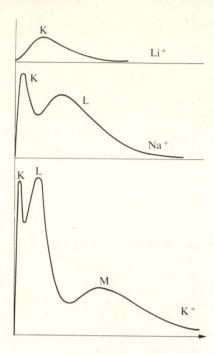

FIG. 2.14. Radial density curves for the ground states of Li$^+$, Na$^+$, and K$^+$. The curves are all drawn to the same scale.

become routine. The results obtained may be represented by tables of numbers showing the forms of the radial factors in the AO's, which differ somewhat from the hydrogen-like factors (polynomial × exponential) listed in Tables 2.2–2.4, or they may be 'fitted' by analytical expressions containing several different exponential terms. Such expressions have been listed very extensively by Clementi and associates (1963, 1964, 1967, 1974). The simplest analytical approximations to the SCF AO's, however, were first suggested by Slater (1930); they are obtained from the functions listed in Tables 2.2–2.4 by keeping only the *highest power* of r in each of the (radial) polynomial factors, replacing the actual nuclear charge Z (in units of e) by an 'effective' nuclear charge Z_e, and modifying the normalizing factors accordingly. The so-called Slater-type orbitals (or STO's) thus have radial factors of the simple form

$$f_n(r) = r^{n-1} \exp\left(-\frac{Z_e r}{n}\right) \tag{2.21}$$

for a shell with principal quantum number n. Distances are again measured in units of a_0. Note also that, for example, a $2p_z$ Slater orbital may be written $N_{2p}(z/r)r\,e^{-Zr/2}$ in which $z/r\,(=\cos\theta$ in Fig. 2.5) is a purely angular factor. The

TABLE 2.6

Normalizing factors for Slater-type orbitals (STO's), where $\zeta = Z_e/n$ is the 'orbital exponent'

1s	2s	2p	3s	3p	3d
$\dfrac{\zeta^{3/2}}{\sqrt{\pi}}$	$\dfrac{\zeta^{5/2}}{\sqrt{(3\pi)}}$	$\dfrac{\zeta^{5/2}}{\sqrt{\pi}}$	$\dfrac{\sqrt{2}\,\zeta^{7/2}}{3\sqrt{(5\pi)}}$	$\dfrac{\sqrt{2}\,\zeta^{7/2}}{\sqrt{(15\pi)}}$	$\dfrac{\sqrt{2}\,\zeta^{7/2}}{3\sqrt{\pi}}$

radial factor is then of the same form as in the Slater 2s orbital. The size of the AO may be varied by adjusting the parameter Z_e—increasing Z_e makes the AO shrink closer to the nucleus, while decreasing Z_e allows the AO to become more diffuse. The modified normalizing factors are given in Table 2.6.

It should be noted that the Slater orbitals possess no nodes. For $n > 1$ they rise up smoothly from the nucleus to a peak and then decay smoothly to zero; they are thus best fitted to describe the outermost part of an SCF AO, outside the region of the nodes, and that is the most important part of an AO in discussing bond formation.

When Z_e is related to the actual nuclear charge Z by writing

$$Z_e = Z - s \qquad (2.22)$$

we call s a 'screening constant'; it describes the self-consistent field in an approximate way as being that of a nucleus of charge Z units (an attraction) diminished by an amount s which roughly represents the repulsive effect of electrons in other orbitals closer to the nucleus than the one considered. The mutual repulsion of the electrons is thus recognized, somewhat less accurately than in the SCF method, as a 'screening' effect. The lithium 2s electron, for example, feels a central field not of the nucleus alone ($Z = 3$) but of a nucleus surrounded by two tightly bound 1s electrons; the effective charge might be expected to be not much greater than the net charge of a helium-like 'core', namely $3 - 2 = 1$; in fact the SCF 2s orbital is more accurately approximated when $Z_c = 1\cdot28 = 3 - 1\cdot72$. The screening effect of electrons in the *same* shell as the one considered is much less complete; in helium, for example, the central charge $Z = 2$ felt by each 1s electron is effectively diminished by $0\cdot31$ because of the presence of the other electron.

Slater gave the following rules for estimating screening constants.†
(i) Divide the AO's into the following subshells:

(1s)   (2s, 2p)   (3s, 3p)   (3d)   (4s, 4p)   (4d)   (4f)   etc

---

† The original rules have been improved upon by Clementi and others. A good discussion is available in Karplus and Porter (1970). Optimum values of $\zeta = Z_e/n$ have been obtained by Clementi and Raimondi (1963).

(ii) The screening constant s for the AO's in a given subshell is then a sum of the following contributions: (a) nothing from any shell outside the one considered; (b) an amount 0.35 from each other electron in the group considered (except in the 1s group where 0.30 is used instead); (c) if the shell considered is an s, p shell, an amount 0.85 from each electron in the next inner shell and 1.00 from all electrons still further in; but if the shell is a d or f, an amount 1.00 from every electron inside it.

As an example, the carbon 2s and 2p AO's would be given the common screening constant

$$s = 3 \times (0.35) + 2 \times (0.85) = 2.75$$

and the effective charge would then be $Z_e = 6 - 2.75 = 3.25$. The actual SCF AO's are fitted still better by taking $Z_c(2s) = 3.216$ but $Z_c(2p) = 3.136$. However, Slater's rules give a fairly accurate picture of the situation, at least for AO's with $n < 4$. They even hold fairly well for the electron configurations of atoms and ions excited states in which some of the subshells are incompletely filled.

An important application of screening constants is to the discussion of trends in AO energies and sizes as we move across or down the periodic table. The orbital energies may be estimated roughly by using (2.17) with Z replaced by the effective charge Z_e. The 'size' may be measured by the value of r which gives the peak in the radial density function for the outermost AO (i.e. the most probable distance of the electron from the nucleus). With the form (2.21) we then need the maximum of $r^{2n}\exp(-2Z_e r/n)$ which occurs at $r = n^2/Z_e$. Since we have been using atomic units, the AO energy and size are

$$E \simeq -\frac{1}{2}\frac{Z_e^2}{n^2}E_h, \qquad r \simeq \frac{n^2}{Z_e}a_0. \qquad (2.23)$$

As we move along the rows of the periodic table, sets of orbitals in the same shell are filling. Thus n is fixed but Z_e, in (2.22), is increasing steadily (Z in steps of 1.0, s in steps of 0.35); the ionization potentials therefore rise along a row and the atomic radii fall. At the end of a row, in the rare-gas configuration, Z_e has reached its maximum value for the shell, but at the beginning of the next row Z_e falls because, while Z increases by 1.0, all the other electrons now become 'inner-shell' electrons and give a much greater screening of the nucleus (by 0.85 each); the trends in the previous row are then repeated. As we move down a column, the elements have a similar outermost electron configuration (e.g. ns^2 for the alkaline earths), but Z_e increases comparatively slowly owing to the efficient screening by inner-shell electrons. Thus, for Be and Mg, with $(2s)^2$ and $(3s)^2$ configurations, the Z_e values are about 1.9 and 3.3, respectively, compared with Z values of 4 and 12. It is then the increasing value of the principal quantum number which dominates, leading to increasing atomic radii and decreasing ionization potentials. In this way observed trends throughout the periodic system can be rationalized quite well using only the basic ideas of the self-consistent field and its simplified representation in terms of screening constants.

In conclusion, we note that the energies of s and p electrons in the same shell are not distinguished at this level of approximation (note that they would be if different values of Z_e were allowed for s and p orbitals), but that the difference (cf. Fig. 2.11) is important in *aufbau* and other considerations. Some idea of orders of magnitude is therefore useful. If the electrons in question are in the valence, or outer, shell of the atom, then the difference $E_p - E_s$ between a p-type and s-type orbital with similar principal quantum number is usually in the region of 2–4 eV for elements in the first column of the periodic table and increases steadily as we move to later columns. For example, the familiar sodium D lines arise from a 3p → 3s transition, with energy 2·10 eV, but in chlorine the corresponding transition has an energy of about 10 eV. Similarly, the difference $E_d - E_p$ is usually a little larger than $E_p - E_s$, so that a d orbital for principal quantum number n has an energy in the same region as an s orbital for quantum number $n + 1$. In potassium, for example, the energies of 4s, 4p, and 3d are nearly equal, and in that order; $E_f - E_d$ is also large, with the result, shown in Fig. 2.11, that 4f, 5d, and 6s are comparable in energy. Useful compilations of orbital energies are available in the literature (see in particular Slater 1955).

Problems

2.1. The work function for metallic sodium is 2·3 eV. For what minimum frequency of radiation would incident photons eject electrons? You may use the wavenumber $(1/\lambda)$ in units of cm^{-1} as a measure of frequency $(v = c/\lambda)$. This unit, widely used in spectroscopy, is a 'permitted non-SI unit'. If the incident radiation is of wavelength 106·7 nm (Argon (I) source) what is the greatest kinetic energy with which an electron can escape? Why should some electrons escape with less energy?

2.2. If all absorbed photons in the experiment of Problem 2.1 were effective in ejecting electrons, how many per second would be ejected from unit area of surface by radiation of intensity 10 W m^{-2}? (Hint: what is the photon energy in joules?)

2.3. Turn back to Problem 1.4. The photoelectron spectrum considered (Fig. 1.5) was produced using Al K_α radiation of wavelength 1487 nm. Estimate the depth of the 1s level (in eV) below the ionization limit for each of the three types of nitrogen atom. At what distance from the nucleus would an extra unit charge (e.g. a nearby positive ion) produce a 'chemical shift' of the observed order of magnitude?

2.4. Is the electron in the hydrogen 1s orbital more likely to be found at a distance from the nucleus greater than 1 bohr or less than 1 bohr? The empirical 'covalent radius' of the hydrogen atom is ~0·03 nm; what is the probability of finding the electron within the corresponding sphere? (Hint: work in atomic units and use equation (2.14)).

2.5. The 'fine structure constant' $e^2/4\pi\varepsilon_0 hc$ is a pure number with the value 1/137·03. What is the velocity of light in atomic units? Hence (knowing $c = 2·998 \times 10^8$ m s^{-1})

infer the value of the atomic unit of time (τ_0, say) in seconds. How many bohrs will a photon travel in 1 sec?

2.6. The virial theorem in classical mechanics states that for particles moving under an inverse-square force law the average potential and kinetic energies are related by $\bar{V} = -2\bar{T}$. Hence, since $E = \bar{T} + \bar{V}$, $\bar{T} = -\frac{1}{2}E$. The same result holds in quantum mechanics.

   Estimate the root-mean-square velocity (using $T = \frac{1}{2}mv^2$) for a 1s electron in (i) the hydrogen atom, (ii) the argon atom, and (iii) the uranium atom. How do these values compare with the velocity of light? (Hint: use atomic units.) These results are important because as the electron velocity approaches that of light the Schrödinger equation must be corrected to take account of the mass variation predicted by relativity theory.

2.7. Propose electron configurations for the atoms Li, B, N, F and Na, Al, P, Cl. Use Slaters' rules to estimate the effective nuclear charges for the valence electrons and hence, using equations (2.23), substantiate the qualitative argument in the text (p. 44) concerning trends in ionization potentials and atomic radii along rows and down columns of the periodic table. How much smaller would these atoms be if the outer electrons were not screened from the full nuclear attraction by the repulsion due to inner shell electrons?

2.8. Use Slater's rules to estimate (i) the 2s orbital energy for the (ground state) lithium atom, and hence the first ionization potential, and (ii) the second ionization potential. Finally, calculate the third ionization potential (for removal of the 1s electron from Li^{2+}) and hence the total electronic energy of the lithium atom. (Experimental values are $I_1 = 5\cdot39$ eV, $I_2 = 75\cdot7$ eV, $I_3 = 122\cdot4$ eV, $E = -203\cdot5$ eV.) Why do you not get the same value of E by adding together the orbital energies ($E = 2E_{1s} + E_{2s}$) calculated for the neutral atom?

2.9. Show that the electron density in the nitrogen atom is spherically symmetrical (i.e. depends only on r), using the orbital forms given in Table 2.2.

   Will a similar result hold (i) for *any* p^n shell, or (ii) only for certain numbers of electrons?

   Can you extend your results to fully and partially occupied d shells using the AO's in Table 2.4?

3
Wave-mechanical principles

3.1. The wave equation

The last chapter was devoted chiefly to a review of the conclusions that follow from the application of wave mechanics to atomic structure. We must now show how these results follow from a wave equation, and thus prepare the ground for a subsequent discussion of molecules. Of course we cannot 'derive' the wave equation, any more than we can derive Newton's equations of motion; such equations, which embody fundamental 'laws of nature', are simply very compact and general expressions of experimental conclusions. We can therefore only give a 'plausibility argument' to show what kind of equation the wavefunction must satisfy if it is to conform to the experimental requirements indicated in §2.1.

Let us consider first a particle moving freely along the x axis; for any kind of wave propagation we can then write $\Psi = \Psi(x, t)$. For a discussion of many types of wave motion see Coulson (1941). The simplest form of wave is sinusoidal, the value of Ψ repeating itself when x increases by λ (the *wave length*) and when t increases by $1/v$ (the reciprocal frequency or *period* of the oscillation); Ψ then depends on x, through a factor $\exp(\pm 2\pi ikx)$ where $k = 1/\lambda$ is the 'wave number' (number of waves per unit length), and on t through a factor $\exp(\pm 2\pi ivt)$. Remember that $\exp(i\theta) = \cos\theta + i\sin\theta$, so $\exp(2\pi ikx)$ is a mixture of sine and cosine terms. In general any combination, or 'mixture', of the four possible products, with constant coefficients, is acceptable. If x is increased by λ, the exponent is increased by $\pm 2\pi i$ and Ψ is multiplied by $\exp(\pm 2\pi i)(=1)$, and similarly for t. If we take any one of the four products, e.g. $\Psi = \exp(2\pi ikx)\exp(2\pi ivt)$, the equation it satisfies can be found by differentiating. Thus, differentiating twice with respect to x and twice with respect to t,

$$\frac{\partial^2 \Psi}{\partial x^2} = -4\pi^2 k^2 \Psi \qquad \frac{\partial^2 \Psi}{\partial t^2} = -4\pi^2 v^2 \Psi$$

and hence Ψ has the property that

$$\frac{\partial^2 \Psi}{\partial x^2} = \frac{1}{u^2}\frac{\partial^2 \Psi}{\partial t^2} \tag{3.1}$$

where $u = \lambda v$ is the velocity of propagation ('phase' velocity). Equation (3.1) is

the classical 'wave equation', satisfied by sound waves, water waves, elec-
tromagnetic waves, etc. according to the interpretation of Ψ.

A little consideration shows that (3.1) cannot be the correct equation for the
Schrödinger wavefunction, because in general $\int \Psi^* \Psi \, dx$ will vary in time and a
normalized wavefunction therefore will not stay normalized. If *only one* choice
of time factor were allowed, *either* $\exp(2\pi ivt)$ *or* $\exp(-2\pi ivt)$ (but not a
mixture), the product $\Psi^* \Psi$ would clearly contain $\exp(2\pi ivt) \times \exp(-2\pi ivt)$
$= 1$ and the normalization would become time-independent, but the differential equation would then have to be *first* order in t (one solution) instead of
second order (two solutions). Let us suppose, therefore, that only *one* time
factor is allowed, $\exp(-2\pi ivt)$, and look for the differential equation satisfied
by

$$\Psi(x, t) = \{A \exp(2\pi ikx) + B \exp(-2\pi ikx)\} \exp(-2\pi ivt)$$

where the factor in braces depends only on position and is the *amplitude* of the
wave. Let us further incorporate the experimentally based assumptions (see
§2.1) that, for a particle moving freely (i.e. in a region of constant potential),

$$k = 1/\lambda = p/h \qquad v = E/h \tag{3.2}$$

and that E and p are related classically;

$$E = \tfrac{1}{2}mv^2 + V = p^2/2m + V. \tag{3.3}$$

To eliminate Ψ we must go to the second derivative in x, but now only to the
first in t:

$$\frac{\partial^2 \Psi}{\partial x^2} = -\frac{4\pi^2}{\lambda^2} \Psi = -\frac{p^2}{\hbar^2} \Psi \qquad \frac{\partial \Psi}{\partial t} = -2\pi iv\Psi = -\frac{Ei}{\hbar} \Psi \tag{3.4}$$

where we use (3.2) and have again introduced 'crossed h' ($= h/2\pi$). We can now
write down, using (3.3), the mathematical identity

$$\left(\frac{1}{2m}p^2 + V\right)\Psi = E\Psi$$

and, on substituting from (3.4), the differential equation

$$\boxed{-\frac{\hbar^2}{2m}\frac{\partial^2 \Psi}{\partial x^2} + V\Psi = i\hbar\frac{\partial \Psi}{\partial t}} \tag{3.5}$$

This is the one-dimensional form of Schrödinger's famous wave equation; it
has been designed to give the experimentally inferred properties for a free
particle (moving in a region where V is constant) but it turns out to be more
generally satisfactory by giving complete agreement with experiment even
when $V = V(x)$ and is *not* constant.

It will be noticed that equation (3.5) can be written down by a very simple recipe starting from the classical expression for the energy as a function of position and momentum (known as the Hamiltonian function and denoted by the symbol H). We take

$$E = H(x, p) = \frac{1}{2m}p^2 + V(x) \tag{3.6}$$

and wherever p appears we replace it by the 'operator' $(\hbar/i)(\partial/\partial x)$ (i.e. differentiate with respect to x and multiply by $\hbar/i$); similarly we replace E by the operator $-(\hbar/i)(\partial/\partial t)$. Then we put a factor Ψ on the right of each term in (3.6) so that the operators have a function to work on. The result is (3.5). This rule for associating operators with dynamical quantities is fundamental to Schrödinger's formulation of quantum mechanics.

For a particle moving in three dimensions, p^2 in (3.6) is a sum of squares of the three momentum components $p_x = mv_x$, etc., while V depends on $x, y,$ and z. The corresponding association is

$$p_x \rightarrow \frac{\hbar}{i}\frac{\partial}{\partial x}, \quad p_y \rightarrow \frac{\hbar}{i}\frac{\partial}{\partial y}, \quad p_z \rightarrow \frac{\hbar}{i}\frac{\partial}{\partial z}, \quad E \rightarrow -\frac{\hbar}{i}\frac{\partial}{\partial t}. \tag{3.7}$$

The three-dimensional form of the wave equation is then

$$-\frac{\hbar^2}{2m}\nabla^2\Psi + V(x, y, z)\Psi = -\frac{\hbar}{i}\frac{\partial\Psi}{\partial t} \tag{3.8}$$

where $\nabla^2\Psi$ is simply a convenient shorthand notation,

$$\nabla^2\Psi = \frac{\partial^2\Psi}{\partial x^2} + \frac{\partial^2\Psi}{\partial y^2} + \frac{\partial^2\Psi}{\partial z^2} \tag{3.9}$$

and ∇^2 is called the Laplacian operator. The operator notation is used very widely and (3.8) is commonly written

$$\hat{H}\Psi = -\frac{\hbar}{i}\frac{\partial\Psi}{\partial t} \tag{3.10}$$

in which

$$\hat{H} = -\frac{\hbar^2}{2m}\nabla^2 + V(x, y, z) \tag{3.11}$$

is called the Hamiltonian operator (being derived from the classical Hamiltonian function (3.6)). In (3.10) the operator is simply provided with a function to work on by writing Ψ on the right of every term—yielding, in explicit form, the original equation (3.8). The addition of a circumflex (^) is commonly used to indicate an operator quantity; it is the convention adopted throughout this book.

3.2. Stationary states

In §2.1 we remarked that in valence theory the most important solutions of the Schrödinger equation would be those which referred to *stationary states*, in which the energy and charge density were independent of time, and in §2.4 we dealt entirely with time-independent wavefunctions—more correctly 'amplitude functions'. Such states have wavefunctions of the special form

$$\Psi(x, y, z; t) = \psi(x, y, z)f(t) \tag{3.12}$$

and the amplitude function then satisfies a simplified form of (3.10) in which the time dependence is eliminated. If we insert (3.12) in (3.10) and require that the equation be satisfied we find, on dividing throughout by Ψ,

$$\frac{\hat{H}\psi}{\psi} = -\frac{\hbar}{i}\frac{1}{f}\frac{\partial f}{\partial t}.$$

Here we have used the fact that $\hat{H}\Psi = f\hat{H}\psi$ since the factor $f(t)$ is treated as a constant in performing the partial differentiations implied in $\hat{H}$; similarly $\partial\Psi/\partial t = \psi(\partial f/\partial t)$. Now the two sides of this equation are entirely independent, one being a function of co-ordinates only and the other of time only; they can only be equal (for all x, y, z and all t) if each is separately equal to the same constant, which we denote by E. The solution of

$$-\frac{\hbar}{i}\frac{1}{f}\frac{df}{dt} = E$$

is simply

$$f(t) = \exp\left(-\frac{iEt}{\hbar}\right) \tag{3.13}$$

while the equation for ψ, namely

$$\boxed{\hat{H}\psi = -\frac{\hbar^2}{2m}\nabla^2\psi + V\psi = E\psi} \tag{3.14}$$

is the *time-independent Schrödinger equation*. Thus, the stationary states are found by solving (3.14) to obtain the amplitude function $\psi(x, y, z)$ and then attaching the time factor (3.13). Now the presence of this latter factor means that differentiating the wavefunction with respect to t is equivalent to multiplying it by $(-iE/\hbar)$, but this is the property (3.4) of the de Broglie waves—in which E is interpreted as the *particle energy*. We conclude that the quantity E in the time-independent equation (3.14) is the *energy* of the system in the stationary state with (time-independent) wavefunction ψ.

Clearly, the wavefunction (3.12), which now becomes

$$\Psi(x, y, z; t) = \psi(x, y, z) \exp\left(-\frac{iEt}{h}\right)$$

yields a charge density function which is independent of time, for

$$P = \Psi\Psi^* = \psi \exp\left(-\frac{iEt}{h}\right) \psi^* \exp\left(+\frac{iEt}{h}\right) = \psi\psi^*$$

which depends only on the spatial co-ordinates x, y, z. All the physical properties of the system, depending on the product $\Psi\Psi^*$, then become stationary and persist indefinitely without change. To get the system from one state to another (i.e. to cause a transition) we must apply some kind of time-dependent disturbance, allowing $\hat{H}$ to change accordingly, and then study the time evolution of ψ by solving the more general equation (3.10). However, in the theory of valence we are nearly always concerned with stationary states, and when we refer to 'the wave equation' or 'the Schrödinger equation' we shall normally mean (3.14) or its many-electron analogue.

3.3. Illustrative examples. The Hartree field

It is obviously important to be able to write down the wave equation quickly for all sorts of problems. The following examples will show how easy this is.

Hydrogen atom. The electron moves in three dimensions so that its kinetic energy is $\frac{1}{2}m(v_x^2 + v_y^2 + v_z^2)$. Its potential energy is $-e^2/\kappa_0 r$, for the nuclear charge is $+e$, the electronic charge is $-e$, and r is the distance between the two. The wave equation is

$$\hat{H}\psi = -\frac{h^2}{2m}\nabla^2\psi - \frac{e^2}{\kappa_0 r}\psi = E\psi. \tag{3.15}$$

If the nuclear charge had been Ze, this would have been

$$\hat{H}\psi = -\frac{h^2}{2m}\nabla^2\psi - \frac{Ze^2}{\kappa_0 r}\psi = E\psi. \tag{3.16}$$

Helium atom. This represents our first application of quantum mechanics to a system containing more than one electron. We simply use the same recipe (3.7), admitting similar terms for all particles. There are now two electrons, 1 and 2 (Fig. 3.1), with co-ordinates (x_1, y_1, z_1) and (x_2, y_2, z_2), which move around a fixed nuclear charge $2e$ at the origin. Using the notation of Fig. 3.1 it follows that

$$T = \frac{1}{2m}(p_{x1}^2 + p_{y1}^2 + p_{z1}^2 + p_{x2}^2 + p_{y2}^2 + p_{z2}^2)$$

$$V = -\frac{1}{\kappa_0}\left(\frac{2e^2}{r_1} + \frac{2e^2}{r_2} - \frac{e^2}{r_{12}}\right).$$

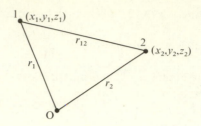

FIG. 3.1. Co-ordinates for the helium atom. The electrons are at points 1 and 2 and the nucleus is at the origin O.

The wave equation is now

$$\hat{H}\psi = -\frac{\hbar^2}{2m}\nabla_1^2\psi - \frac{2e^2}{\kappa_0 r_1}\psi - \frac{\hbar^2}{2m}\nabla_2^2\psi - \frac{2e^2}{\kappa_0 r_2}\psi + \frac{e^2}{\kappa_0 r_{12}}\psi = E\psi \quad (3.17)$$

where ∇_1^2 denotes the Laplacian operator (3.9) in terms of the co-ordinates x_1, y_1, z_1, and ∇_2^2 denotes ∇^2 in terms of x_2, y_2, z_2. This equation is a partial differential equation in all six variables $x_1, y_1, z_1, x_2, y_2, z_2$. There are only six independent variables because the other quantities r_1, r_2, r_{12} may all be expressed in terms of the original six.

Hydrogen molecule ion H_2^+. Let us next write down the wave equation for the simplest of all molecules, the hydrogen molecule ion, where there is only one electron moving in the presence of two fixed attracting centres A and B (Fig. 3.2). We may then write

$$T = \frac{1}{2m}(p_x^2 + p_y^2 + p_z^2) \quad V = -\frac{e^2}{\kappa_0 r_a} - \frac{e^2}{\kappa_0 r_b}.$$

Therefore the wave equation is

$$\hat{H}\psi = -\frac{\hbar^2}{2m}\nabla^2\psi - \frac{e^2}{\kappa_0}\left(\frac{1}{r_a} + \frac{1}{r_b}\right)\psi = E\psi. \quad (3.18)$$

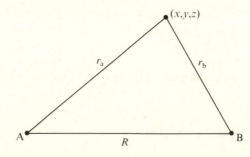

FIG. 3.2. Co-ordinates for the hydrogen molecule ion H_2^+.

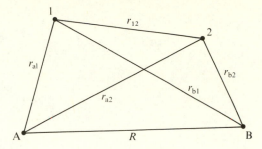

FIG. 3.3. Co-ordinates for the hydrogen molecule H_2.

The validity of supposing that we may take the nuclei as fixed, dealing with the electrons alone, is a question to which we shall return in §3.5.

Hydrogen molecule H_2. There are now two electrons, 1 and 2, with co-ordinates $(x_1 y_1 z_1)$ and $(x_2 y_2 z_2)$, so that, with the notation of Fig. 3.3,

$$T = \frac{1}{2m}(p_{x_1}^2 + \ldots) + \frac{1}{2m}(p_{x_2}^2 + \ldots)$$

$$V = -\frac{e^2}{\kappa_0}\left(\frac{1}{r_{a1}} + \frac{1}{r_{a2}} + \frac{1}{r_{b1}} + \frac{1}{r_{b2}}\right) + \frac{e^2}{\kappa_0 r_{12}}.$$

The wave equation is, as usual,

$$\hat{H}\psi = -\frac{\hbar^2}{2m}(\nabla_1^2 + \nabla_2^2)\psi + V\psi = E\psi. \tag{3.19}$$

After we have written down the wave equation, our subsequent task is always the same, *viz.* to find those particular values of the energy E for which acceptable wavefunctions exist. As we stated in Chapter 2, an acceptable wavefunction is one which is finite, single valued, and continuous, whose gradient is continuous, and which can be normalized.

The Hartree field. In Hartree's treatment of a many-electron atom (§2.5) each electron is imagined to move in an average field provided by the rest of the system, i.e. the nucleus and all the remaining electrons. If we consider an electron in orbital ψ_1, its potential energy function may be denoted by V_1 and the wave equation to determine ψ_1 will thus be

$$\hat{H}\psi_1 = -\frac{\hbar^2}{2m}\nabla^2\psi_1 + V_1\psi_1 = \varepsilon_1\psi_1 \tag{3.20}$$

where we have denoted the *orbital energy* by ε_1, since it is not the electronic energy of the atom (E in the many-electron Schrödinger equation) but rather

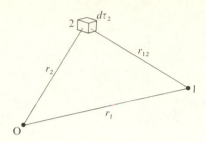

FIG. 3.4. Calculation of potential energy of an electron at point 1 in the field of the charge cloud of a second electron. Volume element $d\tau_2$ contains charge $-e\psi_2^2 d\tau_2$.

the energy of a single electron in a hypothetical 'effective field'—the Hartree field. To calculate V_1 we need to know the field produced by the remaining $N-1$ electrons in orbitals $\psi_2, \psi_3, \ldots, \psi_N$. Now the charge cloud associated with electron 2 in ψ_2 has a density ψ_2^2 electrons/unit volume and the charge in a volume element $d\tau_2$ is thus $-e\psi_2^2 d\tau_2$. The potential energy of electron 1 in the presence of that charge is $e^2\psi_2^2 d\tau_2/\kappa_0 r_{12}$ (Fig. 3.4), and in the presence of the whole charge cloud it is thus

$$\frac{e^2}{\kappa_0} \int \frac{\psi_2^2 \, d\tau_2}{r_{12}}.$$

Contributions such as these are summed, the nuclear term $-Ze^2/\kappa_0 r_1$ is added, and the whole is averaged over all angles to provide the spherically symmetrical potential energy function V_1 which appears in (3.20). In practice this equation has to be solved numerically, but when this has been done we have what was called on p. 37 a first-improved ψ_1. Proceeding in this way for each electron, the whole process described in §2.5 is carried through until further iterations yield no appreciable changes in any of the orbitals. At this stage we have obtained the self-consistent field and the corresponding total electron distribution $P = \psi_1^2 + \psi_2^2 + \ldots \psi_N^2$. Note that if, say, $\psi_2 = \psi_1$ (i.e. ψ_1 is *doubly* occupied), P will contain two identical terms and hence a contribution $2\psi_1^2$ from the *pair* of electrons in ψ_1; if an orbital is doubly occupied we simply double its contribution to the electron density.

3.4. Some exactly solvable problems

Two or three examples will help to familiarize us with the way in which acceptable solutions of the wave equation can be obtained. First, however, taking a one-dimensional case, we may enquire as to the general form of ψ and the origin of 'quantization' of the acceptable energies.

Let us consider a particle moving with the potential energy function shown in Fig. 3.5. The energy diagram, interpreted classically, tells us that there are three important regions: the particle can never be found in (a) or (c)—they are

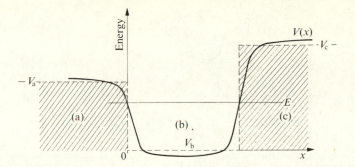

Fɪɢ. 3.5. Classical energy diagram for motion with energy E and potential energy function $V(x)$. Outside the vertical broken lines are the classically 'forbidden' regions (a) and (c). The broken lines indicate a 'potential box', suitable as a rough approximation to $V(x)$.

'classically forbidden' regions (with $V > E$)—and the motion is confined to region (b). The particle therefore moves back and forth between the walls of a 'container' or 'potential box'.

In quantum mechanics, however, this picture is modified; we must calculate ψ and then use ψ^2 to see where the particle is most likely to be. If we consider the potential energy function $V(x)$ to have, in (a), a roughly constant value V_a we can say ψ satisfies the equation (rearranging (3.14))

$$\frac{d^2\psi}{dx^2} = \omega_a^2 \psi \qquad (3.21a)$$

where $\omega_a^2 = 2m(V_a - E)/\hbar^2$ is positive and roughly constant. On making a similar approximation, $V \simeq V_b(\text{constant})$ in region (b), we obtain

$$\frac{d^2\psi}{dx^2} = -\omega_b^2 \psi \qquad (3.21b)$$

where $\omega_b^2 = 2m(E - V_b)/\hbar^2$ is again positive, because here $E > V_b$. Finally in region (c) a similar approximation gives

$$\frac{d^2\psi}{dx^2} = \omega_c^2 \psi. \qquad (3.21c)$$

These differential equations occur very commonly in elementary physics and (as may be verified by differentiating) have solutions

region (a) $\qquad\qquad \psi = A\exp(-\omega_a x) + B\exp(\omega_a x) \qquad (3.22a)$

region (b) $\qquad\qquad \psi = C\sin\omega_b x + D\cos\omega_b x \qquad (3.22b)$

region (c) $\qquad\qquad \psi = E\exp(-\omega_c x) + F\exp(\omega_c x). \qquad (3.22c)$

The wavefunction thus 'wiggles' in a classically allowed region, more rapidly

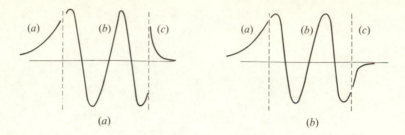

FIG. 3.6. (*a*) Form of the wavefunction in the three regions—exponentially decaying tails in the forbidden regions, wavelike variation in the allowed region. Numerical constants not correctly chosen: ψ discontinuous. (*b*) Numerical constants correctly chosen, giving an acceptable wavefunction.

the greater the energy, but falls or rises exponentially in a forbidden region, more rapidly the greater the value of $V - E$. The acceptability of the wavefunction depends on its being well behaved; it must not become infinite, so in regions (a) and (c) the constants multiplying the *rising* exponential (as $x \to -\infty$ or $+\infty$, respectively) must be put equal to zero. This leaves ψ with *exponentially decaying tails* as we penetrate into a classically forbidden region. Therefore ψ in the three regions must behave in the general way shown in Fig. 3.6 (*a*). To get the complete wavefunction we must join the bits together *smoothly*. This can only be done by choosing the various constants (including E, which appears in each of the ω's) very precisely. In Fig. 3.6 (*a*) the wavelength in the box is about right for a smooth join, which can be achieved by giving the decaying exponential in region (c) a negative coefficient. The pieces then join up smoothly as in Fig. 3.6 (*b*), but clearly this would not be possible if, for example, the wavelength were too long to get at least half a wave into the box. *Quantization of the energy thus arises simply from the condition that ψ be well behaved.*

Of course, the above discussion is qualitative, and if we took account of the variation of V_b, for instance, the wavelength of the oscillation would vary somewhat over region (b), but its main features are quite general. In the first of our solvable examples we consider a box for which all the quantized states can be obtained exactly.

Infinitely deep potential box. We consider the 'rectangular box' in Fig. 3.7 in which the potential energy is infinite in regions (a) and (c) and zero in (b). The exponential tails (with $\omega_a, \omega_b \to \infty$) now fall to zero infinitely rapidly, which means we can set $\psi = 0$ at the boundaries of region (b), which we take to be $x = 0$ and $x = L$. Only the sine function in (3.22*b*) can satisfy these conditions, so we must put $D = 0$, getting

$$\psi = C \sin \omega x \qquad (\omega^2 = 2mE/\hbar^2).$$

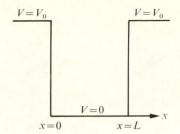

FIG. 3.7. Deep potential box; potential energy taken as zero within the box ($V = 0$), V_0 indefinitely large outside (i.e. $V_0 \to \infty$).

We must also ensure that an integral number of half-waves will fit exactly into the box (Fig. 3.8). This means $L = n \times \frac{1}{2}(2\pi/\omega)$ and therefore fixes the acceptable values of ω ($= \sqrt{(2mE)}/\hbar$) and hence the energy. The acceptable solutions are thus (remember $\hbar = h/2\pi$)

$$\psi_n(x) = C \sin\left(\frac{n\pi x}{L}\right) \qquad E_n = \frac{h^2 n^2}{8mL^2} \tag{3.23}$$

where $n = 1, 2, 3, \ldots$ is a quantum number. It is easy to normalize the wavefunctions (see p. 20) by giving C the value $\sqrt{(2/L)}$.

This system, although hypothetical, is a useful model. It may be used to describe crudely the motion of the 'mobile' electrons in a long polyene chain (length L) and also as a basis for the 'free-electron model' of a metallic conductor.

Generalization to three dimensions is straightforward; it is easily verified that, for a box with sides of length L_1, L_2, L_3, within which $V = 0$, the wave equation is satisfied by a product of three appropriate one-dimensional solutions

$$\psi_{n_1 n_2 n_3}(x, y, z) = C \sin\left(\frac{n_1 \pi x}{L_1}\right) \sin\left(\frac{n_2 \pi y}{L_2}\right) \sin\left(\frac{n_3 \pi z}{L_3}\right) \tag{3.24a}$$

and that the corresponding energy is

$$E_{n_1 n_2 n_3} = \frac{h^2}{8m}\left(\frac{n_1^2}{L_1^2} + \frac{n_2^2}{L_2^2} + \frac{n_3^2}{L_3^2}\right). \tag{3.24b}$$

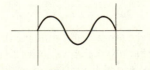

FIG. 3.8. Condition $\psi = 0$ at the boundaries. An integral number of half-waves must fit the box.

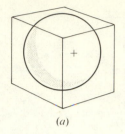
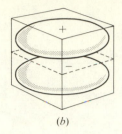

(a)   (b)

FIG. 3.9. Wavefunctions for a three-dimensional box (cube): (a) quantum numbers $n_1 = n_2 = n_3$ $= 1$; (b) $n_1 = n_2 = 1, n_3 = 2$ (z axis vertical). There is a nodal plane indicated by the broken lines.

Every state is thus determined by specifying three quantum numbers. For $L_1 = L_2 = L_3$ the wavefunctions so determined bear some similarity to the AO's discussed in Chapter 2, but refer to a particle bound within a small *cube* rather than a spherical region. Wavefunctions for $n_1 = n_2 = n_3 = 1$, and for $n_1 = n_2 = 1, n_3 = 2$, are shown in Fig. 3.9; they clearly resemble s-type and p$_z$-type AO's.

The hydrogen-like atom. According to (3.16) the wave equation is

$$\nabla^2 \psi + \frac{2m}{\hbar^2}\left(E + \frac{Ze^2}{\kappa_0 r}\right)\psi = 0. \tag{3.25}$$

If we use the fact that with $r = (x^2 + y^2 + z^2)^{1/2}$

$$\frac{\partial r}{\partial x} = \frac{x}{r}$$

it may be verified that this equation is satisfied (see Problem 3.5) by

$$\psi = e^{-kr} \tag{3.26}$$

provided that

$$k = \frac{mZe^2}{\kappa_0 \hbar^2} = \frac{Z}{a_0} \tag{3.27}$$

where a_0 is the Bohr radius (p. 23). The energy is then given by

$$E = -\frac{\hbar^2 k^2}{2m} = -\frac{Z^2 m e^4}{2\kappa_0^2 \hbar^2} = -\frac{Z^2 e^2}{2a_0 \kappa_0}. \tag{3.28}$$

This is the ground state of the atom, and in the particular case of hydrogen, for which $Z = 1$, these are precisely the equations (2.12) and (2.11b) which were given in Chapter 2 without proof. Equation (3.26) gives us, when normalized, the 1s AO of an atom with nuclear charge Ze.

In the same way it may be verified that there is a solution

$$\psi = x \exp(-\tfrac{1}{2}kr)$$

where k is still given by (3.27). By symmetry there are also $y \exp(-\tfrac{1}{2}kr)$ and $z \exp(-\tfrac{1}{2}kr)$. These are the $2p_x, 2p_y, 2p_z$ AO's defined in Table 2.2.

3.5. Need for an effective approximate method of solving the wave equation

Only for certain rather special forms of the potential function V is the wave equation completely solvable in closed terms. For example, it is insolvable for all atoms and ions except those with a single electron, and even more so for molecules. A simple example will show the complexity of the problem. In methane (CH_4) there are five nuclei and 10 electrons, so that the complete wave equation involves a total of $3 \times 15 = 45$ independent variables. A partial differential equation in as many variables as this is quite outside the range of exact solutions, even using present-day computers.

Fortunately we may simplify the problem without trouble, as indeed we have done in earlier sections, by assuming the nuclei fixed and considering only the electronic motion. This is a first step and is valid because, on account of their greater mass (the mass of a proton is about 1836 times as great as the mass of an electron), the nuclei move much more slowly than the electrons. In fact, in picturesque terms, the electrons move so fast that they see the nuclei standing still. The argument was first put in wave-mechanical terms by Born and Oppenheimer (1927), and the conclusion was verified. (See also Born and Huang (1954) and Longuet–Higgins (1961).) It means that the vibrational and rotational motions of a molecule are effectively quite separate from the electronic motions. Corresponding to any given positions of the nuclei, there is a definite energy (or set of energies) for the electronic motion. Further, we can obtain these energies by treating the nuclei as fixed and solving the appropriate wave equation. This is our justification for writing the wave equation for H_2^+ and H_2 in the forms (3.18) and (3.19).

This approximation reduces the complexity of the complete molecular wave equation by removing from it terms involving the motion of the nuclei. It is known as the fixed-nucleus approximation, and is astonishingly accurate. Thus, the error for H_2 has been shown (see Kolos 1964) to be $4 \cdot 95 \, \text{cm}^{-1}$ ($0 \cdot 00062 \, \text{eV}$); this may be compared with a total electronic energy at the equilibrium configuration of about 32 eV and a dissociation energy of $4 \cdot 746$ eV. Quite apart from this accuracy, however, the significance of the approximation itself is of the utmost importance, for the validity of this separation of electronic and nuclear motions provides the only real justification for the idea of a potential curve, or potential energy surface, of a molecule. Thus, using E henceforth to denote the electronic energy supplemented by the nuclear repulsion energy (p. 3), E may be regarded as a function of the relative positions of all the nuclei. In the particular case of a diatomic molecule, E is a function simply of the internuclear distance R and, at least in principle, the wave equation enables us to determine the function $E(R)$. The graph of $E(R)$ against R is the potential energy curve of the molecule, such as was discussed in §1.4. Since $E(R)$ actually contains electronic *kinetic* energy it is a 'potential energy' only in the formal sense that it may be so regarded *in discussing motion of the nuclei*.

This fixed-nucleus approximation certainly simplifies our wave equation, but it is still far too complex. In methane there still remain 30 variables, and even in H_2 there are six. Quite clearly, if any progress is to be made, we must have some technique for finding approximate solutions of the wave equation and corresponding approximate energies. If by some means we knew the true ψ, we could easily determine the true energy. In fact we know neither, so that we are faced with two interlocking problems.

(i) How can we get the most suitable approximate ψ?

(ii) How can we use this approximate ψ to calculate an approximate E?

It is for the solution of these two problems that the variation method has been found to be the most effective technique.

3.6. The variation method

We first return to the original formulation (3.14) of the wave equation

$$\hat{H}\psi = E\psi. \tag{3.29}$$

If we multiply both sides of this equation by ψ^* and integrate over all the co-ordinates involved, represented† simply by $d\tau$, we have the result‡

$$E = \frac{\int \psi^* \hat{H} \psi \, d\tau}{\int \psi^* \psi \, d\tau}. \tag{3.30}$$

If we have normalized ψ the denominator will of course be unity, but (3.30) is valid generally provided ψ satisfies (3.29). This extremely important formula enables us to calculate E *if we know* ψ.

If we do not possess an *exact* ψ, (3.30) will not give us an exact energy. However, as Rayleigh first showed, any 'trial function' ψ can be used to give us an *approximation* to the energy; we denote this by

$$\mathscr{E}(\psi) = \frac{\int \psi^* \hat{H} \psi \, d\tau}{\int \psi^* \psi \, d\tau} \tag{3.31}$$

the notation meaning that $\mathscr{E}$ depends on the form of the *function* ψ—which

---

† We shall adhere to this convention throughout the book. Unless the contrary is stated, $\int \ldots d\tau$ will imply an integration over all the co-ordinates represented in the integrand.

‡ We must be careful to write $\psi^* \hat{H} \psi$ and not $\hat{H}\psi\psi^*$ because $\hat{H}$ is an operator, and (3.29) shows that it is only to operate on ψ.

may have been suggested by physical or chemical intuition. The 'Rayleigh ratio' $\mathscr{E}(\psi)$ is called by mathematicians a 'functional'. The important property of $\mathscr{E}(\psi)$ is that it is always greater than or equal to the exact energy E_1 of the ground state (i.e. the lowest energy):

$$\mathscr{E}(\psi) \geqslant E_1. \tag{3.32}$$

This result is usually called the *variation theorem*. It means simply that however we may vary our approximation to the ground-state wavefunction the *best* choice is the one which gives the lowest value of $\mathscr{E}$.

It is rather ineffective and unnecessarily tedious to guess isolated ψ's, so as far as possible we deal with a whole family of ψ's at the same time. This may be done by choosing a trial function with one or more variable parameters. If we call these $c_1, c_2, \ldots$ then ψ is a function of $c_1, c_2, \ldots$. We now vary the values of $c_1, c_2, \ldots$ until $\mathscr{E}$ is minimized. These values give us the best approximate wavefunction consistent with the type of function we have adopted, and the whole procedure is called the *variation method*. Naturally, the more arbitrary parameters c we are able to introduce, the more flexible is our trial function, and the better our final ψ. If we have enough parameters, we can approach as near as we like to the true ψ, though except in very rare instances we never quite get there. In practice the introduction of additional parameters adds considerably to the labour, so that some compromise is necessary between accuracy and convenience.

The variation method can also be applied to excited states, but the technique is a little more involved except in special cases. We shall be dealing mainly with ground states, which are of the greatest interest chemically.

3.7. Variation method for the ground state of the hydrogen atom

The variation method is so important a tool that it is worth illustrating at once, even though the example that we choose happens to be one where the wave equation can be solved exactly and for which therefore no approximate methods are necessary. We choose the hydrogen atom, for which (see (3.15)) the Hamiltonian is

$$\hat{H} = -\frac{\hbar^2}{2m}\nabla^2 - \frac{e^2}{\kappa_0 r}. \tag{3.33}$$

We are to look for acceptable wavefunctions. In this case it is quite obvious that $\psi \to 0$ at large distances. A plausible hypothesis would be that it has spherical symmetry and that ψ dies away exponentially with r. Let us, in fact, try a hypothetical wavefunction

$$\psi = \mathrm{e}^{-cr} \tag{3.34}$$

with one arbitrary parameter c which, to be physically acceptable, must be

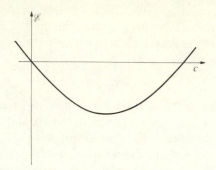

FIG. 3.10. Variation of energy function $\mathscr{E}(c)$ with the parameter c.

positive. It is soon verified (cf. Problem 3.5) that

$$\nabla^2\psi = \left(c^2 - \frac{2c}{r}\right)e^{-cr}.$$

The Rayleigh ratio (3.31) then becomes

$$\mathscr{E}(c) = \frac{\hbar^2}{2m}c^2 - \frac{e^2c}{\kappa_0}.$$

The graph of $\mathscr{E}(c)$ against c is shown in Fig. 3.10. The minimum is easily obtained by requiring $\partial\mathscr{E}/\partial c = 0$, and this gives

$$c = \frac{me^2}{\kappa_0\hbar^2} \qquad \mathscr{E}_{\min} = -\frac{me^4}{2\kappa_0^2\hbar^2}. \tag{3.35}$$

It will be recognized from (2.11a) that $c = 1/a_0$ and $\mathscr{E}_{\min} = -e^2/2a_0\kappa_0$. In this particular case we have managed to obtain the true ground-state wavefunction and energy, as given in (3.26) and (3.28), though this could not be recognized from the analysis that we have made. It arose from the particularly fortunate choice of an exponential function in (3.34).

We might not have chosen so fortunately. For example, we might have tried a Gaussian type of function,

$$\psi = \exp(-cr^2)$$

finding

$$\mathscr{E}(c) = \frac{3\hbar^2}{2m}c - \frac{2e^2\sqrt{2}}{\kappa_0\sqrt{\pi}}c^{1/2}.$$

This has a minimum value when

$$c = 8/9\pi a_0^2$$

and the corresponding result

$$\mathscr{E}_{\min} = -\frac{4}{3\pi}\frac{e^2}{\kappa_0 a_0} = -\frac{0\cdot424e^2}{\kappa_0 a_0}$$

which is about 15 per cent higher than the true value $-0\cdot5e^2/\kappa_0 a_0$.

It would have been possible to introduce greater flexibility in our trial function. We could, for example, have used a combination of two Gaussian functions, putting

$$\psi = \exp(-c_1 r^2) + k\exp(-c_2 r^2)$$

and treated c_1, c_2, and k all as variable parameters. We should then have found (see McWeeny (1953) where Slater orbitals were first approximated by variationally determined combinations of Gaussians) that the error was reduced from 15 per cent to about 1·3 per cent. Gaussian functions are now very widely used in molecular theory since they are particularly easy to handle computationally (see, for example, Boys (1950) and, for a review, Shavitt (1963)).

Finally, we note that if the functional form of the assumed ψ is incorrect it cannot give the *exact* energy, but by including enough parameters it may be able to give an exceedingly good approximation. Another happy circumstance is that the error in the energy is normally much less than that in the wavefunction itself, errors in ψ being 'averaged out' in the integration (3.31).

3.8. The method of linear combinations

There is one particular form of the variation method which is outstandingly convenient to use. It is also important, and is worth a separate description, because it represents one of the most powerful ways in which chemical intuition may be introduced into the finding of a suitable wavefunction.

Suppose first that we have grounds for believing that the true wavefunction ψ has characteristics typical of any two known functions which we shall call ϕ_1 and ϕ_2. It is not necessary that ϕ_1 and ϕ_2 should themselves be solutions of any particular wave equation. For example, if an electric field acts in the x-direction on an electron in an s-type AO, it will tend to pull the electron away from the nucleus preferentially in the negative (remember the electron has a negative charge) x direction and will give a polarity to the atom. We could imagine the true ψ as having at the same time characteristics both of the original s-type orbital ϕ_1 and of some directed orbital ϕ_2 related to the direction of the field, and here perhaps one of the p_x AO's. In a case like this we should naturally try to describe the complete wavefunction as a sum of contributions from each separate orbital; i.e. we might write

$$\psi = c_1\phi_1 + c_2\phi_2 \tag{3.36}$$

where c_1 and c_2 are constants to be chosen so that the energy expression is minimized. In the example mentioned above, c_2 presumably would vanish when the field was zero, but if the field was large c_2 would be relatively more important.

Trial wavefunctions such as (3.36) are particularly well suited to the variation method. If, for simplicity, we use real functions and put (3.36) into (3.31) we obtain

$$\mathscr{E}(\psi) = \frac{c_1^2 \int \phi_1 \hat{H} \phi_1 \, d\tau + 2c_1 c_2 \int \phi_1 \hat{H} \phi_2 \, d\tau + c_2^2 \int \phi_2 \hat{H} \phi_2 \, d\tau}{c_1^2 \int \phi_1^2 \, d\tau + 2c_1 c_2 \int \phi_1 \phi_2 \, d\tau + c_2^2 \int \phi_2^2 \, d\tau}. \tag{3.37}$$

Here, we have used a very important 'symmetry property'

$$\int \phi_1 \hat{H} \phi_2 \, d\tau = \int \phi_2 \hat{H} \phi_1 \, d\tau. \tag{3.38a}$$

characteristic of the operators in quantum mechanics (see Problem 3.9). If the wavefunctions are complex, the corresponding property is

$$\int \phi_1^* \hat{H} \phi_2 \, d\tau = \int (\hat{H} \phi_1)^* \phi_2 \, d\tau \tag{3.38b}$$

and is called 'Hermitian symmetry'. The operator may work on the left-hand function (starred) or the right-hand function (unstarred); it makes no difference.

To make (3.37) look simpler we use the notation

$$H_{rs} = \int \phi_r \hat{H} \phi_s \, d\tau \qquad S_{rs} = \int \phi_r \phi_s \, d\tau \tag{3.39}$$

noting that $H_{rs} = H_{sr}$, and $S_{rs} = S_{sr}$. With this notation (3.37) becomes

$$\mathscr{E} = \frac{c_1^2 H_{11} + 2c_1 c_2 H_{12} + c_2^2 H_{22}}{c_1^2 S_{11} + 2c_1 c_2 S_{12} + c_2^2 S_{22}}. \tag{3.40}$$

If ϕ_1 and ϕ_2 are separately normalized—which is often useful, but by no means necessary—then $S_{11} = S_{22} = 1$. We refer to H_{rs} as the 'matrix element' of $\hat{H}$ with respect to ϕ_r and ϕ_s since such numbers are often set out in a square array or *matrix*, H_{rs} being found at the intersection of the rth row and the sth column. In the same way we could call S_{rs} the matrix element of unity, but it is more usual to refer to it as an *overlap integral*. The interpretation of the integral is shown in Fig. 3.11, where ϕ_1 and ϕ_2 are taken to be two orbitals with ϕ_1 and ϕ_2 negligible in value outside the regions indicated. Obviously, the only appreciable contributions to $\int \phi_1 \phi_2 \, d\tau$ arise from parts of space where ϕ_1 and ϕ_2 are *both* fairly large, i.e. from the region of 'overlap' of the two functions.

Let us now return to (3.40). The only variables are the parameters c_1 and c_2. We ask ourselves the following question: What are the values of c_1 and c_2 which make a function of the general type $c_1 \phi_1 + c_2 \phi_2$ the best possible approximation to the ground-state wavefunction? The answer is that $\mathscr{E}$ must

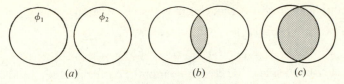

FIG. 3.11. Overlap (S_{12}) between two identical s-type orbitals, ϕ_1 and ϕ_2: (a) $S_{12} \simeq 0$; (b) S_{12} small; (c) $S_{12} \rightarrow 1$.

be a minimum with respect to variation of c_1 and c_2. If we write down the conditions

$$\frac{\partial \mathscr{E}}{\partial c_1} = 0 \qquad \frac{\partial \mathscr{E}}{\partial c_2} = 0$$

we obtain what are known as the 'secular equations'

$$c_1(H_{11} - ES_{11}) + c_2(H_{12} - ES_{12}) = 0$$
$$c_1(H_{12} - ES_{12}) + c_2(H_{22} - ES_{22}) = 0. \tag{3.41}$$

In (3.41) we have written E instead of $\mathscr{E}$, because for these values of c_1 and c_2 $\mathscr{E}$ is the closest approximation to the true energy E. These equations (3.41) may be solved to determine E and also the ratio of the coefficients c_1 and c_2.

This particular form of the variation method is sometimes known as the Rayleigh–Ritz, or Ritz, method; like Rayleigh's method, it was developed long before wave mechanics to deal with a wide variety of problems in classical mechanics and wave motion. It is easily extended to include more than two component functions. Thus with a trial wavefunction ψ compounded out of n functions $\phi_1, \phi_2, \ldots, \phi_n$ in the form of a linear combination

$$\psi = c_1\phi_1 + c_2\phi_2 + \ldots + c_n\phi_n \tag{3.42}$$

there are n secular equations:

$$c_1(H_{11} - ES_{11}) + c_2(H_{12} - ES_{12}) + \ldots + c_n(H_{1n} - ES_{1n}) = 0$$
$$c_1(H_{12} - ES_{12}) + c_2(H_{22} - ES_{22}) + \ldots + c_n(H_{2n} - ES_{2n}) = 0$$
$$\cdots \cdots \cdots \cdots \cdots \cdots \cdots \cdots \cdots \cdots \cdots \cdots \cdots \tag{3.43}$$
$$c_1(H_{1n} - ES_{1n}) + c_2(H_{2n} - ES_{2n}) + \ldots + c_n(H_{nn} - ES_{nn}) = 0.$$

These equations give the ground-state energy and the corresponding coefficients c_r ($r = 1, 2, \ldots n$) to use in (3.42); the resultant wavefunction will be more accurate than the two-term approximation (3.36).

In the case of only two functions it is easy to solve (3.41); division allows us to eliminate c_1 and c_2, obtaining a condition which must be satisfied by E, namely

$$\frac{H_{11} - ES_{11}}{H_{12} - ES_{12}} = \frac{H_{12} - ES_{12}}{H_{22} - ES_{22}}. \tag{3.44}$$

Cross multiplication shows that this may also be written

$$\begin{vmatrix} H_{11} - ES_{11} & H_{12} - ES_{12} \\ H_{12} - ES_{12} & H_{22} - ES_{22} \end{vmatrix} = 0 \qquad (3.45)$$

where the left-hand side is the *determinant* of the coefficients of c_1 and c_2 in the secular equations (3.41). Now (3.44), or (3.45), gives a quadratic equation in E with two roots E_1 and E_2 (both real). These roots are approximations to the energy of the ground state *and the first excited state*. The fact that an excited-state energy and wavefunction arise as byproducts of the ground-state calculation is an extremely useful feature of the Ritz form of the variation method.

With n functions the conclusions are exactly parallel; the secular determinant vanishes for n values of E, and these energies $E_1, E_2, \ldots, E_n$ are all upper bounds to the first n exact energy levels.[†] To obtain the 'mixing coefficients' $c_1, c_2, \ldots, c_n$ for any particular solution (e.g. $E = E_1$) we need only insert that value in the original secular equations and solve them as ordinary simultaneous equations.

The Ritz method is powerful, general, and easy to apply. The functions we mix together are completely at our disposal and may embody any preconceived ideas based on chemical intuition; a 'bad' function will not spoil the calculation, for it will automatically come in with a small coefficient and will thus reject itself; and we always have to deal with equations of exactly the same form (3.43) which can thus be solved using a single standard computer programme. The Ritz method is the principal mathematical tool used throughout this book.

3.9. Mutual repulsion of two energy curves: the non-crossing rule

There is one important general feature of the Ritz-type variation method with just two functions ϕ_1 and ϕ_2 which has very wide significance and merits special attention. Let us suppose, for convenience, that ϕ_1 and ϕ_2 are each separately normalized, so that

$$S_{11} = S_{22} = 1.$$

The energy $\mathscr{E}$ associated with the component ϕ_1 by itself is simply

$$\mathscr{E} = \frac{\int \phi_1 \hat{H} \phi_1 \, d\tau}{\int \phi_1^2 \, d\tau} = \frac{H_{11}}{S_{11}} = H_{11}. \qquad (3.46)$$

It will be convenient to call this E_1^0, where the superscript zero means 'before allowing ϕ_1 and ϕ_2 to mix'; similarly H_{22}, which is the energy associated with

---

[†] More correctly the energies of the first n states of *given symmetry*; if $\phi_1, \ldots, \phi_n$ were all p_x-type functions, for example, no amount of mixing could yield approximations to 1s, 2s, 3s, but we *should* get approximations to $2p_x, 3p_x, \ldots$.

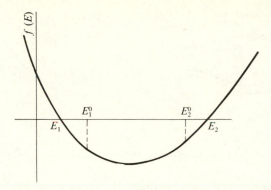

Fig. 3.12. The function $f(E)$. E_1 and E_2 must be displaced outwards from the levels E_1^0 and E_2^0 before interaction.

the component ϕ_2 by itself, is written E_2^0. The secular equations (3.41) become

$$c_1(E_1^0 - E) + c_2(H_{12} - ES_{12}) = 0$$
$$c_1(H_{12} - ES_{12}) + c_2(E_2^0 - E) = 0.$$

If we eliminate c_1/c_2 we find that E is a root of the quadratic equation

$$(E - E_1^0)(E - E_2^0) - (H_{12} - ES_{12})^2 = 0.$$

Let us call the left-hand side of this equation $f(E)$; the graph of $f(E)$ is drawn in Fig. 3.12 for the particular case in which $E_1^0 < E_2^0$. The curve is a parabola whose ordinate has a positive value when E is large, either positive or negative, and which has a negative value when $E = E_1^0$ or E_2^0. Evidently the values of E for which $f(E) = 0$ lie outside the region between E_1^0 and E_2^0. This result is quite independent of the relative magnitudes of E_1^0, E_2^0, H_{12}, and S_{12}. Thus, when we combine two functions ϕ_1 and ϕ_2, we find two possible combinations and two possible energies; one of the original energies has gone downwards, the other has gone up. It is as though the two separate energy levels repelled one another.

A very important application of this law occurs in diatomic molecules. We have already seen in §3.5 that the energy of the molecule depends on the separation R of the two nuclei. This means that E_1^0 and E_2^0, computed for approximate wavefunctions ϕ_1 and ϕ_2, are themselves functions of R. The energies which we obtain when we allow combination of ϕ_1 and ϕ_2 always lie above and below the energies E_1^0 and E_2^0, with the result that, even if the curves of E_1^0 and E_2^0 against R cross one another, the energies in the improved approximation can never cross. Two possible situations are illustrated in Fig. 3.13 which shows the apparent repulsion between the two original energy curves. The fact that the two final curves do not cross one another is referred to as the 'non-crossing rule'. It is of the greatest importance, as we shall see, in correlating molecular states with the products of dissociation. Naturally, it

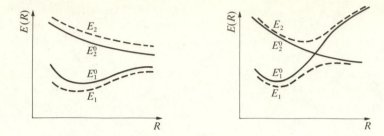

FIG. 3.13. Two instances of the non-crossing rule. The second illustrates an 'avoided crossing'.

is assumed that ϕ_1 and ϕ_2 may legitimately be combined; if they are of different symmetry no combination occurs and there is no non-crossing rule—the E_1^0 and E_2^0 curves may cross and remain unaffected. The importance of symmetry will be discussed later.

The above discussion of linear combinations of two functions may be extended to three or more. Each time that we add a new component (ϕ_{n+1}) we obtain a new set of secular equations with one more energy value. It may be shown (a simple discussion is given by McWeeny and Sutcliffe (1969), pp. 30–33) that between any two consecutive energies in the new series there lies just one of the previous series; thus, the first n approximate energies go down (approaching the exact values) and a new level E_{n+1} appears at the top. This result is in fact the basis of the statement (p. 66) that the *linear* variation method provides valid approximations to excited states as well as to the ground state.

3.10. More physics†

During recent years there have been great advances in the quantum theory of valence; these inevitably bring with them the need for a somewhat broader understanding of the main ideas of quantum mechanics. At the same time there has been an impressive development in the use of experimental techniques borrowed from physics, many of which (e.g. magnetic resonance and electric quadrupole resonance spectroscopy) involve the application of electric and magnetic fields. Again, such techniques bring with them the need for a slightly deeper discussion of ideas such as 'electron spin' which, 20 years ago, required only the briefest of explanations. A detailed account of such matters would be out of place in this book. The reader is referred to Atkins (1970) for an introductory exposition, and to McWeeny (1972) for a slightly more advanced account of basic quantum mechanics. (Both books are addressed primarily to chemists.) A few principles must be noted in passing; we shall deal with these serially for easy reference.

† Some readers may like to defer the study of this section until the need arises. Most of what follows can be understood without it.

(*i*) *Observables and operators.* The Schrödinger equation, written in the form (3.14), exemplifies a number of ideas which may be generalized immediately. The *energy*, expressed in the Hamiltonian form (3.6) (i.e. in terms of particle co-ordinates and corresponding momentum components), is an example of a classically defined *observable*; by suitable observations on a system in a given state it can be assigned a 'measured value'.

In quantum mechanics, we have associated with the energy an *operator* $\hat{H}$, using a certain prescription (3.7), and have then used the equation

$$\hat{H}\psi = E\psi \tag{3.47}$$

to find possible wavefunctions ($\psi_1, \psi_2, \ldots$) and associated energies ($E_1, E_2, \ldots$) characterizing acceptable *stationary states*, in which the energy (and possibly other properties) has definite observable values which do not change in time; these are the 'quantized' atomic and molecular states observed in spectroscopy, the lowest-energy state (ground state) being of particular importance in chemistry.

Such procedures are, in fact, completely general and may be applied to *other* physically observable quantities besides the energy; it is only necessary to take the classical definition of the observable (let us call it L), set up an associated operator ($\hat{L}$) using the same prescription (3.7), and formulate an equation analogous to (3.46), namely

$$\hat{L}\psi = \lambda\psi. \tag{3.48}$$

Equations of this kind are called *eigenvalue* equations; the values of λ for which well-behaved solutions can be found are the *eigenvalues* of $\hat{L}$ and represent the quantized observable values of L, while the associated wavefunctions or *eigenfunctions* describe the corresponding states in the usual way. In an eigenstate the corresponding observable is often said to possess a 'definite' value—the eigenvalue.

A spectroscopic state is often characterized by the values of *several* observable quantities, not just the energy. In such cases the wavefunction of the state must be *simultaneously* an eigenfunction of $\hat{H}$ (for the energy) and of the operators associated with each of the other observables; the corresponding observables then have simultaneously definite values which are called 'constants of the motion', as in classical mechanics. There is, however, a limit to the number of quantities which can be constants of the motion for any given system. It is easy to prove that two quantities can only be simultaneously definite if the order in which their operators are applied successively to the wavefunction ψ is immaterial. For example, with just two quantities, and corresponding operators $\hat{H}$ and $\hat{L}$, we must have

$$\hat{H}\hat{L}\psi = \hat{L}\hat{H}\psi. \tag{3.49}$$

When this is true generally (not just for a particular ψ but for *all* states) we say the operators $\hat{H}$ and $\hat{L}$ *commute* and write $\hat{H}\hat{L} = \hat{L}\hat{H}$. *Quantities are eligible as constants of the motion provided their associated operators all commute with $\hat{H}$ and with each other.*

An important example of the idea of constants of the motion is provided by the hydrogen-like wavefunctions ψ_{nlm} discussed in §2.4. If we set up the operators associated with the classical components of angular momentum of the electron moving around the nucleus (see Problem 3.8), we find ψ_{nlm} is an eigenfunction of the operator associated with the z-component, with eigenvalue $m\hbar$. As $\hbar$ is a natural unit of angular momentum, it is convenient to denote the operator by $\hbar\hat{L}_z$ and in this case

$$\hat{L}_z\psi_{nlm} = m\psi_{nlm}. \tag{3.50}$$

The eigenvalue of the angular momentum operator $\hat{L}_z$ is the integer m, and in this state the electron has m units of angular momentum around the z axis. Similarly, with the square of the magnitude of the angular momentum (in units of $\hbar^2$) we can associate the operator $\hat{L}^2 = \hat{L}_x^2 + \hat{L}_1^2 + \hat{L}_2^2$. This operator can be shown to commute with L_z, and we find

$$\hat{L}^2\psi_{nlm} = l(l+1)\psi_{nlm}. \tag{3.51}$$

Therefore s orbitals describe states of zero angular momentum, p orbitals describe states of angular momentum $\sqrt{(1 \times 2)}\hbar$, d orbitals describe states of angular momentum $\sqrt{(2 \times 3)}\hbar$, and so on. The quantum numbers n, l, m are now seen to specify a set of constants of the motion—the *energy*, the *angular momentum*, and *one component of angular momentum*. Two *different* components cannot simultaneously have definite values, because their operators are found *not* to commute. If we want to specify one component we must of course observe it and this means we must physically define an axis (e.g. by means of an applied magnetic field, as in many spectroscopic experiments). The fact that we usually call such an axis the z axis is purely a matter of convention. Some further properties of angular momentum, which we refer to only occasionally, are summarized in Appendix 2.

(ii) *Properties of eigenfunctions.* The set of eigenfunctions $\psi_1, \psi_2, \psi_3, \ldots$ (usually infinite), obtained by solving the Schrödinger equation (3.47), exhibits certain important mathematical properties. Besides being 'well behaved' (p. 21), and hence individually normalizable, different functions may be assumed† *orthogonal* in the sense that

$$\int \psi_i^* \psi_j \, d\tau = 0 \qquad (i \neq j). \tag{3.25}$$

---

† Solutions with $E_i \neq E_j$ are 'automatically' orthogonal; if E_i and E_j happen to be equal, equally acceptable *mixtures* of ψ_i and ψ_j can be constructed which *are* orthogonal (cf. eqn (2.18) et seq.). It is convenient to work with orthogonal functions whenever possible.

When the functions are both normalized and mutually orthogonal we speak of an *orthonormal set*. Under certain conditions a set of eigenfunctions is *complete* in the sense that an arbitrary function (of the same variables, of course) may be expressed in the form

$$\psi = c_1\psi_1 + c_2\psi_2 + c_3\psi_3 + \ldots, \tag{3.53}$$

where the c's are numerical coefficients, with unlimited accuracy provided a sufficient number of terms is included.

(iii) *Expectation values*. When a quantity L, with associated operator $\hat{L}$, does *not* have a definite value in a given state ψ, the most we can hope to obtain is an 'average' or 'expectation' value. This quantity, usually denoted by $\langle L \rangle$, is defined as the average of the many different values we should find in a large number of observations (always starting with the system in the given state ψ). The prescription for getting $\langle L \rangle$ is very simple:

$$\langle L \rangle = \int \psi^* \hat{L} \psi \, d\tau \tag{3.54}$$

where ψ is the normalized wavefunction describing the state of the system. Of course, if ψ *is* an eigenfunction, satisfying (3.48), we shall find $\langle L \rangle = \lambda$—the average will coincide with the definite value λ, every observation giving the same result. More generally, however, there will be a 'scatter' of the observed values and $\langle L \rangle$ is the most we can hope to predict.

The energy functional $\mathcal{E}(\psi)$ used in §3.6 is now seen to be nothing more than an energy expectation value. If the normalized wavefunction ψ is not an eigenfunction of $\hat{H}$, and cannot therefore describe a stationary state of definite energy, we find

$$\mathcal{E}(\psi) = \int \psi^* \hat{H} \psi \, d\tau = \langle E \rangle. \tag{3.55}$$

The variation theorem (p. 61) is then a statement of the obvious; the average value we should obtain in any series of energy measurements could certainly not be lower than the least!

(iv) *Magnetic dipoles. Electron spin*. In classical physics a charged particle moving in an orbit is equivalent to an electric current flowing round a circuit and therefore sets up a magnetic field; it is equivalent to a *magnetic dipole* and the magnetic moment of the dipole is proportional to the angular momentum of the particle. The same is true in quantum mechanics; with the angular momentum vector $\hbar\mathbf{L}$ (components $\hbar L_x, \hbar L_y, \hbar L_z$) is associated a magnetic dipole of moment

$$\mu_{orb} = \gamma \hbar \mathbf{L} \tag{3.56}$$

where the proportionality constant γ is called the 'magnetogyric ratio' and is

$$\gamma = -e/2m. \tag{3.57}$$

The minus sign shows that the dipole points in the opposite direction to the angular momentum, a result of the negative charge on the electron. Equation (3.56) is also often written

$$\mu_{orb} = -\mu_B \mathbf{L} \tag{3.58}$$

where

$$\mu_B = e\hbar/2m \tag{3.59}$$

is called the 'Bohr magneton' and is the magnetic moment associated with one unit of orbital angular momentum. The small magnetic dipole arising this way gives the so-called 'orbital paramagnetism' of free atoms with a resultant orbital angular momentum. In molecules the orbital angular momentum of the electrons is usually 'quenched' by the strong interaction between the atoms, and for this reason it is only rarely that we need to make much use of angular momentum theory in chemistry.

There is, however, another source of magnetic moment which is evident even for an electron in a state with zero orbital angular momentum; it arises from the *intrinsic* angular momentum of the electron itself, which is called *spin*. Thus, for the single 1s electron in the hydrogen atom, application of a magnetic field reveals two possible states, the electron being found with its magnetic dipole aligned parallel or antiparallel to the field. Comparison with the orbital angular momentum, characterized by quantum numbers l, m_l (temporarily adding a subscript l to the second quantum number) suggests the introduction of spin quantum numbers s, m_s. Just as there are $2l + 1$ values of m_l, so we would expect $2s + 1$ values of m_s. However, as there are only *two* observed states, the implication is that $s = \frac{1}{2}$; we call an electron a 'spin $\frac{1}{2}$' particle. When we observe the angular momentum of a spin $\frac{1}{2}$ particle around some chosen axis (which we call the z axis) we always find, in units of $\hbar$, a value $+\frac{1}{2}$ or $-\frac{1}{2}$. These two states (often depicted as in Fig. 3.14) are commonly called 'spin-up' and 'spin-down' states. We now see the deeper significance of level-and-arrow diagrams like those in Fig. 2.12.

The non-integral quantum number $s = \frac{1}{2}$ is not the only anomalous feature of spin. If we denote the spin angular momentum vector by $\hbar\mathbf{S}$ (the analogue of $\hbar\mathbf{L}$) with components $\hbar S_x, \hbar S_y, \hbar S_z$, then the corresponding magnetic dipole is almost exactly *twice* as strong as we should expect on classical grounds; instead of (3.58) we must write

$$\mu_{spin} = -g\mu_B \mathbf{S} \tag{3.60}$$

field direction is then $-g\mu_B S_z = -g\mu_B m_s$. Now the energy of a dipole in a Dirac who took account of the requirements of relativity theory. Further refinement leads to the value $g = 2 \cdot 0023$ in good agreement with experiment. When a magnetic field is applied to fix a z axis, the component of μ_{spin} in the

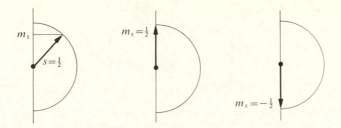

F$_{IG}$. 3.14. Representation of electron spin by a vector. Only two states are observed, corresponding to $m_s = \frac{1}{2}$ (vector parallel to applied field) and $m_s = -\frac{1}{2}$ (vector antiparallel).

field direction is then $-g\mu_B S_z = -g\mu_B m_s$. Now the energy of a dipole in a magnetic field of flux density B is $-B \times$ (component along the field) and the interaction energy of the electron with the field will thus be

$$\Delta E_{\text{spin}} = g\mu_B B m_s \qquad (m_s = \pm\tfrac{1}{2}) \qquad (3.61)$$

The energy separation of two states $m_s = \pm\frac{1}{2}$ (e.g. in the case of the hydrogen atom ground state), owing to imposition of the field, should thus be $g\mu_B B$. This is called the spin 'Zeeman splitting' after Zeeman who first investigated the effect of magnetic fields on atomic energy levels; quantum jumps between such pairs of levels may be induced by radiation of frequency $h\nu = g\mu_B B$ and occur in electron spin resonance (ESR) spectroscopy.

(*iv*) *Spin functions and spin-orbitals.* The magnetic effects of spin are of comparatively small importance in valence theory; but the existence of spin, as we already know from Chapter 2, is of crucial importance in determining electron configurations and therefore has implications for the whole of chemistry. We therefore need to know how to incorporate spin in the wavefunction. The difficulty is that there is no classical quantity corresponding to electron spin; nor are there any 'spin co-ordinates' corresponding to the x, y, z which specify the position of the electron and which appear in the wave function $\psi(x, y, z)$. To introduce spin we therefore go back to basic principles: if there are just two possible spin states, corresponding to $S_z = \pm\frac{1}{2}$, there must be two eigenfunctions of $\hat{S}_z$—which we shall call α and β—which satisfy the equations

$$\hat{S}_z \alpha = \tfrac{1}{2}\alpha \qquad \hat{S}_z \beta = -\tfrac{1}{2}\beta.$$

The eigenvalues are the two possible values we can find (in units of $\hbar$) for the z-component of the spin angular momentum. Very often α and β are written formally as functions of a spin variable s, so that $\alpha(s)$ and $\beta(s)$ look like wavefunctions $\psi(x, y, z)$, and, although it is not strictly necessary, it is even possible to press the analogy further by interpreting s as the value of S_z and regarding $|\alpha(s)|^2 \, ds$, for example, as the probability of finding S_z in the range s

FIG. 3.15. Schematic representation of the two spin functions. $\alpha(s)$ is a function of the 'spin variable' s and is zero unless $s \simeq +\frac{1}{2}$, and similarly for $\beta(s)$, whose peak is at $s = -\frac{1}{2}$. The functions are clearly non-overlapping and are assumed to be normalized.

to $s + ds$. With this interpretation, $\alpha(s)$ must have the form shown in Fig. 3.15(a) (an infinitely narrow 'spike' at $s = \frac{1}{2}$) since in spin state α there is, by definition, zero probability of finding any value of S_z other than $\frac{1}{2}$; similarly $\beta(s)$ must have the form shown in Fig. 3.15(b). The spin functions are clearly non-overlapping, and therefore orthogonal, and we may assume them normalized in the usual way. Thus

$$\int |\alpha(s)|^2 \, ds = \int |\beta(s)|^2 \, ds = 1$$

$$\int \alpha(s)\beta(s) = 0.$$

The spin functions may then be used just like any other wavefunction.

We now verify that the wavefunction for an electron moving in space *and with a given spin* may be regarded as a *product* of an orbital factor ψ and a spin factor α or β. In other words

> $\psi\alpha$ describes an electron in orbital ψ with $S_z = \frac{1}{2}$
>
> $\psi\beta$ describes an electron in orbital ψ with $S_z = -\frac{1}{2}$.

To show that this is so we need only use the definitions

$$\hat{H}\psi = E\psi \text{ (position described by wavefunction } \psi, \text{ energy } E)$$
$$\hat{S}_z\alpha = \tfrac{1}{2}\alpha \text{ (spin described by spin function } \alpha, z \text{ component } \tfrac{1}{2})$$

and remember that $\hat{H}$ acts only on functions of the *spatial* variables (x, y, z) while $\hat{S}_z$ acts only on functions of the spin variable (s). Thus

$$\hat{H}(\psi\alpha) = (\hat{H}\psi)\alpha = E(\psi\alpha)$$
$$\hat{S}_z(\psi\alpha) = \psi(\hat{S}_z\alpha) = \tfrac{1}{2}(\psi\alpha)$$

and $\psi\alpha$ is an eigenfunction of $\hat{H}$ *and* $\hat{S}_z$, simultaneously, with energy eigenvalue E (just as if spin were neglected) and spin eigenvalue $S_z = \frac{1}{2}$. A similar argument applies to the product $\psi\beta$, but in this state $S_z = -\frac{1}{2}$. The one-electron wavefunctions $\psi\alpha$ and $\psi\beta$ are called *spin-orbitals*; to take spin into account we shall have to assign electrons to spin-orbitals instead of orbitals. The exclusion principle, used in discussing electron configurations in the *aufbau* approach, now has a somewhat simpler meaning; 'not more than two electrons per orbital, with opposite spins' evidently means we can assign not more than *one* electron to each available *spin-orbital*. We shall find a deeper significance of this principle in a later section (§5.6).

Problems

3.1. Write down the Schrödinger equation (*a*) for the molecule LiH, and (*b*) for its ion LiH$^+$. (Hint: use a notation similar to that in Fig. 3.3.)

3.2. Hartree supposed each electron of a many-electron atom to move in an average field provided by the nucleus and all other electrons (p. 53). What qualitative form would the potential energy function $V(r)$ take for a lithium 2s electron (*a*) ignoring the 1s electrons, and (*b*) including them by Hartree's method. Try to suggest suitable forms of $V(r)$ (*a*) near the nucleus ($r \rightarrow 0$), and far from it ($r \rightarrow \infty$).

3.3. A particle moves along the x axis, its potential energy function having the form

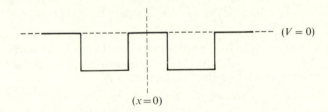

$(x=0)$

Sketch suitable wavefunctions $\psi(x)$ for (*a*) the ground state (lowest energy), (*b*) the first excited state (one node), and (*c*) a highly excited (but still bound) state.

What would be the effect of making the left-hand box much deeper than the right? How could you find the relative probabilities of finding the particle in the left- and right-hand boxes? Can you suggest any chemical application of this model?

3.4. Write down the wavefunctions and energy levels for a particle confined within a cube of side L. What choice of quantum numbers would give a wavefunction resembling a d$_{xy}$ atomic orbital? How many other wavefunctions would have the same energy and what would they look like? How would this degeneracy be affected by compressing the cube slightly in the z direction? (Hint: start from equations (3.24) and ask where the three sine waves in (3.24*a*) have their maximum values for various combinations of n_1, n_2, n_3.)

(Your results shed light on the effect of cubic and tetragonal 'crystal fields' upon the energy levels of d electrons in transition metal ions, Chapter 9.)

3.5. Verify the statement (p. 58) that a wavefunction of the form $\psi = e^{-kr}$ satisfies the Schrödinger equation for a hydrogen-like system, with values of k and E as given in the text. (Hints: you need to use

$$\frac{\partial \psi}{\partial x} = \frac{d\psi}{dr}\frac{\partial r}{\partial x} = \frac{x}{r}\frac{d\psi}{dr}$$

in finding $(\partial^2\psi/\partial x^2)$ and then to add similar results with $x \to y, z$ to get $\nabla^2\psi$. Evaluate the left-hand side of equation (3.25) and set equal to zero the coefficient of each power of r.)

3.6. Show, by the method of Problem 3.5, that the 2p orbital $\psi = x\,e^{-kr/2}$ is also a solution of equation (3.25), verifying that k is given by (3.27) and obtaining the value of E_{2p}.

3.7. Use the variation method to obtain an approximate 1s wavefunction for a hydrogen-like system (see equation (3.16)) in the form $\psi = e^{-cr^2}$. Hence generalize the conclusions given on p. 62. (Hint: use the same procedure for evaluating $\nabla^2\psi$ as in Problem 3.5. The volume element in the integrals is $d\tau = 4\pi r^2\,dr$; and you will need the definite integral

$$\int_0^\infty r^{2n}\,e^{-cr^2}\,dr = \frac{1.3.5\dots(2n-1)}{2^{n+1}}\left(\frac{\pi}{c^{2n+1}}\right)^{1/2}$$

in evaluating the Rayleigh ratio $\mathscr{E}(c)$.)

3.8. The components of angular momentum of a particle moving about the origin are defined, in classical mechanics, by

$$M_x = (yp_z - zp_y), \quad M_y = zp_x - xp_z, \quad M_z = xp_y - yp_x.$$

Write down the quantum mechanical operators associated with these quantities. Note that M has dimensions ML^2T^{-1} (length $\times$ mass $\times$ velocity), which coincide with those of $\hbar$; hence if we write $M_x = \hbar L_x$, etc., the related quantities L_x, L_y, L_z are dimensionless. (Hint: use the basic association given in equation (3.7).)

3.9. Show that the wavefunction $\psi = e^{2\pi i k x}$ is an eigenfunction of the Hamiltonian operator for a free particle ($V = 0$), and is at the same time an eigenfunction of the operator associated with the linear momentum p_x. Show that the corresponding values of energy and momentum are, in this case, related exactly as in classical mechanics.

Verify that $\hat{p}_x$ is a Hermitian operator in accordance with (3.38). (Hint: use integration by parts to 'move' the d/dx.)

3.10. Attach the appropriate time factor (given in (3.13)) to the wavefunction of Problem 3.9 to obtain the wavefunction including the time (in the form (3.12)). Verify that (with k positive) Ψ at time t, evaluated for any point $x' = x + ut$ (i.e. displaced to the right by an amount ut) where $u = E/p_x$, coincides in value with Ψ for $t = 0$ at point x. This means that Ψ is a *travelling* wave moving to the right with a phase velocity u.

(This is the reason for choosing the time factor as we did on p. 48. With the alternative sign, the particle and the wave describing it would have turned out to be moving in opposite directions—making physical nonsense.)

3.11. Verify that an electron in a $2p_x$ AO does not have a definite angular momentum around the z axis, and that the average value obtained from a large number of measurements would be zero.

3.12. Show that, on introducing polar co-ordinates (Fig. 2.5), the operator $\hat{L}_z$ (see Problem 3.8) may be written equivalently as

$$\hat{L}_z = \frac{1}{i}\frac{\partial}{\partial\phi}$$

and hence that a wavefunction of the form (2.15), which depends on ϕ through a factor $e^{im\phi}$, describes a state with m units of angular momentum (i.e. $M_z = m\hbar$) around the z axis. (Hint: from Fig. 2.5, $x = r\sin\theta\cos\phi$, $y = r\sin\theta\sin\phi$, $z = r\cos\theta$. Now assume the given result and use

$$\frac{\partial\psi}{\partial\phi} = \frac{\partial\psi}{\partial x}\frac{\partial x}{\partial\phi} + \frac{\partial\psi}{\partial y}\frac{\partial y}{\partial\phi} + \frac{\partial\psi}{\partial z}\frac{\partial z}{\partial\phi}$$

to show its equivalence to the cartesian form (Problem 3.8).)

4

Qualitative MO theory: Diatomic molecules

4.1. Theories of electrons in molecules

For reasons which are already evident, drastic approximations must usually be made in trying to solve the wave equation for a many-electron molecule, even in a fixed-nucleus approximation. Historically, two extreme approaches emerged: one was based on the first satisfactory calculation, by Heitler and London (1927) of the bond in H_2, the simplest of all molecules; the other was modelled along the lines of the *aufbau* approach to atomic structure, in which the electrons are assigned independently to their own orbitals. The first approach came to be known as *valence bond* (*VB*) *theory*, the second as *molecular orbital* (*MO*) *theory*. In VB theory the wave function is constructed in such a way that the role of the separate *atoms*, and the localized *atomic* orbitals, is emphasized; in MO theory, on the other hand, the electrons are assigned to orbitals which extend over the molecule as a whole, and consequently the sharing and de-localization of the electrons is emphasized. Of course, the variation method tells us that if we refine our first approximations, by adding more and more terms in the variation function, we shall be able to come as close as we wish to the exact ground-state solution. So as both theories are improved and extended their conclusions should converge, and which one we start from should be immaterial. However, the simplicity and efficiency of the first approximation is an important matter, for often it is not feasible to proceed to higher accuracy and we then need to be aware of the relative merits of the different approaches. The first applications of wave mechanics to molecular structure were dominated by the VB theory, but nowadays most calculations are performed in the general framework of MO theory, largely owing to its mathematical simplicity.

We shall develop the MO approach first because of its *conceptual* simplicity and its close resemblance to the one used in discussing atoms. In this chapter we concentrate on the simple qualitative features of the theory, as an introduction to its systematic development in later chapters.

4.2. Molecular orbitals

The molecular orbital theory starts by supposing that the main ideas of the self-consistent-field method for atoms may be applied equally well to molecules. This means that we may adopt a large proportion of §2.5 and enunciate the following set of fundamental principles.

(i) Each electron in a molecule is described by a certain wavefunction ψ, which, because it describes the electron in the *molecule*, is properly called a molecular orbital (abbreviated to MO). These orbitals are essentially polycentric, and not monocentric as in the case of an atom. When the orbitals are normalized, so that $\int \psi^2 \, d\tau = 1$, the interpretation of ψ is that $\psi^2 \, d\tau$ (or $\psi\psi^* \, d\tau$ if ψ is complex) gives the probability of finding the electron in volume element $d\tau$ at the point where ψ is evaluated. Alternatively we may speak (§2.2) of the charge cloud associated with this orbital. The density of the charge cloud is ψ^2 in electrons per unit volume. We may represent the orbital by drawing contours of constant ψ (or ψ^2), or we may draw a 'boundary surface', which is such that effectively all the charge cloud lies within this surface.

(ii) Each ψ is defined by certain quantum numbers, which govern its energy and its shape.

(iii) Each ψ is associated with a definite energy value, which represents very closely the energy required to remove an electron from it by ionization. The total energy of the molecule is the sum of the energies of the occupied MO's, corrected for the mutual interaction between the electrons. For rough purposes, this latter is often neglected. Such a neglect is not really justified because (section (*c*) on p. 38) in calculating the energy of an electron in any one MO we take account of the repulsion between it and all the other electrons. This means that when we add together the various MO energies, we do in fact count each electronic repulsion term twice over. Those terms (see Table 4.2 on p. 91) are usually much larger than the total binding energy and further discussion will therefore be needed (§8.9).

(iv) Each electron has a spin, which, just as in the case of an electron in an atom, has a z component $\pm \frac{1}{2}\hbar$, corresponding to the spin quantum number $m_s = \pm \frac{1}{2}$.

(v) When describing a molecule we determine the allowed orbitals and then adopt the *aufbau*, or building-up, process of §2.5 in which the electrons are fed into the allowed orbitals one at a time, taking due account of the Pauli exclusion principle, so that not more than two electrons have the same ψ—in which case they must have opposed spins.

All this is almost identically word for word the same as for electrons in an atom. However, before it is very much use to us, we must be able to describe the MO's in some detail and show in what ways new features appear on account of their molecular character. What then can we say in general terms about the nature of these MO's?

4.3. LCAO approximation

The most obvious characteristic of an MO for a diatomic molecule is that it is bicentric. It is this which distinguishes it from an AO. The most appropriate physical description is that the electron moves in an orbital which encloses the neighbourhood of both nuclei.

There is a further point to be made here, however. When the electron is in the region of one of the nuclei, the forces on it are those due chiefly to that nucleus and to the other electrons near that nucleus. If we disregard the electron interactions, the wave equation will be similar to that for H_2^+ (p. 52), namely,

$$\hat{H}\psi = -\frac{\hbar^2}{2m}\nabla^2\psi - \frac{e^2}{\kappa_0}\left(\frac{Z_A}{r_a} + \frac{Z_B}{r_b}\right)\psi = E\psi \tag{4.1}$$

where Z_A and Z_B are the charges on the two nuclei (which, as in §2.6, may be replaced by 'effective' nuclear charges to simulate the effect of electron interactions). We notice that when the electron is near nucleus A, the most significant parts of $\hat{H}$ are precisely those terms which would comprise the Hamiltonian of an electron in the field of A alone. The (Z_B/r_b) term, although it is not exactly zero, is relatively small. Thus, in this region, the wave equation approximates that of an electron near an isolated nucleus; and so also must its solution. This means that in the neighbourhood of nucleus A, the MO resembles an AO ϕ_A. Similarly in the neighbourhood of the other nucleus B, the MO resembles ϕ_B. Since the complete MO has characteristics separately possessed by ϕ_A and ϕ_B, it is a natural step to adopt the method of linear combinations described in §3.8 and write

$$\psi = c_A\phi_A + c_B\phi_B. \tag{4.2}$$

This is also conveniently written

$$\psi = N(\phi_A + \lambda\phi_B) \tag{4.3}$$

where N is a normalizing factor. The value of N is fixed (see (2.9)) by requiring

$$\int \psi^2\, d\tau = N^2(S_{AA} + 2\lambda S_{AB} + \lambda^2 S_{BB}) = 1$$

with the notation of (3.39), or

$$N = (1 + \lambda^2 + 2\lambda S_{AB})^{-1/2} \tag{4.4}$$

if we suppose that ϕ_A and ϕ_B are each separately normalized. For many purposes we may omit the normalizing factor, and write (4.3) in the unnormalized form

$$\psi = \phi_A + \lambda\phi_B. \tag{4.5}$$

The constant λ measures the 'polarity' of the orbital and may have any value ranging from $+\infty$ to $-\infty$, according to the nature of the combining atoms. According to the variation method this value of λ must be chosen to minimize the energy functional

$$\mathcal{E}(\psi) = \frac{\int \psi \hat{H} \psi \, d\tau}{\int \psi^2 \, d\tau}. \qquad (4.6)$$

The approximation represented by (4.2), or its various alternative forms, is called the 'linear combination of atomic orbitals' (LCAO) approximation.

We still have to decide which orbitals ϕ_A and ϕ_B of the two atoms may be combined together in the form (4.5). In order that there may be an effective combination between a given ϕ_A and ϕ_B it is necessary

(a) that the energies of ϕ_A and ϕ_B in their respective atoms should be of comparable magnitude,
(b) that ϕ_A and ϕ_B should overlap one another as much as possible, and
(c) that ϕ_A and ϕ_B should have the same symmetry relative to the molecular axis AB.

Unless these conditions are fulfilled, ϕ_A and ϕ_B mix together either only slightly or not at all (e.g. the λ in (4.5) may be small or zero). The proof of these conditions is not difficult and will be given presently; here we note only that the conditions must follow from the secular equations (3.41) which determine the values of the mixing coefficients which make the energy functional (4.6) stationary. With only two functions, ϕ_A and ϕ_B, it is convenient to use the abbreviations $H_{AA} = \alpha_A$, $H_{BB} = \alpha_B$, $H_{AB} = \beta$, $S_{AB} = S$, in which case the equations become

$$(\alpha_A - E)c_A + (\beta - ES)c_B = 0$$
$$(\beta - ES)c_A + (\alpha_B - E)c_B = 0. \qquad (4.7)$$

On eliminating the coefficients c_A and c_B, we obtain (cf. p. 67)

$$(\alpha_A - E)(\alpha_B - E) - (\beta - ES)^2 = 0 \qquad (4.8)$$

whose roots, E_1 and E_2, determine the values of E for which the equations (4.7) are compatible; these represent the energies of two allowed states.

Quantities analogous to α_A, α_B, β and S appear in all applications of LCAO MO theory, and it is important that we understand their physical meaning. In full,

$$\alpha_A = \int \phi_A \hat{H} \phi_A \, d\tau \qquad \alpha_B = \int \phi_B \hat{H} \phi_B \, d\tau \qquad (4.9)$$

while

$$\beta = \int \phi_A \hat{H} \phi_B \, d\tau \qquad S = \int \phi_A \phi_B \, d\tau. \qquad (4.10)$$

We recognize, from (3.31), that (since ϕ_A is assumed normalized) α_A represents approximately the energy the electron would have if it were constrained to move in orbital ϕ_A located on centre A, with a similar interpretation for α_B. This is not exactly the same as the energy E_A of an electron in ϕ_A on nucleus A *alone*, because $\hat{H}$ defined in (4.1) includes also an attraction term for nucleus B; in the terminology of §3.10 (iii), it is the expectation value of the energy of an electron confined to ϕ_A. However, contributions to the α_A integral arise mainly from the region of space where ϕ_A is large (i.e. near nucleus A) and this is just the region where (cf. p. 80) H is close to that for a free atom. Thus $_A$ or α_B) is roughly the energy of an electron in ϕ_A on atom A (or ϕ_B on atom B) modified slightly by the presence of the second atom; since α_A and α_B contain coulombic interaction terms they are usually called *coulomb integrals*. Generally speaking, α values increase in magnitude, for orbitals of given type, as we proceed from left to right across the periodic table (i.e. towards atoms with greater electronegativity or 'electron attracting power'). The integrals in (4.10), however, depend jointly on ϕ_A and ϕ_B; it is clear that both β and S will be large in magnitude only when both AO's have a region of 'overlap' (Fig. 3.11) for otherwise their product will be negligible at any given point and corresponding contributions to the integrals will be extremely small. It is usual to call quantities such as S *overlap integrals* and quantities such as β *bond integrals* (or, in older terminology, *resonance integrals*).†

The solutions of the secular equations are somewhat different for homonuclear molecules ($\alpha_A = \alpha_B$) and heteronuclear molecules ($\alpha_A \neq \alpha_B$), and we shall deal with the two cases separately. First, however, we note the qualitative significance of the conditions for combination, stated above. Condition (*a*) implies, for example, that while there may be considerable mixing between the valence orbitals of two atoms there is unlikely to be appreciable mixing between valence orbitals and those of the inner shells, the latter lying very much lower in energy. Condition (*b*) is often called the 'principle of maximum overlap' and, as we shall see, implies that the bigger the overlap the stronger the bonding. Condition (*c*) will be discussed more fully in later sections, but it is immediately clear, from (4.7) that *no mixing* of ϕ_A and ϕ_B will occur if for any reason $S = 0$ and $\beta = 0$, for then there will be two separate equations, with solutions $E = \alpha_A$, $\psi = \phi_A$ and $E = \alpha_B$, $\psi = \phi_B$, respectively. This will happen if the two orbitals do not overlap at all (in the sense of Fig. 3.11), or it may

† The alternative name 'resonance integral', although still commonly used, is misleading because it suggests a mechanical analogy which has been a source of much misunderstanding. If we were able to start an electron in ϕ_A and to compel its wavefunction at future times to be $\phi = c_A\phi_A + c_B\phi_B$, then solution of (3.10) would show an oscillatory behaviour, the electron moving back and forth between ϕ_A and ϕ_B. This is reminiscent of the behaviour of two coupled pendulums, where the amplitude of one grows as that of the other diminishes. Whereas mechanical resonance is an observable phenomenon, however, no such oscillation occurs in a quantum-mechanical *stationary state* of the kind under discussion; ψ^2 is time independent and the electron in no sense oscillates between the two atoms. The stationary states are obtained by solving (3.14), not (3.10).

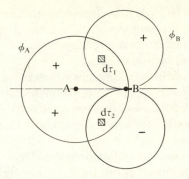

FIG. 4.1. Overlap between orbitals of different symmetry type. Here ϕ_A is of s type, ϕ_B of p_x type, AB being taken as the z axis.

happen because of certain symmetries in ϕ_A and ϕ_B which enable us to separate each of the integrals S_{AB} and β_{AB} into two parts of equal magnitude and opposite sign. When this occurs, we can say that ϕ_A and ϕ_B are of the wrong symmetry type to be combined together in a MO. An example of this occurs if ϕ_A is an s-type AO, and ϕ_B is a p_x-type AO (Fig. 4.1) where the x axis is directed perpendicular to the molecular axis AB, which is usually taken (by convention) as the z axis. It is clear from Fig. 4.1 that the integral S_{AB} $= \int \phi_A \phi_B d\tau = 0$, on account of the difference in sign in the two lobes of a p_x orbital; to every small volume element $d\tau_1$ there is a corresponding element $d\tau_2$ such that the values of the integrand in $d\tau_1$ and $d\tau_2$ are equal and opposite. We can therefore say that the integral vanishes *by symmetry*; in fact $S_{AB} = 0$, whatever the detailed forms of ϕ_A and ϕ_B, when they are of s and p_x types respectively. It is evident at once that a p_x on one atom cannot be combined with a p_y or a p_z or an s on the other atom. Allowed combinations for the s-type, p-type, and some of the d-type orbitals are shown in Table 4.1.

TABLE 4.1
Allowed combinations of AO's in LCAO method

ϕ_A	Allowed ϕ_B	Forbidden ϕ_B
s	s, p_z, d_{z^2}	p_x, p_y, $d_{x^2-y^2}$, d_{xy}, d_{yz}, d_{xz}
p_z	s, p_z, d_{z^2}	p_x, p_y, $d_{x^2-y^2}$, d_{xy}, d_{yz}, d_{xz}
p_x†	p_x, d_{xz}	s, p_y, p_z, $d_{x^2-y^2}$, d_{z^2}, d_{xy}, d_{yz}
d_{xz}†	p_x, d_{xz}	s, p_y, p_z, $d_{x^2-y^2}$, d_{z^2}, d_{xy}, d_{yz}
$d_{x^2-y^2}$	$d_{x^2-y^2}$	s, p_x, p_y, p_z, d_{z^2}, d_{xy}, d_{yz}, d_{xz}
d_{z^2}	s, p_z, d_{z^2}	p_x, p_y, $d_{x^2-y^2}$, d_{xy}, d_{yz}, d_{xz}

† Allowed combinations for p_y and d_{yz} are obtained by interchanging x and y in each of these rows.

When the integral $\int \phi_A \phi_B \, d\tau$ is identically zero, we say that the wave-functions ϕ_A and ϕ_B are *orthogonal*. It is often a matter of the greatest importance to know whether two given functions are or are not orthogonal. The following rules will help us to decide.

(i) If ϕ_A and ϕ_B are orbitals of different symmetry (e.g. behaving differently under reflection across a plane of symmetry) they are orthogonal.

(ii) If ϕ_A and ϕ_B are solutions of a given wave equation they are necessarily orthogonal when the corresponding energies E_A and E_B differ. This is the property referred to in §3.10 (ii).

(iii) Any two approximate wavefunctions found by the Ritz method of linear combinations (§3.8) will be orthogonal if they are built up out of the same given set of functions.

In later chapters it will appear that there are occasions when a given AO ϕ_A is associated with two or more AO's (ϕ_B) on centre B. Possible combinations must be from among those labelled 'allowed' in Table 4.1. It should be noted that the five fundamental d orbitals (Fig. 2.9) $d_{xy}, d_{yz}, d_{zx}, d_{z^2}, d_{x^2-y^2}$ form the appropriate set when, as is conventional, the molecular axis is taken to define the z direction.

We can easily illustrate the conditions (p. 81) for effective combination of AO's in the LCAO method by a consideration of HCl. In the first place the condition (*a*) of comparable energies tells us that the 1s, 2s, 2p, 3s orbitals of Cl will not appreciably combine with any orbital of the H atom, for their energies (with the possible exception of 3s) are much too low. The only other possibilities among the next group are the chlorine $3p_x$, $3p_y$, and $3p_z$. Now the lowest AO of hydrogen is the 1s, and so, according to Table 4.1, the symmetry condition allows an MO to be formed only by combination of H(1s) and Cl($3p_z$). As will be seen later, a small amount of Cl(3s) enters the combination, but this does not seriously affect the description just given. There are possible MO's of higher energy which do not concern us at the moment. It follows from this discussion, and by application of the *aufbau* procedure, that the ground state of HCl could be described as

$$(1s)^2(2s)^2(2p)^6(3s)^2(3p_y)^2(3p_z)^2\{H(1s) + \lambda Cl(3p_z)\}^2 .$$

where the unlabelled orbitals, 1s, 2s, ..., $3p_z$, are all AO's on the chlorine atom.

There is a particular significance in the LCAO relation between an MO in a molecule AB and its component AO's ϕ_A and ϕ_B. This is seen most clearly when the two electrons that occupy the MO $\psi = \phi_A + \lambda \phi_B$ are valence electrons contributing to the bonding between A and B. Normally one of these electrons comes from each atom, and in these atoms they would occupy orbitals ϕ_A and ϕ_B. Thus the MO ψ, which will accommodate two electrons, is correlated with the two AO's into which we might expect the molecular electrons to go if the atoms were separated to infinite distance. We could therefore speak loosely of 'the electron originally in state ϕ_A' and 'the electron originally in state ϕ_B' being paired to form molecular electrons in state ψ $(=\phi_A+\lambda\phi_B)$ for the molecule AB. We could, in fact, imagine ourselves bringing the atoms A and B towards one another, but without allowing interaction, and then pairing together suitable electrons in A and B. This rather crude, and certainly not completely accurate, description of molecule formation reveals a link between MO theory and both the Lewis shared-electron-pair bond and the Langmuir octet theory.

4.4. The hydrogen molecule ion H_2^+

The LCAO approximation takes a particularly simple form for homonuclear diatomics, i.e. molecules where both atoms are of the same kind, as in H_2, O_2, N_2, etc. It will be convenient to begin with the simplest of all such systems, H_2^+. On account of the fact that there is only one electron, this problem has the same central significance for molecules as the H atom has for atomic structure.

We are to form MO's from pairs of AO's using the principles of §4.3. For the system H_2^+ we take ϕ_A and ϕ_B as 1s AO's on the two alternative centres, and suppose

$$\psi = c_A\phi_A + c_B\phi_B \qquad (4.11)$$

noting that in the secular equations (4.6) the quantities α_A and α_B must be equal, for the two ends of the molecule are indistinguishable and an electron would have the same energy in ϕ_A (on the left-hand nucleus) as in ϕ_B (on the right). On putting $\alpha_A = \alpha_B = \alpha$, equation (4.7) has solutions

$$E_{\pm} = \frac{\alpha \pm \beta}{1 \pm S}. \qquad (4.12)$$

On substituting each of these results, in turn, back into (4.6) we find $c_A = \pm c_B$ and the corresponding MO's are thus

$$\psi_+ = N_+(\phi_A + \phi_B) \qquad \psi_- = N_-(\phi_A - \phi_B) \qquad (4.13)$$

where N_+ and N_- are normalizing factors. The energies (4.12) may also be written in the form

$$E_{\pm} = \alpha \pm \frac{\beta - \alpha S}{1 \pm S}. \qquad (4.14a)$$

In approximate work S is sometimes neglected and then

$$E_{\pm} \simeq \alpha \pm \beta. \qquad (4.14b)$$

Equations (4.14a) and (4.14b) show how an original pair of degenerate levels, of energy α, is split into two, one lying higher and the other lower than before. It also shows that the extent of the splitting is dominated by the value of the bond integral β. The lower energy MO is called the *bonding* MO, referring to the molecular ground state in which a bond is present; the other MO, in which

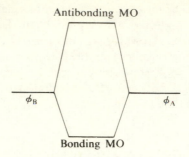

FIG. 4.2. Energy level diagram for molecular orbital formation. The energy of an electron in the bonding MO is *lower* than in the original AO (ϕ_A or ϕ_B); for the antibonding MO it is higher.

the electron would have a higher energy than in the hydrogen atom† is the *antibonding* MO. The situation is indicated schematically in Fig. 4.2, the characteristic asymmetry of the splitting arising whenever $S > 0$.

To proceed further, we give ϕ_A and ϕ_B the explicit form (2.12) corresponding to a 1s AO for the hydrogen atom; β and S may then be calculated for any chosen internuclear distance R. In this way the electronic energy is obtained, and so, by adding the nuclear Coulomb energy $e^2/\kappa_0 R$, we determine the energy curve for the molecule. The results are shown in Fig. 4.3; on the left is the energy curve for the MO ψ_+ and on the right is the curve for ψ_-. The first of these is the energy curve of the molecule in its ground state.

Several features emerge from these energy curves. In the first place the curve for ψ_+ shows a minimum at a nuclear separation a little over $2a_0$ (approximately 10^{-10} m). This shows that the ion H_2^+ must be stable, and indicates a dissociation energy of about 1.77 eV compared with the experimentally determined value of 2.77 eV. It also shows how the type of potential curve for a molecule which has been found experimentally (cf. §1.4) can be calculated theoretically. It is true that H_2^+ is not a chemical species, but is well known from work with the discharge tube and the mass spectrometer. Any criticism that we have been using approximate wavefunctions and therefore getting only approximate energies is met by the observation (p. 61) that better wavefunctions would in fact give us a lower energy and therefore greater stability.

In the second place the final energy curve for ψ_- shows no minimum, so that the molecule would be unstable in such a state breaking up spontaneously into $H + H^+$ with emission of energy. Potential energy curves of these two kinds are called attractive and repulsive respectively; they generally correspond to MO's of bonding and antibonding type.

---

† More correctly (see p. 80) the reference energy α is that of the electron in a hydrogen atom polarized by the presence of a second proton.

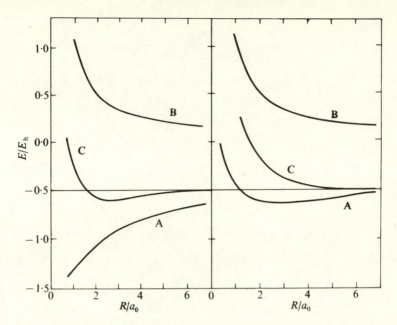

FIG. 4.3. Energy curves for the ion H_2^+. The left-hand diagram is for the bonding MO, and the right-hand diagram for the antibonding MO: curves A, purely electronic energy; curves B, nuclear repulsion energy; curves C, resultant energy curve of the molecule.

The distinction between bonding and antibonding molecular orbitals is so fundamental that it is worth further discussion. Fig. 4.4 shows contours of constant ψ^2, i.e. charge-cloud density, or probability. The plane of the paper is any plane through the nuclei A and B. Even though more accurate wavefunctions would change the shapes of the contours slightly, these diagrams are substantially correct. Fig. 4.5 shows the value of ψ^2 at points along the nuclear axis compared with ϕ_A^2, the electron density for a hydrogen atom A. Clearly, when a second proton is brought to position B the charge density divides itself symmetrically between the two nuclei; in the bonding state there is a *build-up* of electron density in the 'bond region' while in the antibonding state there is a decrease.

Now there is an important theorem in quantum mechanics, already anticipated in §1.5, which states that once we have calculated $P = \psi^2$ the forces which act on the nuclei can be computed classically; the forces which operate are exactly those which would act if the positive nuclei were embedded in a distribution of negative charge of density P electrons/unit volume. This is a very valuable result for it provides a very simple electrostatic picture of the origin and nature of the bonding in a molecule. In the early literature it was sometimes stated that bonding resulted from the decrease in kinetic energy when the electron moved from a smaller region (atom) into a larger region

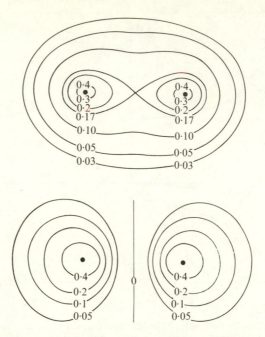

Fig. 4.4. Electron density in the ion H_2^+. The upper diagram is for the electron in the bonding MO, and the lower diagrams is for the antibonding MO ($1\sigma_g$ and $1\sigma_u$ are the conventional names of these MO's).

(molecule). This interpretation is erroneous, for it can be proved that in a molecule the kinetic energy must be *greater* than in its constituent atoms. The kinetic energy decreases only in the early stages of the approach of two atoms (i.e. long before the bond is established). The nuclei repel one another but they are both attracted towards the regions where the negative electronic charge is spread most densely; equilibrium results when the attractions and repulsions come into balance. In the bonding state (MO ψ_+) there is enough charge between the nuclei to hold them together; in the antibonding state (MO ψ_-) there is not, and the nuclei experience a net repulsion at all distances.

Before leaving this calculation on H_2^+, which gives both a satisfactory qualitative interpretation of the bond and a fairly good value of the dissociation energy, it is natural to enquire whether more exact results are available. In fact, an *exact* MO was obtained for this simple one-electron system by Burrau (1927) some time *before* the LCAO calculation was performed. See also Bates *et al.* (1953) and Power (1973). Burrau introduced the co-ordinates $r_a \pm r_b$ and was then able to separate and integrate the Schrödinger equation by direct numerical methods. He achieved results in virtually complete agreement with experiment.

The variation method allows us to refine the simple LCAO approximation

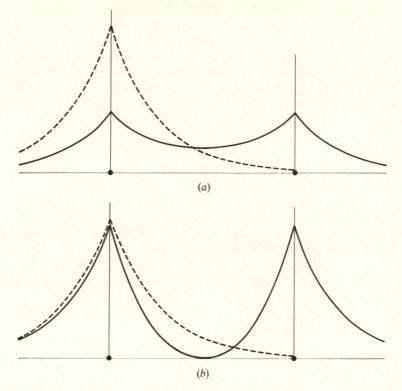

FIG. 4.5. Electron density along the internuclear axis in H_2^+. The upper curve compares ψ_+^2 (bonding density) with ϕ_A^2 for a single hydrogen atom; the lower curve shows ψ_-^2 (antibonding density).

obtaining results quite close to those of Burrau with wavefunctions of fairly simple form containing adjustable parameters. The LCAO calculation was first performed by Pauling (1928); refinements are due to Dickenson (1933), Coulson (1937), and many others. A useful bibliography is given by Richards *et al.* (1971). A first step is to allow the individual AO's to shrink closer to the nuclei as the electron is bound by *two* protons (compared with one in the H atom); a second step is to introduce an axial polarization of each AO to allow for attraction of the electron towards the second proton. Alternatively, we may discard the LCAO form (i.e. $\phi_A \pm \phi_B$) altogether and use, for example,

$$\psi = \{1 + c_2(r_a - r_b)^2\} \exp\{-c_1(r_a + r_b)\} \tag{4.15}$$

where c_1 and c_2 are chosen to minimize the computed energy. This function was used by James (1935) and found to give extremely good results. There is nowadays little difficulty in getting virtually exact results. For a review see Kolos (1968). The fact that a function such as (4.15) may be much closer to an exact wavefunction than one based on an LCAO approximation should remind us that the AO's we use have *no objective significance*; we use them as convenient 'building bricks' in constructing approximate molecular wavefunctions, and they are convenient because they allow us to visualize the electron density in the molecule by mentally breaking it down into contributions associated with the more familiar *atomic* orbitals. It is often said that in molecule formation the AO's on the constituent atoms coalesce or combine to form MO's, but such language is full of dangers unless we recognize it simply as a convenient, but inaccurate, shorthand. An orbital is a mathematical function, the solution of a one-electron Schrödinger equation, describing the state of a particle; the LCAO method represents an attempt to obtain reasonable approximate solutions for a complicated system (molecule) in terms of known solutions for its component parts (atoms).

A final comment is that just as we obtained the lowest two MO's by using the LCAO forms $\phi_A \pm \phi_B$ and taking ϕ_A and ϕ_B to be 1s AO's of atoms A and B, so we could obtain other molecular orbitals by taking ϕ_A and ϕ_B to be 2s, $2p_x$, ... orbitals. These represent excited states (see, for example, Bates *et al.* 1953), some of which are stable and others unstable, dissociating into a proton and an *excited* hydrogen atom (in state 2s, $2p_x$, ... rather than in the 1s ground state). Each pair of AO's leads to two MO's, corresponding to the alternative choice of signs, with a raising or lowering of the energy as in Fig. 4.2.

4.5. The hydrogen molecule

We shall find that much of what we have said about H_2^+ may be carried over to other diatomic molecules. The case of H_2 is particularly simple. It differs from H_2^+ solely in that there are two valence electrons instead of one. Presumably these will both go into an MO similar to the H_2^+ orbital $\phi_A + \phi_B$, and in the ground state of H_2 this MO will therefore be doubly occupied, the electrons having opposed spins. The chief difference between the two molecules is that there are now *two* electrons contributing binding energy, although this is somewhat offset by the energy of their mutual repulsion. This repulsion may be taken into account by Hartree's method (p. 37), essentially as in the case of a single atom. The corresponding MO SCF calculation (Coulson (1938) is the first example of such a calculation) gives an energy curve similar in general appearance to that of H_2^+ but shifted so that the minimum lies at an internuclear separation $1{\cdot}40a_0$ (0·074 nm), somewhat less than the value of $2{\cdot}00a_0$ for H_2^+. By accident, the binding energy of H_2 (4.75 eV) is about twice that for H_2^+ (2·79 eV), but this has no special significance. When comparing H_2^+ and H_2 we must remember that there is a large mutual electron repulsion in H_2 not found in H_2^+, and there is also a considerable change in the nuclear Coulomb repulsion, 13·6 eV in H_2^+ and 19·3 eV in H_2. There is no reason why a similar 'accident' should occur with other molecules, and indeed it seldom does. It would not even be the case for H_2^+ and H_2 if we calculated the energies

TABLE 4.2
Energy terms in H_2 calculated by MO theory

Energy of each electron in its MO	$-16\cdot2\,eV$
Repulsion between electrons	$17\cdot8\,eV$
Nuclear Coulomb repulsion	$19\cdot3\,eV$
Energy of two H atoms	$-27\cdot2\,eV$
Calculated dissociation energy $D_e = -27\cdot2 - (-2 \times 16\cdot2 - 17\cdot8 + 19\cdot3)$	$3\cdot6\,eV$

Taken from the MO SCF calculations of C. A. Coulson (1938).

of both at the same internuclear distance and not at their respective equilibrium values. These facts become clear when we consider Table 4.2 which shows the calculated values of the separate energy terms for H_2. The calculated dissociation energy is about $1\cdot1\,eV$ too small, which is about the best that can be done by the Hartree SCF approximation for molecules. This error is a little larger than in the corresponding atomic problem (the ground state of helium), though of the same order of magnitude.

In the last line of Table 4.2 the electronic repulsion 17.8 eV has to be subtracted and not added because, as noted on p. 38, this repulsion has already been counted twice in adding the separate MO energies. This table emphasizes once again how much smaller the dissociation energy is than most of the other energy terms.

The hydrogen molecule will be discussed in more detail in a later section, where various calculations are compared, but it should be noted that the self-consistent MO's on which Table 4.2 is based are formally similar to those defined in (4.15) and are therefore not of LCAO form. When an LCAO approximation is used, the results are broadly similar but rather less accurate.

As in the case of H_2^+, it is of interest to examine the form of the charge cloud. Here there are two electrons in ψ, giving a density $P = 2\psi^2$, and if for simplicity we use the LCAO approximation (4.13) to the bonding MO we obtain

$$P = 2N_+^2(\phi_A^2 + \phi_B^2 + 2\phi_A\phi_B).$$

The normalizing factor may be evaluated by requiring $\int \psi^2 \, d\tau = 1$, which gives $N_+^2 = 1/2(1+S)$. Hence

$$P = \frac{\phi_A^2 + \phi_B^2 + 2\phi_A\phi_B}{1+S}. \tag{4.16}$$

If we integrate P over all space we obtain the total charge, namely two electrons, but the contributions to P come from three regions: the first (i.e. ϕ_A^2) term in (4.16) is the electron density in a single hydrogen atom (A) multiplied by $1/(1+S)$ and since $\int \phi_A^2 \, d\tau = 1$ the corresponding charge contribution is

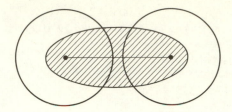

FIG. 4.6. Schematic breakdown of charge cloud into 'orbital' and 'overlap' regions. The overlap region is shown shaded.

$1/(1+S)$; similarly, the second term gives a contribution $1/(1+S)$ associated with the region of atom B; the final term gives a contribution $2S/(1+S)$ (since $\int \phi_A \phi_B \, d\tau = S$), arising from the 'overlap region' where ϕ_A and ϕ_B both have substantial values. This gives us a very graphic way of describing the bond; we may say that in the molecule $1/(1+S)$ electrons are associated formally with each atom, while $2S/(1+S)$ have 'gone into the bond'. The three regions concerned are indicated schematically in Fig. 4.6 and the numbers of electrons associated with these regions are described as their 'electron populations'. This term was introduced by Mulliken (1955), but the quantities referred to had been used previously for various purposes (see, for example, McWeeny (1952) for applications in crystallography). The populations in this case add up to 2,

$$1/(1+S) + 1/(1+S) + 2S/(1+S) = 2$$

and the 'bookeeping' is therefore correct. This idea will be developed later; here we note only that the charge cloud has the full symmetry of the molecule, both atoms having identical populations as we should expect, and that the positive overlap population indicates a 'piling up' of density in the bond region. The two electrons are thus 'shared' between the nuclei and the attraction of the nuclei towards the high-density region of the charge cloud leads to a typical electron-pair bond.

Again we must remember that terms in an LCAO approximation have no objective significance, relating only to our efforts to construct wavefunctions which can be readily visualized, though not of high accuracy. If we wish to avoid such an approximation, we can simply plot the electron density—either along a given axis, as in Fig. 4.5, or in the form of a contour map, as in Fig. 4.4. To verify that the electron density between the nuclei is enhanced by bonding, we may subtract the densities corresponding to two non-bonded hydrogen atoms located at the positions of the nuclei. The resultant 'difference density' (Fig. 4.7), commonly used by crystallographers, shows clearly how density is built up in the bond region and depleted elsewhere. As the numerical values show, the actual movement of charge (perhaps 0·1–0·3 of an electron flowing into the bond region) is rather smaller than the population values would

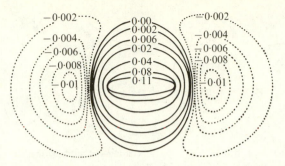

FIG. 4.7. Difference-density contours in H_2. The solid lines denote excess electron density; the dotted lines denote reduced density.

suggest; for example, with $S = \frac{1}{2}$, the formal overlap population would be $\frac{2}{3}$ electron; this is merely because the latter refer to large mutually overlapping regions and fail to take account of the point-by-point variation of the density.

4.6. Homonuclear diatomic molecules

Molecular orbital calculations, using a modified form of Hartree's original SCF techniques, have now been made for a wide variety of diatomic molecules. Typical electron density maps may be found in Wahl (1966) and Bader (1970). Some of these results are reproduced in Chapter 6. Nevertheless, a satisfactory qualitative discussion can still be based on the principles formulated many years ago by Lennard-Jones (1929) and by Mulliken (1932). Thus, when we bring two identical atoms together, the AO's ϕ_A and ϕ_B are replaced by MO's whose LCAO forms are $N_{\pm}(\phi_A \pm \phi_B)$; one of these will usually be bonding and the other will be antibonding, the difference in energy between the two being determined by the appropriate bond integral β. At large distances the energies of both MO's tend to the energy of the AO ϕ_A. Thus two 2s AO's yield two MO's which could be written

$$\psi_{\pm} = N_{\pm}[\phi_A(2s) \pm \phi_B(2s)] \tag{4.17}$$

with a corresponding energy diagram exactly like that in Fig. 4.2. We may picture such orbitals in exactly the same way as for *atomic* orbitals (Fig. 2.3) by sketching boundary contours to indicate the regions where ψ is large. The formation of the MO's in (4.17) from their constituent AO's is indicated in Fig. 4.8. Similar diagrams for p-type AO's, pointing along and transverse to the molecular axis, are shown in Figs. 4.9 and 4.10 respectively. In every case the value of ψ at any point is obtained by adding (or subtracting) the values of ϕ_A and ϕ_B at that point, and subsequently multiplying by the constant N_+ (or N_-). It is important to note that the different 'lobes' in any one MO diagram are all parts of the same MO; they normally alternate in sign and are separated by nodal planes on which $\psi = 0$, exactly as in the case of the AO's (Fig. 2.7).

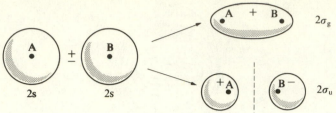

FIG. 4.8. Combination of two 2s AO's to yield bonding and antibonding MO's (denoted by $2\sigma_g$ and $2\sigma_u$). The nodal plane is indicated by the broken line.

The MO's in Figs. 4.4 and 4.8–4.10 have been named according to a standard convention, which has some similarity to that used for AO's. The symbols σ and π are analogous to s and p and refer to the *symmetry* of the MO; a σ orbital is unchanged by rotation around the molecular axis, i.e. has axial symmetry; a π orbital, which has positive and negative lobes separated by a single nodal plane, changes sign on rotation by half a turn; an orbital constructed from d_{xy} AO's on centres A and B would have *two* nodal planes, breaking it into four lobes, and would change sign on rotation through a *quarter* turn. Another mathematical property of such $\sigma, \pi, \delta, \ldots$ orbitals is that they describe states in which the electronic angular momentum around the bond axis has magnitude $0, 1, 2, \ldots$ in units of $\hbar$ (cf. §3.10(i)); as in the case of s, p, d, $\ldots$ AO's, behaviour under rotation around an axis is thus linked with a corresponding angular momentum component; the greater the number of angular nodes the greater is the angular momentum. Actually it is the combinations such as $\pi_x \pm i\pi_y$ (cf. p. 407) which represent the appropriate 'travelling-wave' solutions with definite angular momentum; to be correct we should thus use labels ± 1 instead of x, y on our MO's.

The subscript g or u (*gerade*, even, or *ungerade*, odd) is also a symmetry symbol and applies when the molecule has a *centre of symmetry*, being

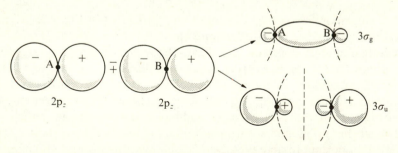

FIG. 4.9. Combination of two $2p_z$ AO's to yield bonding and antibonding MO's (denoted by $3\sigma_g$ and $3\sigma_u$). The z axis is taken pointing to the right; note the signs on the lobes and the ways in which they combine. The nodal surfaces are indicated by broken lines.

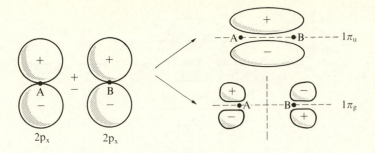

FIG. 4.10. Combination of two $2p_x$ AO's to yield bonding and antibonding MO's (denoted by $1\pi_u$ and $1\pi_g$), x axis taken vertically. Note the signs on the lobes. The nodal surfaces are indicated by broken lines. Each MO is degenerate, its partner being similarly constructed from $2p_y$ AO's.

unchanged in appearance under the operation of *inversion* (in which every point is interchanged with a corresponding point diametrically opposite across the centre of symmetry). If a wavefunction is left unchanged by such an operation it is of g type; if it changes sign it is of u type. For example, s and d AO's are g type, while p AO's are u type, the centre of symmetry being the nucleus. Reference to Figs. 4.9 and 4.10 shows that the g or u classification depends both on the nature of the AO's combined and on whether they are combined with a + or a − sign. For example, in a σ MO the bonding combination is g type, whereas in a π MO the bonding combination is u type.

Finally, the number (1, 2, 3, ...) which preceeds the symmetry symbol simply puts the MO's of that particular symmetry in ascending energy order. If we are to adopt an *aufbau* approach to molecular structure we shall need to know the energy order of all the available orbitals, just as we did in the case of atoms (cf. Fig. 2.11). The sequence most usually found is

$$1\sigma_g < 1\sigma_u < 2\sigma_g < 2\sigma_u < 1\pi_u < 3\sigma_g < 1\pi_g < 3\sigma_u < \ldots$$

and the way in which the MO energies are related to those of the constituent AO's is shown in Fig. 4.11. It should be noted that every π type MO is doubly degenerate; for example, $1\pi_u$ stands for the pair $1\pi_{xu}$, $1\pi_{yu}$ which could be constructed from p_x and p_y AO's, respectively. Also, the $1\pi_u$ and $3\sigma_g$ MO's have rather similar energies and in some molecules (O_2, O_2^+, F_2, F_2^+) may change places; for *second*-row diatomics the σ_g is always below the π_u.

The energy ordering of the MO's was first inferred by Mulliken (1932) from a study of molecular spectra. His predictions have in the main been verified by direct calculation of MO's, using the SCF method which (with large computers) has since become feasible. There is, however, another striking experimental verification of the order (and indeed of the remarkable applicability of the orbital picture, which, as already noted, is itself an

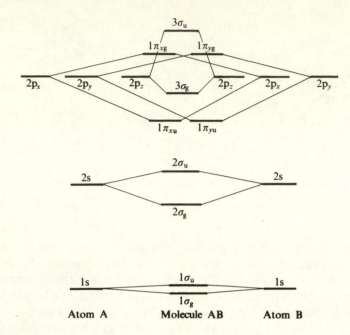

FIG. 4.11. MO energy levels in a homonuclear diatomic moleculé AB related to those of the separate atoms A and B (schematic).

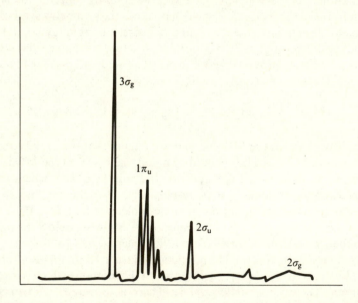

FIG. 4.12. Photoelectron spectrum of N_2 showing ionization peaks for removal of electrons from the four MO's indicated. The 'structure' is due to vibrational effects. (Courtesy of W. C. Price.)

approximation); this rests upon the photoelectric effect (p. 16) which has been adapted by Turner and Price for the accurate measurement of ionization potentials—a technique basically similar to that already sketched in §1.5 (iii). A useful review has been given by Hollander and Jolly (1970), and a comprehensive account by Turner *et al.* (1970). If I is the energy needed to remove an electron from a given MO in the molecule, and this is achieved by an incident photon of energy hv, then the kinetic energy of the ejected electron will be

$$T = hv - I.$$

Accurate measurement of the kinetic energy of ejected electrons when a homogeneous beam of light falls on a sample generally gives a series of peaks corresponding to I values for a whole range of MO's. Since the corresponding orbital energy is simply $-I$, such experiments give very direct information on the energy-level diagram. An example is given in Fig. 4.12 where the peaks are labelled with the names of the MO's from which the electrons are ejected. The 'fine structure' which splits some peaks is associated with the vibrational energy levels of the ionized species and proves to be a valuable guide to the assignment.

The positions of the MO energy levels are frequently discussed by means of a *correlation diagram*. Such a diagram shows how the orbital energies change as a function of the internuclear distance R. When R is very large the MO's must turn into AO's of the separate atoms, with corresponding AO energies; when $R \to 0$, the MO's must turn into the AO's of a 'united atom' in which the two nuclei have coalesced. The two ways of looking at the behaviour of the MO's are called the 'separated-atom' and 'united-atom' viewpoints. At intermediate distances the AO's must be mixed to give MO's, those which are closest in energy mixing most heavily, but as $R \to 0$ the *symmetry* of each MO is unchanged and it must therefore turn into a united-atom AO of the same symmetry type. A typical correlation diagram for a homonuclear diatomic molecule is shown in Fig. 4.13. To give one example of a correlation, we note (Fig. 4.14) that combination of the two $2p_x$ AO's of the separated atoms (extreme right) will yield $1\pi_u$ and $1\pi_g$ MO's which, as $R \to 0$ must pass into p_x and d_{xz} AO's, respectively, on the united atom. The straight lines connecting the orbitals on the two sides of Fig. 4.13 simply indicate such correlations and suggest the directions in which the levels move as the internuclear distance changes; at the distance indicated by the vertical broken line the sequence of energies is the one shown in Fig. 4.11. It should be noted that the non-crossing rule (p. 67) must not be violated in constructing such diagrams; if a crossing occurs it must be between states of *different* symmetry type. There are no exceptions to the rule *so long as symmetry is preserved*. However, as the example of H_2^+ shows, assignment of an electron to a *separated-atom* MO may not give a correct description of the *physical* situation; the H_2^+ electron cannot remain half on one nucleus and half on the other as $R \to \infty$. We see later that this is a defect of the MO description, not of the correlation diagram.

It is now a relatively simple matter to use the *aufbau* approach to describe simple homonuclear diatomics. We have seen that in H_2 the two electrons occupy the bonding MO, i.e. the electron configuration is $(1\sigma_g)^2$. Similarly, He_2^+ would have a ground state $(1\sigma_g)^2(1\sigma_u)$ and He_2 would be $(1\sigma_g)^2(1\sigma_u)^2$. In fact, He_2 is an unstable molecule in its ground state and stability is achieved only in the excited states (spectroscopically observed) in which one electron has been promoted to a higher-energy MO.

The instability of He_2 illustrates a rather general principle that when a bonding orbital and its antibonding 'partner' are both fully occupied there is no resultant bonding, or even a small degree of antibonding (i.e. repulsion). By 'partners' we mean MO's constructed from the same pair of AO's. The reason is clear on energetic grounds from Fig. 4.2 since the resultant energy will clearly be *greater* than that of two separate He atoms. However, it is also

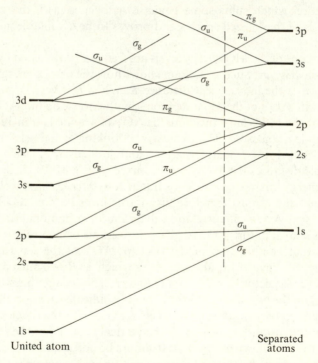

FIG. 4.13. Correlation diagram for a homonuclear diatomic molecule, relating united-atom levels (left) to those of the separated atoms (right). Symmetry of the LCAO's (σ_g, σ_u, π_g, ...) must be preserved as the internuclear distance changes; there can be no mixing between orbitals of different symmetry. The vertical broken line, nearer the separated atom limit, corresponds to the energy level sequence shown in Fig. 4.11. The diagram is purely schematic; actual calculations of energy against internuclear distance show more complicated behaviour.

informative to derive the charge density, $P = 2\psi_+^2 + 2\psi_-^2$ where ψ_+ and ψ_- are the MO's in question, given by (4.13). The result is easily found to be

$$P = \frac{2}{1-S^2}[\phi_A^2 + \phi_B^2 - 2S\phi_A\phi_B]. \qquad (4.18)$$

Thus (cf. the discussion of H_2, following equation (4.16)) an amount of charge $4S^2/(1-S^2)$ has been 'drained away' from the overlap region, as indicated by the negative sign with which the $\phi_A\phi_B$ term enters. For He_2, S is small and the antibonding effect is not large. When S is neglected, the density reduces to $2\phi_A^2 + 2\phi_B^2$ which is exactly that of two non-interacting helium *atoms*, as if each had retained two electrons in its own AO. More accurately, the corresponding density is slightly reduced, in the region of overlap, and the nuclei then experience a net repulsion.

We can now proceed to other diatomics, noting that superposition of bonding and antibonding pairs leads to no net bonding and that for inner shells in particular (with very small values of S) the MO description is precisely equivalent to leaving the atomic densities undisturbed. We frequently indicate

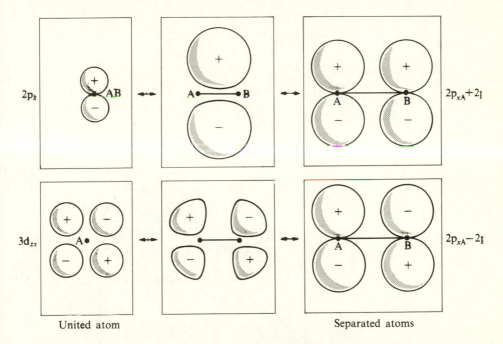

United atom   Separated atoms

Fig. 4.14. Examples of correlation between orbital forms in the united-atom and separated-atom limits (homonuclear diatomic). The upper diagram shows how the π_u combination of two separated-atom p orbitals (right) correlates with a united-atom p orbital (left). The lower diagram shows how the π_g combination (right) goes to a united-atom d orbital (left). The united-atom AO's are relatively compact owing to the increased nuclear charge when A and B coalesce.

TABLE 4.3

Electron configurations in ground states of homonuclear diatomic molecules

		No. of valence electrons		Bond number	Dissociation‡ energy (eV)	Spectroscopic ground state
	Electron configuration†	Bonding	Antibonding			
H_2^+	$1\sigma_g$	1		$\frac{1}{2}$	2·793	$^2\Sigma_g^+$
H_2	$1\sigma_g^2$	2		1	4·748	$^1\Sigma_g^+$
He_2^+	$1\sigma_g^2 1\sigma_u$	2	1	$\frac{1}{2}$	2·6	$^2\Sigma_u^+$
He_2	$1\sigma_g^2 1\sigma_u^2\ (=KK)$	2	2	0		$^1\Sigma_g^+$
Li_2	$KK\,2\sigma_g^2$	2	0	1	1·14	$^1\Sigma_g^+$
Be_2	$KK\,2\sigma_g^2 2\sigma_u^2$	2	2	0		$^1\Sigma_g^+$
B_2	$KK\,2\sigma_g^2 2\sigma_u^2 1\pi_u^2$	4	2	1	3·0	$^3\Sigma_g^-$
C_2	$KK\,2\sigma_g^2 2\sigma_u^2 1\pi_u^3$	6	2	2	6·36	$^1\Sigma_g^+$
N_2^+	$KK\,2\sigma_g^2 2\sigma_u^2 3\sigma_g 1\pi_u^4$	7	2	$2\frac{1}{2}$	8·86	$^2\Sigma_g^+$
N_2	$KK\,2\sigma_g^2 2\sigma_u^2 3\sigma_g^2 1\pi_u^4$	8	2	3	9·90	$^1\Sigma_g^+$
O_2^+	$KK\,2\sigma_g^2 2\sigma_u^2 3\sigma_g^2 1\pi_u^4 1\pi_g$	8	3	$2\frac{1}{2}$	6·77	$^2\Pi_g$
O_2	$KK\,2\sigma_g^2 2\sigma_u^2 3\sigma_g^2 1\pi_u^4 1\pi_g^2$	8	4	2	5·21	$^3\Sigma_g^-$
F_2^+	$KK\,2\sigma_g^2 2\sigma_u^2 3\sigma_g^2 1\pi_u^4 1\pi_g^3$	8	5	$1\frac{1}{2}$		$^2\Pi_g$
F_2	$KK\,2\sigma_g^2 2\sigma_u^2 3\sigma_g^2 1\pi_u^4 1\pi_g^4$	8	6	1	1·34	$^1\Sigma_g^+$

† The order of the orbitals does not necessarily correspond to increasing energy (see text).
‡ Best values currently available, but some uncertainties remain.

the inner shells simply by giving their names K, L, M, ..., and thus Li_2 with two valence electrons outside the inner shells could be written

$$Li_2[(1\sigma_g)^2(1\sigma_u)^2(2\sigma_g)^2] \qquad \text{or} \qquad Li_2[KK(2\sigma_g)^2]$$

the KK indicating the two undisturbed K shells. Similarly, F_2 would have the configuration

$$F_2[KK(2\sigma_g)^2(2\sigma_u)^2(1\pi_u)^4(3\sigma_g)^2(1\pi_g)^4]$$

where $(1\pi_u)^4$, for example, stands for the *degenerate pair* of doubly occupied MO's $(1\pi_{xu})^2(1\pi_{yu})^2$. Since $2\sigma_g$ and $2\sigma_u$ are bonding and antibonding partners, and likewise $1\pi_u$ and $1\pi_g$, it follows that out of the 14 valence electrons only the $(3\sigma_g)^2$ pair actually provides bonding; we say there is a *single bond* of σ type.

In the nitrogen molecule there are four fewer valence electrons and we obtain the configuration

$$N_2[KK(2\sigma_g)^2(2\sigma_u)^2(1\pi_u)^4(3\sigma_g)^2].$$

This is of special interest because the molecule is usually said to possess a *triple bond*. In MO language there is a single bond of σ type $(3\sigma_g)^2$, supplemented by two bonds of π type arising from the electron pairs $(1\pi_{xu})^2(1\pi_{yu})^2$. The triple bond thus differs from a superposition of three axially symmetric single bonds, the π contributions lying *off* the axis (Fig. 4.10). To investigate the form of the π density we write the $1\pi_{xu}$ MO in LCAO form

$$\psi_{\pi x} = x\{f(r_a) + f(r_b)\}$$

i.e. a combination of two $2p_x$ AO's pointing transverse to the bond (z) axis. On taking a similar expression for the $1\pi_{yu}$ MO, the π density $P_\pi = 2\psi_{\pi x}^2 + 2\psi_{\pi y}^2$ becomes

$$P_\pi = 2(x^2 + y^2)\{f(r_a) + f(r_b)\}^2.$$

Now $x^2 + y^2 = r^2$, where r is the distance of point (x, y, z) from the z axis, and P_π is therefore axially symmetric but zero on the bond axis itself ($x = y = 0$). The π density in a triple bond can thus be visualized as a 'sheath' of charge, enveloping and strengthening the central σ bond.

Table 4.3 summarizes the results for a number of molecules, showing also a 'bond number' which indicates the number of bonds formed and is defined as one-half the excess number of bonding electrons over antibonding electrons. This table indicates why bond numbers in excess of 3 do not occur and shows a rough proportionality between bond number and dissociation energy.

The last column of Table 4.3 shows the spectroscopic notation for the various states. These involve (*a*) a raised prefix numeral 1, 2, 3, ..., (*b*) a capital Greek† symbol Σ, Π, Δ, ..., and (*c*) a final subscript, u or g. The first of these

† Σ, Π, Δ, ... are the Greek capitals corresponding to σ, π, δ, ...

tells us the spin multiplicity $2S + 1$ for total spin S. Any pair of electrons which completely fills an MO contributes nothing because, by the Pauli principle, their spins must necessarily be opposed. If there is no resultant spin, the multiplicity is 1. However, if there is one unpaired electron, its spin may be $+\frac{1}{2}$ or $-\frac{1}{2}$ and the state is a doublet. If there are two unpaired electrons, their spins may be parallel, giving a triplet ($S = 1$), or they may be antiparallel, giving a singlet. Singlets and triplets are more fully discussed in §5.6. The question whether the singlet or triplet is lower, is usually settled by an appeal (p. 40) to Hund's rules.

The main symbol in the spectroscopic notation denotes the resultant symmetry character, and again depends on contributions from each electron (0 for a σ-type MO, ± 1 for a π-type MO, etc.). It indicates the component of total angular momentum around the nuclear axis (§3.10). If the total value is 0, the state is a Σ state; if it is 1, the state is Π, etc.

Finally the subscript u or g describes the behaviour of the complete wavefunction† with reference to inversion (p. 95) through the mid-point of the bond; while the superscript ± 1 refers to reflection through a plane containing the nuclei. Since a u-type orbital is multiplied by -1 on inversion, while a g-type orbital is unchanged, the total wavefunction will necessarily be g unless there is an odd number of u orbitals occupied, in which case it must be u.

Reference to the last column of Table 4.3 will show these rules in operation. It may be stated that precisely similar descriptions apply to heteronuclear diatomics, with the single exception of the u and g character. On account of the lack of any centre of symmetry in such a molecule, the u and g classification disappears.

We conclude with a few general comments. First, it will be noted that the terms single, double, and triple bond refer to the number of bonding electron pairs which effectively contribute to the bond, and that the type of bond depends on the σ or π type of such a pair. The characteristic patterns of the electron configurations are

$$\text{single bond } \sigma^2, \qquad \text{double bond } \sigma^2\pi^2, \qquad \text{triple bond } \sigma^2\pi^4.$$

No further possibilities exist for first-row atoms, but further variety is encountered higher in the periodic table. Thus in some strong homonuclear metal–metal bonds (Cotton 1969) there may be quadruple bonds with pattern $\sigma^2\pi^4\delta^2$.

Secondly, the principle of maximum overlap (p. 82) suggests that, since π-type AO's (transverse to the bond) appear to overlap less than σ-type AO's (along the bond), π bonds should be weaker than σ bonds. This conjecture is supported in many cases by actual calculation. For instance, at a

---

† In §5.1 we show that this may be represented as a product of orbital factors; its behaviour may thus be inferred from that of the individual orbitals.

carbon–carbon distance of 0·154 nm, the p_σ and p_π overlap integrals are roughly in the ratio 2:1; the relative weakness of the π bond is reflected in the high reactivity of the ethylene double bond and the acetylene triple bond, for it is easier to disengage the less tightly bound π electrons and link them up with other approaching atoms. However, at shorter distances, owing to the form of the p AO's, the situation may be reversed by a *decreasing* σ overlap. This occurs in N_2 for example (Mulliken 1952). Later, however, we find the situation is less simple than this argument would suggest (§6.2).

Finally, we note that in filling sets of degenerate orbitals (in this case the π_x, π_y pairs) Hund's rules may be applied exactly as in the atomic *aufbau* procedure. It is often useful to indicate the filling of the MO's in a level diagram of the kind already shown in Fig. 4.11. Such a diagram is drawn for the molecule O_2 in Fig. 4.15, where the levels for the separate atoms are shown on the right and the left and related to those for the molecule, shown in the centre. The figure reveals at once how in the O_2 molecule all the levels up to $1\pi_u$ are completely filled and how the two parallel spins of the degenerate $1\pi_g$ level lead to a triplet ground state ($S = 1$). It was one of the earliest triumphs of MO theory to account so neatly for the observed triplet ground state of O_2 and its related paramagnetism.

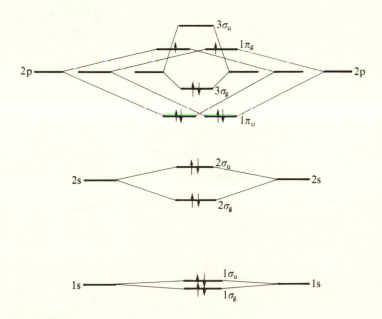

FIG. 4.15. Occupation of MO's in the molecule O_2. Energy levels for degenerate orbitals are shown separately (e.g. $1\pi_u$ includes the two MO's $1\pi_{xu}$ and $1\pi_{yu}$) and the spins of the electrons in the $1\pi_g$ pair are parallel coupled. The molecule is paramagnetic.

4.7. Heteronuclear diatomic molecules

The molecules we have considered so far, in which the atoms A and B are identical, exhibit 'homopolar' binding. Each atom has the same polarizing effect on the electron distribution, with the result that there is no net shift of electron density from one atom to the other. Much more commonly, however, we encounter heteropolar binding, in which atoms A and B are different. Nevertheless, much of our previous discussion will still apply. Thus, in heteronuclear diatomics we can still apply the LCAO approximation (§4.3), writing for the MO an expression

$$\psi = c_A \phi_A + c_B \phi_B \tag{4.19}$$

as in (4.11). But no longer does $c_A = \pm c_B$; nor is the energy given by the simple expression (4.12), and we are obliged to solve the quadratic secular equation (4.8). The conditions (*a*)–(*c*) of §4.3, *viz.* nearly equal energies, maximum overlapping, and the same symmetry around the bond axis, generally allow us to infer quite easily which AO's ϕ_A and ϕ_B may be compounded together. It is still true that MO's are of $\sigma, \pi, \ldots$ type, and the total number of such orbitals, i.e. their degeneracy, is the same as in the

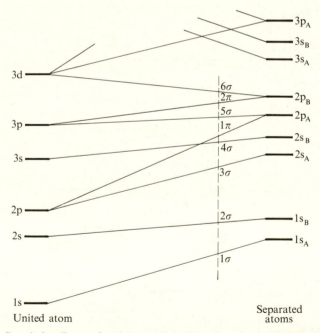

FIG. 4.16. Correlation diagram for a heteronuclear diatomic molecule (cf. Fig. 4.13). Only σ and π symmetry types occur, the g or u classification being lost. The vertical broken line corresponds to the sequence of the MO energies in the NO molecule. (Adapted from R. S. Mulliken, *Rev. mod. Phys.* **4**, 1 (1932).)

homonuclear case. The non-crossing rule of §3.9 is still operative to keep the energies of MO's with the same symmetry apart and to correlate the united-atom and separated-atom parentages of given molecular levels. The corresponding diagram, similar to that of Fig. 4.13 for homonuclear molecules, is shown in Fig. 4.16. It is clear that the chief differences between the two diagrams are (a) the lack of the u, g property (p. 94), which follows from the absence of a centre of symmetry in all these molecules, and (b) the fact that in the separated atoms the energies of similar orbitals are different. For example, in NO the 3σ MO correlates with the oxygen 2s and the 4σ with the nitrogen 2s; on account of its greater nuclear charge the oxygen level is lower than the nitrogen level. However, in O_2 both 3σ and 4σ correlate with the same atomic energy at infinite separation of the nuclei.

To understand the nature of the correlation diagram we go back to the secular equations (4.7) and (4.8) which determine the energies of the bonding and antibonding MO's, which we shall now denote by ψ_1 and ψ_2 respectively, and the corresponding coefficients in (4.19). The solutions for homonuclear and heteronuclear bonding are radically different and it is useful to contrast the two cases.

Case (i) $\alpha_A = \alpha_B$. This occurs in a homonuclear molecule when ϕ_A and ϕ_B are identical AO's, e.g. the two 2s AO's in N_2. We have already found (§4.4) that in such a situation, with $\alpha_A = \alpha_B = \alpha$,

$$E = \frac{\alpha \pm \beta}{1 \pm S} \qquad \psi = \frac{\phi_A \pm \phi_B}{\sqrt{\{2(1 \pm S)\}}}. \qquad (4.20)$$

The correlation between the bonding and antibonding levels (upper and lower signs, respectively) and those in the unbonded atoms is indicated in Fig. 4.17 (a). When $S \ll 1$, the splitting of the AO levels is 2β, and since the magnitude of β (here assumed negative) is larger the larger the overlap, we obtain a basis for the principle of maximum overlap (p. 82). The forms of ψ_1 and ψ_2 are *independent of* β, being determined by symmetry, and in each MO the two AO's occur with equal weight at all internuclear distances.

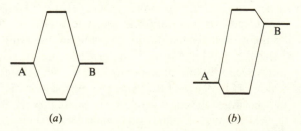

(a)　　　　　　　　　(b)

FIG. 4.17. Modification of AO energy by mixing (a) homonuclear case and (b) heteronuclear case (atom A more electronegative than B).

Case (ii) $\alpha_A \neq \alpha_B$. This occurs in a heteronuclear molecule, where the AO's to be combined may differ somewhat in energy, e.g. the 2s AO's in NO. Instead of solving the quadratic equation (4.8) directly, we may obtain the general character of the results by a simple approximation. The effect of interaction arises from the term $(\beta - ES)^2$, whose neglect leads to the solutions $E = \alpha_A$ or $E = \alpha_B$, corresponding to uncombined AO's ϕ_A and ϕ_B, respectively. Let us suppose $E = \alpha_A$ is the lower of the two AO energies (i.e. atom A is more electronegative than atom B) and suppose the bonding MO has an energy close to α_A. We can then substitute $E \simeq \alpha_A$ everywhere in (4.8) except in the factor $\alpha_A - E$, which would vanish if we did, to estimate how much E is changed by the interaction. This yields

$$(\alpha_A - E)(\alpha_B - \alpha_A) - (\beta - \alpha_A S)^2 = 0$$

which gives, on rearranging, the bonding MO energy

$$E_1 = \alpha_A - \frac{(\beta - \alpha_A S)^2}{\alpha_B - \alpha_A}. \qquad (4.21a)$$

The lower of the AO energies (α_A) is thus pushed down by approximately $\beta^2/(\alpha_B - \alpha_A)$ (a little less if S is not neglected) to give the energy of the bonding MO. A similar argument shows that the upper level (α_B) is pushed *up* by the same amount (for $S = 0$, or by a little less when S is not neglected). In fact, the upper level has

$$E_2 = \alpha_B + \frac{(\beta - \alpha_B S)^2}{\alpha_B - \alpha_A}. \qquad (4.21b)$$

The situation is summarized in Fig. 4.17 (*b*). Whereas the shifts in the homonuclear case were roughly $\pm \beta$, they are now proportional to β^2 and are also diminished in proportion to the separation of the AO energies. The effects of interaction are thus of 'second order' by comparison with the homonuclear case, though clearly they may be large for AO's of similar energy $\alpha_A \simeq \alpha_B$. When α_A and α_B are very close, our simple approximation will of course break down.

We now have a clear basis for the qualitative rules (*a*) and (*b*) of §4.3 (p. 81). First, whether a molecule is homonuclear or heteronuclear, the energy lowering is greater the greater the magnitude of the bond integral β, and since this is the case when overlap is large we obtain the maximum overlap criterion for efficient bonding. Secondly, in constructing MO's from different types of AO we need only 'pair' AO's of about the same energy, for with different AO's (4.21a) applies and the larger the energy difference the weaker the interaction.

The forms of the AO's follow on substituting each energy, in turn, into (4.7). The first equation gives for the ratio of the AO coefficients (λ in the notation of equation (4.3)) in the bonding MO

$$\lambda = \frac{c_B}{c_A} = -\frac{(\beta - \alpha_A S)}{(\alpha_B - \alpha_A)} \qquad (4.22a)$$

whilst for the antibonding MO we obtain

$$\frac{1}{\lambda} = \frac{c_A}{c_B} = \frac{\beta - \alpha_B S}{\alpha_B - \alpha_A}. \qquad (4.22b)$$

Instead of mixing with equal weight at all distances, as in the homonuclear case ($c_B/c_A = \pm 1$), the relative weights will vary strongly with β and S and hence with internuclear distance. We are therefore in a position to discuss the correlation diagram of Fig. 4.16. For this purpose it is convenient to write the bonding MO as

$$\psi_1 = N_1 \left(\phi_A - \frac{\beta - \alpha_A S}{\alpha_B - \alpha_A} \phi_B \right) \qquad (4.23a)$$

where N_1 is a normalizing factor, and the antibonding MO as

$$\psi_2 = N_2 \left(\phi_B + \frac{\beta - \alpha_B S}{\alpha_B - \alpha_A} \phi_A \right). \qquad (4.23b)$$

Since β and the α's are normally negative and S is positive, ϕ_A and ϕ_B are combined with the same sign in (4.23a) but with opposite sign in (4.23b), giving the expected node in the antibonding MO. What is more important, however, is that when the fractions in (4.22a) and (4.22b) are less than 1 (which is implicit in our approximations) the following conclusion may be drawn.

In a heteronuclear molecule, a bonding MO leans towards the more electronegative atom (A, say) (being predominantly ϕ_A), while its antibonding partner leans towards the less electronegative atom (being predominantly ϕ_B).

At the actual internuclear distance (β large), both AO's may occur with comparable weight, but we have now established what happens in the separated-atom limit ($\beta \to 0$).

It is now a simple matter to sketch the way in which the orbital forms change as we pass between the united atom and the separated atoms. Some examples are shown in Fig. 4.18 which is constructed by pairing similar AO's (e.g. $2p_{zA}$, $2p_{zB}$), with a $+$ or a $-$ sign but with emphasis on the A orbital in the bonding combination and the B orbital in the antibonding combination. It should be noted again (cf. p. 34) how the increasing number of nodes, arising on combination, leads to AO's of higher quantum number on the united atom. The imperfections of the diagram in Fig. 4.16 must also be stressed; it is based on *pairwise* combination of AO's (i.e. the assumption that, say, $2s_A$ and $2s_B$ lie closer in energy than $2s_A$ and $2p_B$) whereas in reality *all* AO's of any given symmetry (e.g. σ or π) should be allowed to mix in the LCAO approach. This means that the diagram is grossly oversimplified and that the lines connecting

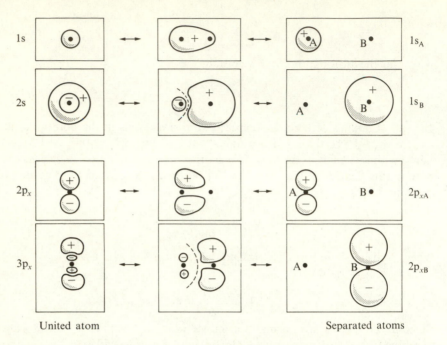

FIG. 4.18. Examples of correlation between orbital forms in the united-atom and separated-atom limits (heteronuclear diatomic). The separated-atom AO's on the right yield, by taking a bonding or an anti-bonding combination, an MO (centre) and finally a united-atom AO (left). Note how antibonding combinations lead to AO's of higher principal quantum number by introducing extra nodes.

separated and unit atom levels, far from being straight, may show avoided crossings (as in Fig. 3.13) owing to heavy interaction at certain distances among *several* orbitals of the same σ or π character. Nevertheless, such diagrams can often correctly suggest the sequence of the MO energies in a molecule. When they fail, more extensive AO mixing may need to be considered, and this leads to the concept of 'hybridization' which is taken up in later sections.

As a concrete example of the use of a correlation diagram we may consider the molecule NO. Fig. 4.16 does in fact give a fairly satisfactory interpretation of the structure of NO, which may be described as follows:

$$N[1s^2 2s^2 2p^3] + O[1s^2 2s^2 2p^4] \rightarrow NO[KK(3\sigma)^2(4\sigma)^2(5\sigma)^2(1\pi)^4(2\pi)].$$

On account of its single 2π electron this molecule possesses an unpaired electron, whose spin will give rise to paramagnetism. The broken guideline in Fig. 4.16 indicates the approximate position of this molecule on the diagram.

Although the correlation diagram is very useful, we must not press it too far. It may give a correct account of the change in an individual MO as the atoms separate, but it may nevertheless fail to suggest the correct electron configurations of the dissociation products. Consider, for example, the four-electron molecule LiH. The MO description of the molecule is $(1\sigma)^2(2\sigma)^2$, as suggested by Fig. 4.16. According to this diagram, as we separate the nuclei the lowest energy orbital should go over into the 1s orbital of the lithium atom, and the second orbital should go over into the 1s orbital of hydrogen. This might suggest LiH $\to$ Li$^+$ + H$^-$, a situation only found in solution where solvation energies favour charged species rather than neutral ones. The true situation with LiH is that the mutual repulsion of the two electrons in the 2σ orbital is sufficiently great to force them on to different nuclei, leading to

$$\text{LiH} \to \text{Li}[(1s)^2(2s)] + \text{H}[(1s)].$$

In other words the occupation numbers of the orbitals must change, to give an electron configuration $[(1s_A)^2(2s_A)(1s_B)]$ instead of $[(1s_A)^2(1s_B)^2]$. We should have had no similar difficulty with the dissociation of the three-electron system LiH$^+$, where the assignment $[(1s_A)^2(1s_B)]$ correctly describes Li$^+$ + H. In fact, the unreliable description of dissociation is a defect of the MO theory itself, rather than of the correlation diagram, as will become clear in Chapter 5.

4.8. Types of molecular orbital

It follows from §§4.6 and 4.7 that electrons in diatomic molecules may be regarded as occupying four types of orbital. These are (*a*) inner shell, (*b*) valence-shell non-bonding, (*c*) valence-shell bonding, and (*d*) valence-shell antibonding. It will be convenient to illustrate these four types in terms of two particular examples, HCl and O$_2$.

In HCl all the chlorine K- and L-shell electrons are of type (*a*), quite localized around the chlorine nucleus. If the *z* direction is from Cl towards H, then the valence-shell bonding orbital (type (c)) is compounded from Cl($3p_z$) and H(1s), both being of σ character. However, the Cl($3p_x$) and Cl($3p_y$) orbitals cannot form any LCAO with hydrogen, for the lowest AO in hydrogen which possesses π character is the H(2p) whose energy is so much greater than that of Cl($3p_y$) that no effective combination occurs. As a result, even though they are in the valence shell the electrons in these AO's remain centred round their original nucleus and merely suffer a small polarization; they are (type (*b*)) valence-shell non-bonding electrons. Two electrons with opposed spins in an orbital of this kind are called lone-pair electrons. The chlorine 3s electrons, although their AO has the same σ symmetry as the bonding MO, are considerably lower in energy than the 3p electrons and therefore also may be regarded as undisturbed by bonding; they provide a third lone pair†. This accounts for all the electrons in HCl.

---

† More accurately (§6.2) there is slight mixing of 3s and $3p_z$ to give a 'hybrid' AO, but this is still non-bonding and localized on the chlorine atom.

In the same way most of the electrons in O_2 can be allotted without difficulty to their proper type by reference to Fig. 4.15 and Table 4.3; we must remember that if there are two low-lying orbitals such as $2s_A \pm 2s_B$ it is immaterial whether or not we replace them by the AO's $2s_A$ and $2s_B$ from which both were composed (p. 98). Chemical language usually favours the latter description, describing the filled AO's as lone pairs. The only new point about O_2 is the top degenerate level $1\pi_g$ (i.e. $1\pi_{xg}$, $1\pi_{yg}$). This would be expected to be antibonding. If it were, the bond energy of O_2^+ in which there is only one of these electrons should be greater than that of O_2 in which there are two. Such is indeed the case, the appropriate values being 6·48 and 5·08 eV. We may contrast this with the situation in N_2, whose configuration (see Table 4.3 and p. 101) is similar to that in Fig. 4.15 except for the omission of the antibonding $1\pi_g$ electrons and for which the dissociation energy of the neutral molecule is greater than that of the positive ion (7·38 eV for N_2, 6·35 eV for N_2^+).

Problems

4.1. At the equilibrium internuclear distance ($R = 1·40$ bohr) in the molecule H_2, the overlap integral S between two hydrogen 1s AO's is about 0·75. Evaluate the normalizing factors for the bonding and antibonding MO's given in equation (4.13). Then use 1s functions of the form given in equation (1.12) to evaluate ψ^2, for each MO, at points along the internuclear axis. Finally, plot curves comparing the charge density in the molecule with that for two non-interacting hydrogen atoms, (*a*) with both electrons in the bonding MO, and (*b*) after exciting one electron to the anti-bonding MO.

4.2. Estimate the orbital and overlap populations (p. 92) for the two situations ((*a*) and (*b*)) referred to in Problem 4.1.

4.3. Label the rows in Table 4.1 with the symbols $\sigma, \pi, \delta, \ldots$ to indicate the symmetry of the LCAO MO's. In which cases will the bonding MO be of g type, and when will it be u type?

4.4. Give a qualitative discussion of the electronic structures of the 3rd row diatomic molecules Na_2, P_2, S_2, along the lines of that given in the text for 2nd row molecules. Would the bonds in P_2^+ and S_2^+ be stronger or weaker than those in the neutral molecules?

4.5. Derive equation (4.18) for the charge density in the hypothetical molecule He_2.

4.6. Use Fig. 2.5 to show that the *complex* MO's $\psi_{\pi, \pm 1} = \psi_{\pi x} \pm i\psi_{\pi y}$, formed by combining degenerate MO's of (real)π_x and π_y type (Fig. 4.10), depend on the angle ϕ through a factor $e^{\pm i\phi}$. Hence show (cf. Problem 3.12) that the complex MO's describe states in which the electron has ± 1 unit of angular momentum around the internuclear axis. (As in the case of atomic orbitals, the real and complex wavefunctions represent equally acceptable 'standing wave' and 'travelling wave' solutions of the Schrödinger equation, respectively. The complex forms are used in discussing spectroscopic states. See Appendix 2.)

4.7. Calculate the energies of the σ-type bonding and antibonding MO's formed from the Li 2s AO and the H 1s AO in lithium hydride, (neglecting any 2p participation) assuming overlap $S = 0.47$ and

$$\alpha_{2s} = -0.23 \text{ hartree}, \qquad \alpha_{1s} = -0.39 \text{ hartree}, \qquad \beta = -0.21 \text{ hartree}.$$

Hence construct a diagram like that shown in Fig. 4.17 (*b*). (Hint: solve the quadratic equation (4.8).)

4.8. Repeat the calculation of Problem 4.7, but using the approximations (4.21*a, b*). Compare the two sets of results.

4.9. Use the data of Problem 4.7 to obtain, using equation (4.22*a*), an approximate bonding MO for the molecule LiH. Normalize the MO and use an analysis parallel to that on p. 92 to describe the bond charge density in terms of orbital and overlap populations.

5

Electron-pair wavefunctions

5.1. Two-electron wavefunctions

In previous chapters we have tried to obtain a qualitative description of the electronic structures of atoms and diatomic molecules using an independent particle model (p. 35), in which each electron is considered as if it moved *by itself* in some effective field (the 'self-consistent field'). We have therefore only needed to consider a *one*-electron Schrödinger equation, whose solution leads to the required one-electron wavefunctions or *orbitals*. However, atoms and molecules are generally *many*-electron systems, and it is now time to go a little deeper and show how a molecular wavefunction may actually be calculated. In so doing we shall obtain a firmer basis for the model used so far, and shall find how it may be improved and linked with other methods of constructing approximate wavefunctions. Most chemical bonds are provided by *pairs* of electrons, as we have already seen, and in this chapter we shall be concerned largely with the simplest of all 'electron-pair' bonds—that in the hydrogen molecule.

First, we need a very general and important result. If a system comprises two *non-interacting* parts, A and B, its Hamiltonian operator takes the form

$$\hat{H} = \hat{H}_A + \hat{H}_B \tag{5.1}$$

where $\hat{H}_A$ and $\hat{H}_B$ refer to the separate parts. Suppose system A has possible states $1, 2, \ldots i, \ldots$ with wavefunctions $\Psi_1^A, \Psi_2^A, \ldots \Psi_i^A, \ldots$ and similarly for system B. We shall often use a capital psi when we wish to remind ourselves that the wavefunction may be a *many*-electron function and not just an orbital. Then an allowed wavefunction for the whole system (AB) is

$$\Psi_{ij}^{AB} = \Psi_i^A \Psi_j^B \tag{5.2}$$

in which the two factors of the product describe part A in state i and part B in state j respectively, and the corresponding energy is $E = E_i^A + E_j^B$.

This result is easily verified from the properties

$$\hat{H}_A \Psi_i^A = E_i^A \Psi_i^A, \quad \hat{H}_B \Psi_j^B = E_j^B \Psi_j^B \tag{5.3}$$

which define the stationary states of the two systems respectively. Thus, assuming (5.1),

$$\hat{H}\Psi_{ij}^{AB} = (\hat{H}_A\Psi_i^A)\Psi_j^B + \Psi_i^A(\hat{H}_B\Psi_j^B)$$

since $\hat{H}_A$ works on the co-ordinates describing part A of the system, not affecting the function Ψ_j^B, and *vice versa*. Consequently, from (5.3)

$$\hat{H}\Psi_{ij}^{AB} = E\Psi_{ij}^{AB}, \quad E = E_i^A + E_j^B \qquad (5.4)$$

which verifies the assertion we wanted to prove; Ψ_{ij}^{AB} *is an eigenstate with energy E*. It is, of course, not surprising that the energy is the sum of the energies of the two independent subsystems. We also note that, in the language of §3.10, Ψ_{ij}^{AB} is a simultaneous eigenfunction of $\hat{H}_A$ and $\hat{H}_B$ describing part A with definite energy E_i^A and part B, simultaneously, with definite energy E_j^B.

Let us apply the above result to the system of two hydrogen atoms, each in its ground state. The Hamiltonian operator is (p. 53)

$$\hat{H} = \left(-\frac{\hbar^2}{2m}\nabla_1^2 - \frac{e^2}{\kappa_0 r_{a1}} - \frac{e^2}{\kappa_0 r_{b1}}\right) + \left(-\frac{\hbar}{2m}\nabla_2^2 - \frac{e^2}{\kappa_0 r_{a2}} - \frac{e^2}{\kappa_0 r_{b2}}\right) + \frac{e^2}{\kappa_0 r_{12}}$$
$$(5.5)$$

and provided (see Fig. 3.3) A (with electron 1) and B (with electron 2) are sufficiently far apart this becomes

$$\hat{H} = \left(-\frac{\hbar^2}{2m}\nabla_1^2 - \frac{e^2}{\kappa_0 r_{a1}}\right) + \left(-\frac{\hbar^2}{2m}\nabla_2^2 - \frac{e^2}{\kappa_0 r_{b2}}\right) = \hat{H}_A + \hat{H}_B \qquad (5.6)$$

where $\hat{H}_A$ refers to an isolated hydrogen atom A and $\hat{H}_B$ similarly to B. If we use ϕ_A and ϕ_B to denote the two corresponding ground-state wavefunctions (i.e. atomic orbitals, both of 1s type), the theorem above tells us that[†]

$$\Psi(1, 2) = \phi_A(1)\phi_B(2) \qquad (5.7)$$

is a two-electron wavefunction for the whole system, the corresponding energy being

$$E = E_A + E_B = 2E_{1s} = -E_h$$

i.e. twice the energy of an isolated hydrogen atom in its ground state or -1 hartree. This result can be made as accurate as we please by putting the atoms far enough apart.

Suppose, however, that the internuclear distance is small. In that case (5.5) cannot be simplified except by simply *neglecting* the electron interaction term $e^2/\kappa_0 r_{12}$. The approximate Hamiltonian is then

$$\hat{H} = \hat{H}(1) + \hat{H}(2) \qquad (5.8)$$

[†] Instead of showing all the electronic co-ordinates it is more convenient to abbreviate them to single numbers; $\Psi(1, 2)$ is a function of all the co-ordinates of electrons 1 and 2. Similarly $\hat{H}(1)$, for example, will denote an operator which works on the co-ordinates of electron 1.

where

$$\hat{H}(1) = -\frac{\hbar^2}{2m}\nabla^2(1) - \frac{e^2}{\kappa_0 r_{a1}} - \frac{e^2}{\kappa_0 r_{b1}}$$

and is seen to be the Hamiltonian for the *hydrogen molecule ion* (cf. (3.18)) consisting of electron 1 in the field of *both* nuclei A and B. $\hat{H}(2)$ is a similar Hamiltonian for electron 2. Thus, with *neglect of electron interaction*, the wavefunction could be written

$$\Psi(1, 2) = \psi(1)\psi(2) \tag{5.9}$$

where ψ denotes a wavefunction for H_2^+ (i.e. a *molecular* orbital) and the energy would be $E = 2E(H_2^+)$.

We now see how neglect of electron interaction leads (in either case) to approximate wavefunctions of *product form*, each factor being an *orbital* describing one electron. The orbitals may describe electrons on separate atoms (being AO's as in (5.7)) or, if the atoms are close together, may describe electrons 'shared' over two or more centres (being MO's as in (5.9)). The whole MO picture rests upon the approximation of omitting 'r_{12} terms' from the Hamiltonian, but if we could calculate *exact* wavefunctions they would show no trace of orbitals! At the same time we see how orbital forms of *approximate* wavefunctions may be set up using simple physical considerations; such forms, supplied with adjustable parameters, may then be systematically refined by the variation method (§3.6).

5.2. Some hydrogen molecule calculations

The earliest calculations on the molecule H_2 provide an excellent example of the variational refinement of a simple orbital approximation; starting from the function (5.7), which describes two separate H atoms, many modifications may be distinguished. In general every mathematical improvement at the same time embodies some aspect of chemical intuition and experience. It is convenient to use atomic units throughout (see §2.3).

Function (a). Let us take the function (5.7), with normalized AO's

$$\phi_A(1) = \pi^{-1/2}\exp(-r_{a1}), \quad \phi_B(2) = \pi^{-1/2}\exp(-r_{b1}). \tag{5.10}$$

We know that this function will be exact in the dissociation limit ($R \to \infty$), but we can also use it as a variation function in computing an upper bound to the energy at *any* internuclear distance. For this purpose we must employ the full Hamiltonian (*including* the electron interaction term) and calculate

$$E = \int \Psi^*\hat{H}\Psi \, d\tau.$$

It should be noted that here $d\tau = d\tau_1 d\tau_2$ indicates integration over all co-ordinates and that Ψ is already normalized

$$\int \Psi^2 \, d\tau = 1.$$

Since $\hat{H}$ in (5.5) may be written

$$\hat{H} = \hat{H}_A + \hat{H}_B - \frac{1}{r_{b1}} - \frac{1}{r_{a2}} + \frac{1}{r_{12}} \tag{5.11}$$

and we have already noted that (5.7), namely

$$\Psi(1, 2) = \phi_A(1)\phi_B(2) \tag{5.12}$$

is an eigenfunction of $\hat{H}_A + \hat{H}_B$, with eigenvalue $2E_{1s}$, it follows easily that

$$E = 2E_{1s} + Q \tag{5.13a}$$

where Q derives from the last three terms in (5.11) and is called a 'Coulomb integral' because it has a purely electrostatic interpretation. In accordance with our convention (§1.4) of including nuclear repulsion terms along with the purely electronic energy in considering the geometry dependence of the energy we thus obtain a total energy (in fixed nucleus approximation)

$$E = 2E_{1s} + Q + 1/R = -E_h + (Q + 1/R) \tag{5.13b}$$

where $1/R$ represents the Coulomb repulsion between the two nuclei. Q contains the Coulomb repulsion between two charge clouds of density ϕ_A^2 and ϕ_B^2 (from the $1/r_{12}$ term), the attraction between nucleus B and a charge cloud ϕ_A^2 ($-1/r_{b1}$ term), and a similar attraction between nucleus A and a charge cloud ϕ_B^2 ($-1/r_{a2}$ term). Only if these last terms (the only negative ones) are sufficiently large to outweigh the other Coulomb terms, will there be any net binding. In fact, the energy curve computed using (5.13b) does exhibit a minimum, $Q + 1/R$ becoming slightly negative at $R \simeq 1\cdot7a_0$. The depth of the minimum (curve a in Fig. 5.1) is, however, only about $0\cdot25$ eV, compared with the observed $4\cdot75$ eV.

Function (b). The wavefunction (5.12), although mathematically acceptable, is not *physically* satisfactory because it treats the two electrons in an unsymmetrical fashion; there is a much higher probability of finding electron 1 near centre A and electron 2 near centre B ($\phi_A^2(1)$ and $\phi_B^2(2)$ then both being large) than of finding 1 near B and 2 near A ($\phi_A^2(1)$ and $\phi_B^2(2)$ then both being small). Since, however, two situations which differ by interchange of identical electrons are *indistinguishable*, they should have the same probability of occurrence. This may be ensured by taking, instead of (5.12),

$$\Psi(1, 2) = N_\pm [\phi_A(1)\phi_B(2) \pm \phi_B(1)\phi_A(2)] = N_\pm [\Phi_1(1, 2) \pm \Phi_2(1, 2)] \tag{5.14}$$

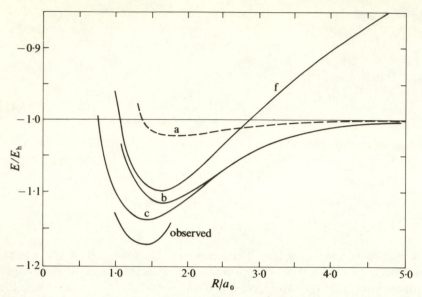

FIG. 5.1. Energy curves for H_2. Curves a, b, and c represent the first three approximations described in the text (functions (a), (b), and (c)), while f refers to function (f).

for, taking either sign, the probability function $|\Psi(1, 2)|^2$ is then unchanged when the electrons change places, the co-ordinates indicated by 1 and 2 being switched. Another way of arriving at this conclusion is to say that the function Φ_2 is just as acceptable as Φ_1, since it also gives the limiting energy $E = E_A + E_B = 2E_{1s}$, and therefore both should be combined with equal weight in a variation function

$$\Psi(1, 2) = c_1\Phi_1 + c_2\Phi_2.$$

The secular equations (3.41) indeed lead to the conclusion $c_1 = \pm c_2$. The wavefunctions and energies are found to be

$$\Psi_+(1, 2) = \frac{[\phi_A(1)\phi_B(2) + \phi_B(1)\phi_A(2)]}{\sqrt{\{2(1 + S^2)\}}}, \quad E_+ = \frac{H_{11} + H_{12}}{1 + S^2} \quad (5.15a)$$

$$\Psi_-(1, 2) = \frac{[\phi_A(1)\phi_B(2) - \phi_B(1)\phi_A(2)]}{\sqrt{\{2(1 - S^2)\}}}, \quad E_- = \frac{H_{11} - H_{12}}{1 - S^2} \quad (5.15b)$$

where

$$H_{11} = \int \Phi_1^* H \Phi_1 \, d\tau = H_{22}$$

$$H_{12} = \int \Phi_1^* H \Phi_2 \, d\tau = H_{21}$$

The denominators in (5.15a, b) arise from overlap of the functions Φ_1 and Φ_2, and S is the usual overlap integral between the AO's ϕ_A and ϕ_B.

Both H_{11} and H_{12} are negative quantities and the solution (5.15a), with $+$ signs, therefore refers to the state of lowest energy. The various integrals may be evaluated, exactly as for Function (a), and the resultant ground-state energy plotted as a function of R (Fig. 5.1, curve B). This curve is much more satisfactory; it gives a minimum energy at about the same internuclear distance as before, but the binding energy is increased to $3.14\,\text{eV}$, much closer to the observed $4.75\,\text{eV}$. The improvement arises mainly from the presence of H_{12} (H_{11} and H_{22} giving the energy for function (a)). The second solution (5.15b), however, leads to a curve (not shown) which rises smoothly for $R \rightarrow 0$ and gives no minimum; it describes an *excited* state in which the molecule would dissociate spontaneously into two separate atoms. The function (5.15a), due to Heitler and London (1927) who later developed a similar approach for polyatomic molecules, is described as an 'electron-pair' or 'valence bond' function; the generalized theory, in which every component of the wavefunction describes an allocation of electrons to *atomic* orbitals, is called valence bond (VB) theory. It is the presence in (5.15a) of two terms, differing by 'exchange' of the electrons, that leads to the appearance of the quantity H_{12}, and the resultant binding energy is often said to be largely 'exchange energy'. However, we must remember this is not a *physical* quantity—it is associated only with a particular way of constructing an approximate wavefunction.

Function (c). The next improvement is again suggested by physical considerations (Wang 1928). The functions ϕ_A and ϕ_B in (5.15a) correspond to unit nuclear charge ($Z = 1$), but an electron in the field of *two* protons might be expected to feel an enhanced attraction corresponding to an effective charge $Z_e > 1$—indeed if the protons were to coalesce, giving the united-atom limit (He), the appropriate value would be $Z_e \simeq 1.65$ (§2.6). This suggests that instead of (5.10) we should try the normalized AO's

$$\phi_A(1) = \sqrt{(c^3/\pi)}\exp(-cr_{a1}) \qquad \phi_B(2) = \sqrt{(c^3/\pi)}\exp(-cr_{b2}) \qquad (5.16)$$

where the effective charge, denoted by c, may vary with internuclear distance but for any value of R should be chosen to minimize the usual energy functional. When this is done, we obtain curve c in Fig. 5.1. This is lower at all points than curves a and b, since extra flexibility in the wavefunction must lead to a lowering of the energy, and gives a binding energy of $3.76\,\text{eV}$ for $c = 1.166$, a considerable improvement on the previous value of $3.14\,\text{eV}$. The dependence of the optimum effective charge $Z_e (= c)$ on internuclear distance is shown in Fig. 5.2 (a).

Function (d). As in the case of the ion H_2^+ (p. 89), a further improvement would arise from admission of 'polarization' of each AO by the presence of a second nucleus. An axially symmetric distortion around the bond (z axis) may

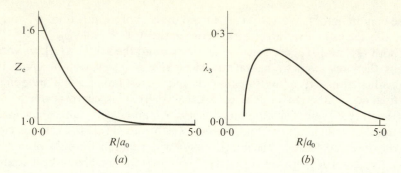

FIG. 5.2. Variation of parameters in H_2 wavefunction with internuclear distance: (*a*) optimum value of effective nuclear charge (Z_e); (*b*) optimum value of ionic character (λ).

be introduced on replacing the spherical ϕ_A, used so far, by

$$\phi_A(1) = \lambda_1 \exp(-c r_{a1}) + \lambda_2 z_a \exp(-c r_{a1}) \qquad (5.17)$$

and similarly for $\phi_B(2)$ (Rosen 1931). On minimizing the energy by variation of the ratio λ_2/λ_1, as well as c, the binding energy improves from 3·76 eV to 4·02 eV.

Function (*e*). Not only is the charge cloud near each hydrogen likely to be contracted and polarized, but there is also the chance, however small, that both electrons should simultaneously be near the same nucleus. This is not at all unreasonable when we remember that the negative ion H^- exists and is quite stable. Now if both electrons were on A, they would have a wavefunction

$$\phi_A(1)\phi_A(2)$$

where, at least approximately,

$$\phi_A(1) = \sqrt{(c'^3/\pi)} \exp(-c' r_{a1}).$$

Since we must allow equal weights to the chances of both being on A or both being on B, we introduce (Weinbaum 1933) what may be called an 'ionic' wavefunction

$$\Psi_{ion} = \phi_A(1)\phi_A(2) + \phi_B(1)\phi_B(2). \qquad (5.18)$$

If we call the previous wavefunction† Ψ_{cov}, in which the subscript implies a purely covalent character for the wavefunction, then the complete molecular wavefunction is taken to be

$$\Psi = \Psi_{cov} + \lambda \Psi_{ion}. \qquad (5.19)$$

The energy of this wavefunction depends on $\lambda_1, \lambda_2, \lambda, c, c'$, and R. A lot of

† That is the VB function (5.15*a*) but with the AO's 'shrunk' and 'polarized' as in (5.17).

tedious calculation is required to show which values for these quantities lead to the lowest energy. In the end it appears that the binding energy increases to 4·10 eV. When Ψ_{cov} and Ψ_{ion} are individually normalized, $\lambda \simeq \frac{1}{4}$, showing that at the equilibrium distance the ionic character of the bond is nothing like so pronounced as the covalent character. At large distances the ionic character would be expected to be much smaller than $\frac{1}{4}$. Figure 5.2 (*b*) shows how λ_3 varies.†

An interpretation of (5.19) is that there is mixing between the 'purely covalent' structure H—H, with a wavefunction Ψ_{cov} or $\Psi(\text{H—H})$, and the two 'ionic' structures H^+H^-, H^-H^+ with wavefunctions $\Psi(H^+H^-)$, $\Psi(H^-H^+)$. Provided that we are not tempted thereby to suppose that the purely covalent and purely ionic structures exist, there is no harm in this description. The mixing of the various structures represented in (5.19) is still frequently described (cf. p. 82) as 'resonance', but it must be stressed that terms such as 'covalent–ionic resonance' are purely formal and merely indicate that the wavefunction may be approximated by a mixture such as (5.19) in which the terms have a certain pictorial interpretation. Such interpretations may indeed be useful, but they must not be regarded as having any objective significance.

Function (*f*). All the preceding wavefunctions (*a*)–(*e*) have been of the VB type, in so far as they are constructed so as to recognize various possible allocations of electrons to *atomic* orbitals (or AO's distorted by contraction and/or polarization). Let us now return to the MO approach outlined in §4.3 in a more quantitative way and ask whether it can be described in similar language.

The MO description of the H_2 ground state (§4.5) assigns two electrons to the bonding *molecular* orbital (normalized)

$$\psi = \frac{\phi_A + \phi_B}{1 + S} \tag{5.20}$$

to give the two-electron wavefunction (5.9). The energy calculated using this function is shown as curve (f) in Fig. 5.1. Evidently the MO function represents a backward step from even the first approximation‡ of VB type; it gives a slightly worse binding energy (2·70 eV) and it apparently goes completely wrong for large values of R, i.e. it does not describe correctly the dissociation of the molecule into its constituent atoms. This is a serious and rather general defect of MO theory; to see what it means in physical terms we insert (5.20) into (5.9) and obtain

$$\Psi(1, 2) = \frac{\phi_A(1)\phi_A(2) + [\phi_A(1)\phi_B(2) + \phi_B(1)\phi_A(2)] + \phi_B(1)\phi_B(2)}{(1 + S)^2}. \tag{5.21}$$

---

† Coulson and Fischer (1949). These calculations were actually made neglecting the polarization effect, Function (*d*). This neglect would not be very likely to change the curve significantly.

‡ Discounting Function (*a*) which lacks the necessary symmetry.

TABLE 5.1
Results of some calculations on H_2

Type of wavefunction	D_e eV	R_e nm
Heitler–London (Function (b)†)	3·14 eV	0·0869 nm
Heitler–London, with screening (Function (c))	3·78	0·0743
Heitler–London, with screening and polarization (Function (d))	4·04	0·074
Heitler–London-plus-ionic, with screening and polarization (Function (e))	4·12	0·0749
Molecular orbital (Function (f))	2·70	0·085
Molecular orbital, with screening (Coulson 1937)	3·49	0·0732
Molecular orbital SCF (Coulson 1938)	3·62	0·074
James–Coolidge (1933), without r_{12} terms	4·29	0·0740
James–Coolidge (1933) (Function (g), 13 terms)	4·72	0·0740
Kolos–Wolniewicz (1968) (Function (g), 100 terms)	4·7467	0·074127
Observed	4·7467	0·07412

† Functions defined in the text.

Evidently in 'AO language' the wavefunction can be regarded as a mixture of covalent and ionic structures of the kind distinguished in Function (e); on normalizing the structures this may be written in the form (5.19) but with $\lambda = 1$. This should be compared with the optimum λ value which varies with internuclear distance as in Fig. 5.2 (b). Evidently, with $\lambda = 1$, the MO function for the dissociation products is a mixture of two *equally weighted* terms representing the situations

(i) H—H,   (ii) H^-H^+ or H^+H^-.

This is physically unreasonable as a description of dissociation into two neutral hydrogen atoms; the Heitler–London function (b) for structure H—H gives the correct dissociation limit, while the presence of the ionic terms in (5.21) leads to an energy half way between that for two hydrogen atoms and that for the hydrogen negative ion and a bare proton.

Function (g). Finally, we take a step which does not depend on physical or chemical intuition, but rather on purely mathematical considerations. James and Coolidge (1933) were the first to introduce new co-ordinates, $\lambda_1 = (r_{a1} + r_{b1})/R$ and $\mu_1 = (r_{a1} - r_{b1})/R$ for electron 1 (similarly for electron 2), and r_{12} for the interelectronic distance. They used as variation function a mixture of terms such as

$$\lambda_1^m \lambda_2^n \mu_1^j \mu_2^k r_{12}^p \exp\{-\alpha(\lambda_1 + \lambda_2)\} \tag{5.22}$$

with m, n, j, k, p as integers and α a variable parameter. They obtained virtually complete agreement with the experimentally determined energy curve using a 13-term function. Both experiment (Herzberg 1970) and theory (Kolos and Wolniewicz 1968) have subsequently been extended and refined, and agreement in values of the binding energy (with due allowance for vibrational and other, much smaller, corrections) has been pressed to astonishingly high accuracy (to within 1 part in 10^6). Our main reasons for mentioning such calculations are (i) that they demonstrate fairly conclusively the validity of Schrödinger's equation as the correct starting point for calculations of electronic behaviour, giving it a role similar to that of Newton's laws in classical dynamics, and (ii) that they show no trace of either orbitals or exchange energy! Such concepts relate only to our simplified mathematical models for describing electrons, not to the electrons themselves. Table 5.1 summarizes the results of some representative calculations on H_2.

5.3. The valence bond (VB) method

The wavefunction (c) of §5.2 is the prototype function of valence bond (VB) theory, developed during the late 1920's and early 1930's by Heitler and London, Slater, Pauling, and others. The standard reference for this approach and its qualitative applications is Pauling (1960). Although now largely superseded (at least in its original form) by MO theory and its variants, the VB approach has left a legacy of well-established concepts which are still widely used in describing chemical bonds. We therefore summarize the essentials of the method and indicate how it is applied to other diatomic molecules. In subsequent chapters the discussion will be extended somewhat and used in the description of polyatomic molecules.

The essential feature of the VB description of H_2 was the selection of two electrons in AO's ϕ_A, ϕ_B around A and B, and the formation of a two-electron wave function $\phi_A(1)\phi_B(2) + \phi_B(1)\phi_A(2)$. This process is referred to as pairing the electrons, and the method is sometimes called the method of electron pairs. When electron spin is taken into account (§5.6), we shall find that the bonding function requires the spins to be antiparallel coupled or 'paired' to give a zero resultant spin; the electron in ϕ_A must therefore be free to be coupled with that in ϕ_B, and we can only form a bond by using electrons not already paired in the original atoms. There is, however, no reason why, if there are two—or three or more—unpaired electrons in one atom, we should not pair these with corresponding electrons in another atom. In this way we shall obtain multiple bonds and, just as in MO theory (Chapter 4), the different bonds are in large measure independent, the total charge cloud being obtained by superimposing those for the individual bonds.

If we are not sure which electrons to pair with which, we can make use once more of the criterion of maximum overlap (§4.3). This criterion still applies because, as follows from (5.15a), the substantial binding occurs only when H_{12}

is numerically large, and further analysis shows this to be possible only if there are regions where the product $\phi_A \phi_B$ is large. This can occur only if the orbitals concerned, ϕ_A and ϕ_B, overlap to a considerable extent. The pairing of ϕ_A and ϕ_B will therefore yield a strong bond if ϕ_A and ϕ_B have considerable overlap.

A few examples will show at once the way in which the VB approach is applied.

(i) The Li—Li bond is a single bond, since an isolated Li atom is described in terms $(1s)^2(2s)$; the bond arises from the pairing together of the two previously unpaired 2s electrons of the separate atoms.

(ii) The N≡N bond is a triple bond since an N atom is

$$(1s)^2(2s)^2(2p_x)(2p_y)(2p_z).$$

We may therefore pair together the two $2p_z$ AO's in which the z direction is along the nuclear axis to form a σ bond, and the $2p_x$ and $2p_y$ AO's to form two π bonds, just as in §4.6 (p. 101). Nothing can be done with the 2s electrons because they are already paired internally in the separate atoms.

(iii) The bonds in F_2, Cl_2, ... are single. In the case of Cl_2, for example, only one 3p AO on each atom ($3p_z$ say) is singly occupied, and taking the z axis along the bond the AO's may be paired to form a σ bond.

(iv) Two normal He atoms on approaching each other will not form a bond at all, since there are no unpaired electrons available in either atom. If any sort of bond is to be formed, it can only be after one atom has been excited. This is in complete agreement with a great deal of spectroscopic evidence concerning the reaction of helium atoms.

In this way much of the information obtained from the MO method in Table 4.3 (p. 100) may be verified again in the VB theory. However, there are certain difficulties, one at least of which must be mentioned. Since an oxygen atom has the configuration

$$O[(1s)^2(2s)^2(2p_x)^2(2p_y)(2p_z)]$$

we should have expected that mutual coupling of the two sets of unpaired electrons would lead to a double bond. Further, the criterion of maximum overlapping would suggest that one of these should be a σ-type bond and the other a π-type bond. Now the bond is generally regarded as double, but this description does not show us that such a state has to be a triplet, with a pair of parallel spins, rather than the expected singlet; nor is it clear that the π bond should involve $2p_y$, when $2p_x$ could equally claim the odd electron. Only a lengthy and mathematically complicated discussion can resolve such ambiguities, and the detailed implementation of VB theory is in fact beset by severe computational difficulties.

In the case of a heteronuclear diatomic molecule AB further complications arise since the polarity of the bond cannot be recognized without admitting the ionic structures corresponding to A^+B^- or A^-B^+. Even with just one electron-pair bond we should therefore need to use a wavefunction (cf. p. 118)

$$\Psi = c_1\Psi_{AB} + c_2\Psi_{A^+B^-} + c_3\Psi_{A^-B^+} \tag{5.23}$$

where one or other of the ionic structures would appear with a heavy weight (large coefficient) in order to introduce the polarity.

An example clarifies the discussion. Let us consider HCl. The electrons to be paired together to form the bond are the hydrogen 1s electron and the chlorine $3p_z$. If, for the moment, we leave out of account all the other electrons of chlorine, the covalent wavefunction (un-normalized) is

$$\Psi_{HCl} = \phi_H(1)\phi_{Cl}(2) + \phi_H(2)\phi_{Cl}(1) \tag{5.24}$$

and the ionic wavefunction is

$$\Psi_{H^+Cl^-} = \phi_{Cl}(1)\phi_{Cl}(2). \tag{5.25}$$

We see immediately that the other possible ionic function $\Psi_{H^-Cl^+}$ is likely to be unimportant, for the data on atomic ionization potentials and electron affinities show that it requires about $9.6\,\text{eV}$ to take an electron from H and give it to Cl, but about $12.3\,\text{eV}$ for the reverse operation. Since the Coulomb forces between H^+ and Cl^- and between H^- and Cl^+ will be very similar at equal separations of the nuclei, we conclude that the structure H^+Cl^- is nearly $3\,\text{eV}$ more stable than H^-Cl^+. The criterion of comparable energies (cf. p. 106)† for effective combination in a linear variation function is not satisfied, so that the weight of H^-Cl^+ will be expected to be less than that of H^+Cl^-. Unless we are trying to make rather accurate calculations the function for H^-Cl^+ may therefore be neglected. Nevertheless, especially with multiple bonding, large numbers of ionic structures must normally be included in dealing with polar molecules and, even at a qualitative level, the discussion of bonding lacks the simplicity and elegance of the MO treatment.

In closing this brief summary of the VB approach we must again refer to our use of the word 'structure' in describing the composition of various wavefunctions. When we speak of a structure, we mean a certain way of pairing orbitals together, the two AO's being symmetrically involved as in $(5.15a)$.‡ For example, we have just been discussing the so-called covalent structure obtained by pairing ϕ_A and ϕ_B, and the ionic structures in which both electrons are on A or on B. The various wavefunctions are the ones involved in (5.23). Now the important thing about these structures is that they do not exist except in our imagination! What was said for the hydrogen molecule is true generally; the structures themselves have no objective reality, and it is quite inaccurate to

---

† Note that a similar argument applies whether we are combining AO's to form a one-electron wavefunction (MO) or VB structures to form a many-electron wavefunction.

‡ With this definition we should not, therefore, call (5.12) a structure, but we should do so with (5.24) and (5.25). The interpretation of structures in terms of pairing of *spins* is taken up in §5.6.

speak of any kind of resonance between two or more structures as if it implied that each structure existed for a fraction of the time indicated by its weight in the wavefunction. There is no such thing as an independent ionic or covalent structure, for the simple reason that their wavefunctions are not eigenfunctions for the allowed stationary states. The main value of the VB method is that it selects as component structures wavefunctions which do have a simple pictorial appeal. Thus if we could allow two atoms to approach, keeping the various AO's entirely unaffected except that an electron in one atom was paired with an electron in the other, then the resultant orbital wavefunction would describe the covalent (non-polar) situation conventionally indicated by the symbol A–B. Similarly the ionic structures obtained by changing the occupation of the AO's (i.e. transferring electrons) would correspond pictorially to situations represented by the conventional chemical symbols A^+B^- and A^-B^+.

It is of course possible to associate an energy formally with any structure, for if the wavefunction for the structure is Ψ, this energy is simply the 'expectation value' (§3.10) which can be defined whether Ψ is an eigenfunction or not:

$$E = \frac{\int \Psi^* H \Psi \, d\tau}{\int \Psi^* \Psi \, d\tau}. \tag{5.26}$$

Thus we can speak of the energy of a pure covalent structure A–B, or of a pure ionic structure A^-B^+, etc. Like the structures themselves, these energies are fictitious. The only energy that does possess a real observable character is the energy associated with the exact wavefunction. From the very nature of the variation method this energy will lie below any of the components E_{cov} or E_{ion}. With the usual care about the meaning of our words we could say that the superposition of structures corresponds to a 'resonance', as a result of which the energy is lowered. The amount of lowering below that of the lowest-energy component is termed the 'resonance energy'. This resonance energy, of course, has no absolute meaning and is necessarily related to the basic structures we have chosen. If we change our choice of structures we alter the resonance energy—a very unsatisfactory feature of the theory.

The difficulties and defects of the VB method should by now be apparent. Fortunately, however, as we shall find in the next section, the VB and MO approximations must come into agreement as each is refined; we may therefore use whichever theory is mathematically more convenient and interpret the results using the concepts and language drawn from either approach. For the most part we shall employ the MO description, but in dealing with electron-pair bonds it is usually immaterial which kind of approximate wavefunction we have in mind.

5.4. Equivalence of MO and VB methods

Let us consider in more detail the wavefunctions (b) and (f) of §5.2, which are the prototype functions of VB and MO theory, respectively. To simplify the notation a little we shall use, for example, $a(1)$ instead of $\phi_A(1)$ to indicate the AO for electron 1 on centre A, and denote the Heitler–London (VB) function (5.15a) simply by

$$\Psi_{HL} = ab + ba \qquad (5.27)$$

where it is understood that the first factor of each product is a function of the position of electron 1 and the second factor is a function of the position of electron 2. The normalizing factor has been dropped but can always be added later. The corresponding MO function follows from (5.9) and (5.20) as (again without normalizing)

$$\Psi_{MO} = (a+b)(a+b) = (ab + ba) + (aa + bb) \qquad (5.28)$$

where the ionic terms (e.g. $aa = a(1)a(2)$ means both electrons on centre A) are too prominent as we have noted in the discussion on p. 120. A function of type (e), which (neglecting contraction or polarization of the AO's) may be called 'Heitler–London-plus-ionic', is a refinement of the simple VB function (5.27) and is

$$\Psi_{HLPI} = (ab + ba) + \lambda(aa + bb). \qquad (5.29)$$

At the equilibrium distance λ is small and Ψ_{HL} is a better approximation to (5.29) than Ψ_{MO}, the disparity increasing as R increases (see Fig. 5.2 (b)).

Since Ψ_{HLPI} represents a refinement of the VB function (5.27), it is natural to ask whether Ψ_{MO} could be refined in a similar way, i.e. by adding terms corresponding to different assignments of electrons to orbitals—but this time to *molecular* orbitals. The only other MO at our disposal in this simple approximation is the antibonding combination $(a-b)$; so we might try to improve (5.28) by adding a term corresponding to a different electron configuration—both electrons in the antibonding MO, say. The result, which is said to admit *configuration interaction* (CI), is

$$\Psi_{MOCI} = (a+b)(a+b) + k(a-b)(a-b) \qquad (5.30)$$

where increasing the coefficient k increases the CI, $k = 0$ giving the simple MO function (5.28). It is not difficult to show that there is no interaction with a configuration $(a+b)(a-b)$ which lacks the necessary symmetry in a and b.

Comparison of (5.30), after expansion, with (5.29) shows that Ψ_{HLPI} and Ψ_{MOCI} are identical (apart from an irrelevant normalizing factor) provided that

$$\lambda = \frac{1+k}{1-k}. \qquad (5.31)$$

Now in each case we choose k and λ (§3.6) so as to make the energy function stationary; consequently we shall obtain precisely the same wavefunction by either method. In this way, therefore, both the MO and VB methods become equivalent if we carry them beyond the simplest forms. This is quite generally

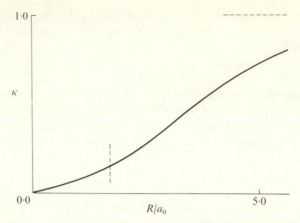

FIG. 5.3. Configuration interaction in H_2. The coefficient κ measures the importance of the 'excited' configuration. The vertical broken line corresponds to the equilibrium distance; the horizontal broken line is the limit for $R \to \infty$. (From C. A. Coulson and I. Fischer, *Phil. Mag.* **40**, 386 (1949).)

true however complicated the molecule may be, provided we consider *all possible* VB structures and *all possible* MO electron configurations. There is thus no real conflict between MO and VB theories; they offer somewhat different first approximations to molecular wavefunctions but converge to the same results as they are systematically refined. The importance of this conclusion can hardly be overemphasized.†

We have already seen (Fig. 5.2 (*b*)) how the coefficient λ, which measures the importance of the ionic structures, varies with R for the H_2 molecule. Equation (5.31) allows us to infer the corresponding value of k. More significance attaches to this coefficient, however, if we normalize the component configurational wavefunctions in (5.30), writing

$$\Psi_{\text{MOCI}} = \frac{(a+b)(a+b)}{2(1+S)} - \kappa \frac{(a-b)(a-b)}{2(1-S)} \tag{5.32}$$

where

$$\kappa = -\frac{1-S}{1+S}k. \tag{5.33}$$

κ now measures the weight of the excited configuration $(a-b)(a-b)$ which interacts with the basic configuration $(a+b)(a+b)$. The graph of κ against R is shown in Fig. 5.3. At the equilibrium distance κ is quite small, but as the internuclear distance increases the configuration interaction, measured by κ,

† In earlier editions of this book the conflict is apparent, but by the 1970's it was recognized (Coulson 1970) that 'In the early 1930's valence theory was confused by a conflict between MO and VB approximations. But during the 1940's a reconciliation between the two methods was found …'.

increases up to unity. This is a serious situation, warning us that even at moderate distances the simple MO method is quite unreliable. For example, if we wished to follow the whole course of a chemical reaction by calculating the energy of the compound system at all stages of the reaction, we should have to choose the VB method rather than the MO method if we wanted anything like reliable results from a simple wavefunction.

This failure of the MO description† at large internuclear distances illustrates one big distinction between the two theories. In a single sentence, we can say that the MO method underestimates *electron correlation* and the VB method overestimates it. By this we mean that the presence of the Coulomb repulsion $e^2/\kappa_0 r_{12}$ between the two electrons of a bond must actually tend to keep them apart. There is, for example, an almost zero chance that both electrons should be simultaneously in exactly the same small region, and if the first electron is momentarily in the neighbourhood of one nucleus there is a greater chance of the second electron being away from this nucleus and in the neighbourhood of the other. We say that their motions are 'correlated' so as to keep them apart. The mathematical formulation of this idea is discussed in Appendix 1. Such correlation is completely absent from the simple MO wavefunction $(a+b)(a+b)$, since the distribution of electron 2 is entirely independent of the instantaneous position of electron 1. When electron 1 is near A, electron 2 has an equal probability of being there also. Indeed, if we expand the MO function as in (5.28) and write it as

$$\Psi_{MO} = ab + ba + aa + bb$$

it suggests, as we have already noted, an equal probability of $\frac{1}{4}$ for the four situations ab, ba, aa, and bb. These ratios are independent of the nuclear separation, so that we should be led to infer that when the molecule dissociated there was a chance of 2 in 4 that it separated as two H atoms and 1 in 4 that it separated as $A^- B^+$ (or $A^+ B^-$). It is precisely the omission of any reference to electron correlation that allows the possibility of $A^- B^+$ and $A^+ B^-$. Electron correlation has the effect of diminishing the weight of the component $aa + bb$. In contrast, the simple VB wavefunction $ab + ba$, although it correctly describes the dissociation, overemphasizes electron correlation, for it insists (by disallowing aa and bb) that at all internuclear distances the electrons should be associated with different centres.

We now also see why the correlation‡ diagrams for heteronuclear diatomics, such as Fig. 4.16 in which we follow the behaviour of individual MO's as we change from the united atom through the actual molecule to the separated atoms, must be used with great care. The assignment of electrons in pairs to the

---

† In its simple form, i.e. without the admission of CI which is in general computationally expensive. The simple form is valid for all values of R only in one special case: that of two interacting closed-shell systems (e.g. He---He).

‡ Note that the word correlation is used here in a less specialized sense.

MO's may give a good one-configuration wavefunction at, say, the normal equilibrium distance, but as the MO's change into AO's on dissociation the double occupation may change, electrons being forced by their mutual repulsion into different AO's. This is why in LiH (p. 109) one of the two bonding electrons $(2\sigma)^2$ goes to each atom, although the correlation diagram suggests that both should go to the hydrogen nucleus.

Finally, it is instructive to look at the charge densities described by the various approximate wavefunctions, for these offer considerable insight into the nature of the chemical bond (§2.2) and, again, should come into agreement as MO and VB theories are progressively improved. The simple MO function (5.28) leads directly to the charge-density function discussed in §4.5; with the present notation for the orbitals it becomes

$$P = \frac{a^2 + b^2 + 2ab}{1 + S} \tag{5.34}$$

where, as usual, S denotes the overlap integral for the AO's a and b.

With the simple VB function (5.27) the argument is slightly less straightforward.† We must remember that $P\,d\tau$ is the probability of finding an electron (no matter which) in a given volume element $d\tau$, while $|\Psi(1,2)|^2\,d\tau_1\,d\tau_2$ is that of finding electron 1 in $d\tau_1$ and electron 2 in $d\tau_2$ *simultaneously*. To obtain the probability of electron 1 in $d\tau_1$ and electron 2 *anywhere*, however, we merely add the probabilities for all possible positions of the second electron, electron 1 always being kept in a fixed $d\tau_1$. The probability of electron 1 being found in $d\tau_1$ is thus, using the normalized function given in (5.15a),

$$d\tau_1 \int |\Psi(1,2)|^2\,d\tau_2 = d\tau_1 \int \frac{[a^2(1)b^2(2) + b^2(1)a^2(2) + 2a(1)b(1)a(2)b(2)]}{2(1+S^2)}\,d\tau_2$$

$$= \frac{[a^2(1) + b^2(1) + 2Sa(1)b(1)]}{2(1+S^2)}\,d\tau_1.$$

In other words, dropping the label 1 on the electron co-ordinates, $(a^2 + b^2 + 2Sab)/2(1+S^2)$ is the probability per unit volume of finding electron 1 at the point where the function is evaluated. It is easily verified that the probability of electron 2 being found at a given point is given by exactly the same expression—this being true generally owing to the indistinguishability of different electrons. Thus, with two electrons, we simply double the above result to obtain

$$P = \frac{a^2 + b^2 + 2Sab}{1 + S^2} \tag{5.35}$$

as the required charge-density function.

---

† For a more complete discussion see Appendix 1.

The interpretation of densities such as (5.34) and (5.35) has been discussed already in §4.5; the total charge, in this case two electrons, may be mentally divided into amounts q_a, q_b, q_{ab}, called the 'electron populations' of orbital and overlap regions. We now see that, at the level of the charge density, the simple MO and VB functions merely give slightly different estimates of the orbital and overlap populations. Both approximations give a charge density which may be written

$$P = q_a a^2 + q_b b^2 + q_{ab}(ab/S) \qquad (5.36)$$

where (ab/S) is a *normalized* overlap density (integration of ab over all space giving S). However, according to MO theory, the piling up of the electron density in the bond region is measured by (comparing (5.36) and (5.34))

$$q_{ab} = \frac{2S}{1+S} \qquad (5.37)$$

while in VB theory, from (5.35), we obtain

$$q_{ab} = \frac{2S^2}{1+S^2}. \qquad (5.38)$$

With $S \simeq 0.75$ these estimates become 0.857 and 0.720 respectively. Both MO and VB theory therefore agree that chemical bonding is accompanied by a substantial migration of charge into the bond region, i.e. the region in which the AO's of the constituent atoms begin to overlap. Although there was, for many years, fierce controversy as to the relative merits of these two extreme approaches (AO's *versus* MO's), the conflict is now seen to have been more apparent than real.

If we refine both approaches, using either of the equivalent forms (5.29) or (5.30), we must of course obtain exactly the same charge density. A simple calculation, similar to that for the Heitler–London function, shows that the resultant overlap population is

$$q_{ab} = \frac{2(S^2 + 2\lambda S + \lambda^2 S^2)}{(1+S^2) + 4\lambda S + \lambda^2(1+S^2)}. \qquad (5.39)$$

For $\lambda \to 0$ this goes over into (5.38) for VB theory excluding ionic structures, while for $\lambda \to 1$ it goes over into (5.35) for MO theory excluding CI.

5.5. The Coulson–Fischer wavefunction

From what has been said so far, it is clear that even the simplest electron-pair bond can be described in a variety of equivalent, or nearly equivalent, ways. However, common to all orbital descriptions is the central idea of the *overlap*

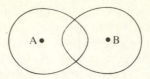

Fig. 5.4. Modified orbitals for use in a generalized valence bond description of H_2.

of a pair of atomic orbitals, one on each atom, and the enhancement of electron density in the overlap region. In qualitative valence theory we want the simplest possible pictorial scheme for describing the bond and, if possible, for following the whole process of making and breaking of bonds. The MO theory, as we have seen, may be adequate for molecules in near-equilibrium geometry but may be completely misleading (without CI) in bond-breaking situations; the simplest VB wavefunction would be preferable in such cases, even without the admission of ionic structures.

There is, however, another possibility, which always presents the wavefunction as a single covalent structure involving two strongly overlapping orbitals and which gives a very graphic description of what happens when a bond is broken; this approximation was introduced by Coulson and Fischer (1949) and may be called a generalized valence bond (GVB) approximation. During recent years GVB theory has been widely developed (see, for example, Goddard *et al.* 1973) as a means of preserving the simple chemical concept of localized electron-pair bonds. The idea is to use a pair of overlapping orbitals, A and B say, which in the dissociation limit go over into the AO's a and b, respectively, but which at intermediate distances are allowed to adjust their shapes in order to give the best possible wavefunction.

As the atoms approach we might make orbital a 'lean' towards the other atom (giving better overlap) by mixing in a bit of b, and *vice versa* for orbital b, to give two egg-shaped orbitals (Fig. 5.4)

$$A = a + \mu b, \quad B = b + \mu a \tag{5.40}$$

where μ is a small positive coefficient, given the same value in each case to preserve symmetry. The simple covalent function of Heitler–London type (5.27) is then

$$AB + BA = (1 + \mu)(ab + ba) + 2\mu(aa + bb)$$

or, apart from a normalization factor,

$$\Psi_{CF} = (ab + ba) + \lambda(aa + bb) \tag{5.41}$$

which coincides exactly with (5.29); the ionic terms have now been incorporated in what was formally a single covalent structure! The relationship between λ, which measures the ionic character, and μ, which measures the 'degree of distortion' of the AO's according to (5.40), is simply

$$\lambda = \frac{2\mu}{1+\mu}, \quad \mu = \frac{\lambda}{2-\lambda}. \tag{5.42}$$

In other words the best possible approximate wavefunction we can construct from two AO's a and b can be written as a single covalent VB structure looking exactly like the Heitler–London function (5.27) except that the AO's have been distorted $(a \rightarrow A, b \rightarrow B)$ to secure more efficient overlapping. This GVB function retains its validity and accuracy, besides its conceptual simplicity, at all internuclear distances and is therefore satisfactory for the discussion of potential energy curves. The behaviour of μ is also illuminating; for large R, $\mu \rightarrow 0$ and gives the Heitler–London function, but when R becomes somewhat less than the equilibrium value (approaching the united-atom situation) the optimum values of λ and μ rapidly approach 1 to give the MO wavefunction with both electrons in the bonding orbital $(a + b)$. In this sense we may say the simple MO function is best near the united-atom limit, and the simple VB function is best near the separated-atom limit. The Coulson–Fischer (GVB) function is flexible enough to yield both limits correctly; wavefunctions of this type will be referred to again in later chapters.

5.6. Singlets and triplets: Pauli's exclusion principle

So far, no mention has been made of electron spin during our discussion of the chemical bond. This in itself should serve to warn us that the very small magnetic interactions arising from electron spin have nothing whatever to do with chemical bonding. Indeed the Hamiltonian we have started from in §5.1 contains nothing but kinetic energy terms and Coulombic interactions. Spin terms appear only as small corrections, so far neglected; they are significant in spectroscopy, where extremely high resolution of energy levels can be achieved, but not in the rest of chemistry. Molecules are held together by *electrostatic* forces, which are several orders of magnitude larger, and the origin of these forces lies in the charge density (§2.2); of course the density must be calculated by solving the wave equation, but the actual interpretation of the bond requires no mysterious non-classical concepts. The electron-pair bond is frequently described in terms of the 'pairing' or 'coupling' of spins and is associated with a 'spin-coupling energy', and we must now try to understand how such a misleading terminology ever came to be adopted.

Let us go back to the Heitler–London calculation (p. 116) in which the products $\phi_A(1)\phi_B(2)$ and $\phi_B(1)\phi_A(2)$ describe electrons 1 and 2 simultaneously on A and B or on B and A, respectively. These were combined to give one wavefunction *symmetric* under exchange of the electrons $(1, 2 \rightarrow 2, 1$ leaving the function unchanged), and one wavefunction *antisymmetric* $(1, 2 \rightarrow 2, 1$ changing the sign only); the former gave a fairly good description of the ground state. If now we include spin the choice of functions becomes wider; instead of ϕ_A we have *spin-orbitals* $\phi_A\alpha$ and $\phi_A\beta$, and instead of ϕ_B we have $\phi_B\alpha, \phi_B\beta$. The products we can combine, representing different allocations of electrons 1 and 2 to centres A and B, each electron with two choices of spin ($m_s = \pm\frac{1}{2}$), are thus

$$\phi_A(1)\alpha(1)\phi_B(2)\alpha(2), \qquad \phi_A(1)\alpha(1)\phi_B(2)\beta(2),$$
$$\phi_A(1)\beta(1)\phi_B(2)\alpha(2), \qquad \phi_A(1)\beta(1)\phi_B(2)\beta(2),$$
$$\phi_B(1)\alpha(1)\phi_A(2)\alpha(2), \qquad \phi_B(1)\alpha(1)\phi_A(2)\beta(2),$$
$$\phi_B(1)\beta(1)\phi_A(2)\alpha(2), \qquad \phi_B(1)\beta(1)\phi_A(2)\beta(2).$$

We could use these eight functions in a variation calculation, calling them $\Phi_1, \Phi_2, \ldots, \Phi_8$, evaluating the quantities H_{11}, H_{12}, etc. and solving to find the energies and mixing coefficients as outlined on p. 65. This may be avoided, however, by a simple symmetry argument coupled with the fact that (since spin does not appear in the Hamiltonian (5.5) and can therefore make no difference to the resultant energies) the wavefunction must factorize into (spatial factor) $\times$ (spin factor), the two alternative spatial factors being the functions already found in (5.15a, b). Thus

$$\Psi = N_{\pm}[\phi_A(1)\phi_B(2)\pm\phi_B(1)\phi_A(2)] \times [\text{spin factor}] \qquad (5.43)$$

and since the spin factor must be some combination of the products $\alpha(1)\alpha(2)$, $\alpha(1)\beta(2), \beta(1)\alpha(2), \beta(1)\beta(2)$ it is clear that, on multiplying out, this will yield some combination of the spin-orbital products listed above. The symmetry argument now runs as follows. If we make the electrons change places, by giving 1 the position (x_2, y_2, z_2) and spin (s_2) originally associated with 2 and *vice versa*, $|\Psi|^2$ must be completely unchanged, for the new situation is physically indistinguishable from the old (the electrons being indistinguishable) and must therefore appear with exactly the same probability. This means that if we make the interchange $1, 2 \rightarrow 2, 1$ in both orbitals and spin functions, Ψ can at most change sign; the wavefunction must be symmetric or antisymmetric under interchange of electronic variables. The behaviour of the spatial factor in (5.43) is already evident; the ground-state function ($+$ sign) is symmetric for interchange of *spatial* variables, while the second function (describing an excited state) is antisymmetric. The spin factor must therefore possess similar properties and clearly there are four possible choices,

$$\alpha(1)\alpha(2), \quad \alpha(1)\beta(2)+\beta(1)\alpha(2), \quad \beta(1)\beta(2), \quad \alpha(1)\beta(2)-\beta(1)\alpha(2)$$

the first three being symmetric and the fourth antisymmetric. If we normalize these spin functions, and use $\Phi_+(1, 2)$ and $\Phi_-(1, 2)$ to denote the two space wavefunctions (which appear in normalized form in (5.15a, b)), the possible space-spin products may be written as follows.

SPACE FACTOR	SPIN FACTOR (i)	SPIN FACTOR (ii)
$\Phi_+(1, 2) \times$	$[\alpha(1)\beta(2) - \beta(1)\alpha(2)]$	$\alpha(1)\alpha(2)$ $\beta(1)\beta(2)$ $[\alpha(1)\beta(2) + \beta(1)\alpha(2)]$
$\Phi_-(1, 2) \times$	$\alpha(1)\alpha(2)$ $\beta(1)\beta(2)$ $[\alpha(1)\beta(2) + \beta(1)\alpha(2)]$	$[\alpha(1)\beta(2) - \beta(1)\alpha(2)]$

The product functions associated with column (i) spin factors are essentially different from those associated with column (ii). This difference is one of symmetry. If we interchange 1 and 2 in every part of the function, we find that the four product functions involving column (i) are all multiplied by -1, those based on column (ii) being unchanged. We say that the first are antisymmetrical with respect to interchange of the electrons and the second are symmetrical. Now it may be shown from the general principles of quantum theory that a symmetry property of this kind is unchangeable (i.e. 'a constant of the motion'), e.g. symmetrical wavefunctions always remain symmetrical.

There appears to be no way of deciding *a priori* whether nature requires symmetrical or antisymmetrical functions. However, experiment shows (as first inferred by Heisenberg (1926)) that for electrons, protons, neutrons, and all particles with half-integer spin (i.e. $n + \frac{1}{2}$, with n an integer) only antisymmetrical wavefunctions occur, and for alpha particles, photons, certain types of meson, and all particles with integer spin (including zero) only symmetrical states are found. We must therefore reject column (ii) of the wavefunctions in the table above and accept only column (i). This shows us that the ground-state wavefunction $\phi_A(1)\phi_B(2) + \phi_B(1)\phi_A(2)$ can have only one possible spin factor, but that the excited-state function can have three. If we apply a strong magnetic field (§3.10) and include small spin-field interactions in the Hamiltonian, the three energies are slightly separated, leading us to the statement that the ground-state function (5.15a) gives a singlet and the other function (5.15b) gives a triplet. As electrons cannot change their spin states (except very slowly, through small spin–spin and spin–orbit interactions) no direct transitions occur between the states with orbital factors (5.15a) and (5.15b).

The interpretation of the spin functions is of considerable interest; $\alpha(1)\alpha(2)$ and $\beta(1)\beta(2)$ obviously correspond to 'parallel spins', both 'up' or both 'down' as in Fig. 5.5, giving a *total* spin z component of† $M_S = +1$ or $M_S = -1$, and the other two functions, with one 'up' and one 'down' both have $M_S = \frac{1}{2}$ $-\frac{1}{2} = 0$. It may be shown (using the eigenvalue 'test' provided by (3.48)

† Capital letters are used to denote the quantum numbers for *total* angular momentum. Thus s, m_s are replaced by S, M_S and l, m_l by L, M_L.

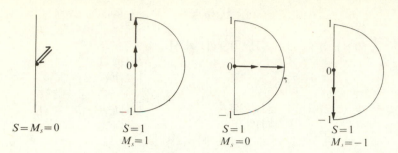

$S = M_s = 0$ $S = 1$ $S = 1$ $S = 1$
                        $M_s = 1$ $M_s = 0$ $M_s = -1$

Fig. 5.5. Vector diagram for coupling of two spins. Antiparallel coupling ('pairing') gives only one state ($S = 0$). Parallel coupling gives a resultant spin $S = 1$ and on applying a magnetic field (vertical direction) the quantized component along the field may take three possible values $M_S = 1, 0, -1$. The triplet is resolved by a field into three states of slightly different energy.

together with the general properties of angular momentum operators) that the three functions of the symmetric family all correspond to the same *total spin*, whose squared magnitude (in units of $\hbar^2$) has the value $S(S+1)$ with $S = 1$, while the single antisymmetric function has $S = M_S = 0$. We speak of 'parallel and antiparallel coupling' of the two $s = \frac{1}{2}$ spins to a resultant total spin $S = 1$ or $S = 0$, respectively. The whole situation is summarized pictorially in Fig. 5.5, which may be compared with Fig. 3.14. The antiparallel coupling is also often described as spin 'pairing', and we now see that the wavefunction which describes the chemical bond is the one with *paired spins*; this does not imply any kind of *physical* coupling, but merely arises from the antisymmetry principle which requires the symmetric space function (Φ_+) to be associated with an antisymmetric spin function. Van Vleck (1935) described this result nicely by saying that the spin coupling was only an *indicator* of chemical bonding. What really determines the bonding, however, is the form of the space factor, and the charge density to which it leads, and we must be careful not to assume that whenever spins are paired there is a chemical bond. For a wavefunction of the form (5.43) the charge density depends *only on the spatial factor*, and in computing the density the spin factor may be ignored. The argument runs along the same lines as that on p. 128 (see also Appendix 2). If we are interested in the probability of finding an electron in a given *spatial* volume element, *irrespective of its spin*, we simply sum (integrate) over all possible spin values. The probability function P, derived from (5.43) will thus contain $\int [\text{spin factor}]^2 \, ds_1 \, ds_2$ which (when the spin factor is normalized) simply reduces to unity. This is the reason why, in the earlier sections of this chapter, we were able to concentrate purely on the spatial part of the wavefunction.

The statement that every electronic wavefunction must be antisymmetric in every pair of electrons, is the *antisymmetry principle*, and is in fact a more fundamental form of Pauli's exclusion principle. Pauli's form relates to an independent-particle model and thus applies very directly to a molecular-orbital-type wavefunction where electrons are assigned to individual MO's. In the MO description of the hydrogen molecule, for example, the bonding MO ψ gives rise to two molecular *spin*-orbitals, $\psi\alpha$ and $\psi\beta$. With one electron in $\psi\alpha$ and the other in $\psi\beta$ an antisymmetric wavefunction is

$$\begin{aligned}\Psi(1, 2) &= \psi(1)\alpha(1)\psi(2)\beta(2) - \psi(2)\alpha(2)\psi(1)\beta(1) \\ &= \psi(1)\psi(2)[\alpha(1)\beta(2) - \beta(1)\alpha(2)]\end{aligned}$$

in which the spins are paired to give a singlet function with $S = 0$ and a symmetric space function with a doubly occupied bonding MO. If, however, we tried to put both electrons into the *same* spin orbital, $\psi\alpha$ say, we could not avoid obtaining a symmetric wavefunction

$$\Psi(1, 2) = \psi(1)\alpha(1)\psi(2)\alpha(2)$$

which is unacceptable. With this type of wavefunction the antisymmetry principle therefore allows not more than one electron in any given spin-orbital, or in other words with a given set of quantum numbers (including m_s). This result is quite general. If N electrons are assigned to N spin-orbitals $\psi_1, \psi_2, \ldots, \psi_N$ (spin factors included) we may construct an antisymmetric function by starting from the product $\psi_1(1)\psi_2(2)\ldots\psi_N(N)$, interchanging electrons in all possible ways, affixing a minus sign for every interchange, and summing all $N!$ terms. If, however, ψ_1 and ψ_2 are the same this will lead to a term $-\psi_1(j)\psi_1(i)\ldots\psi_N(k)$ for every term $\psi_1(i)\psi_1(j)\ldots\psi_N(k)$ (differing only by interchange of i, j) and the result will vanish identically, i.e. *an antisymmetric wavefunction cannot be constructed if any spin-orbital appears twice*. An antisymmetric wavefunction formed in this way, from a single spin-orbital product, is usually called a 'one-determinant' wavefunction since the sum of products is equivalent to the expansion of a determinant.

The considerations of this section clarify some of the statements made in Chapter 4. It should now be clear, for example, why (§4.7) the O_2 electron configuration

$$O_2[KK(2\sigma_g)^2(2\sigma_u)^2(1\pi_u)^4(3\sigma_g)^2(1\pi_g)^2]$$

leads to a triplet or singlet, according as the spins of the electrons in the two $1\pi_g$ MO's are parallel or antiparallel. It should be noted that the singly occupied MO's (e.g. $1\pi_{gx}$ and $1\pi_{gy}$) need not have equal energy. Indeed, it follows at once that whenever we have all MO levels doubly filled, except two which contain one electron each, the resulting *open-shell* state must be either a singlet or a triplet, according to the spins of the two electrons in these orbitals. With three 'odd electrons' (cf. the N atom) we find a doublet or a quartet level, and with four unpaired electrons (cf. the important case of an excited C atom) we have a singlet, triplet, or quintet level.

Finally, it must be emphasized that, without actual calculation, spin-coupling arguments cannot tell us which possible coupling leads to the lower total energy of the molecule. Thus, with two singly occupied orbitals, Hund's rules provide *empirical* evidence for the triplet to be lower, corresponding to parallel coupling, but a proper theoretical discussion is not easy. The fact that Heitler and London found the *singlet* coupling of electrons in ϕ_A and ϕ_B to give the lower energy should warn us that reliable predictions are not possible without detailed consideration of the *orbital* part of the wavefunction. It is true that the lower energy of the singlet state (5.15a) arises because the 'exchange integral' H_{12} is negative when ϕ_A and ϕ_B overlap strongly, whereas for the orthogonal orbitals used in MO theory H_{12} is essentially positive (Slater 1960) and thus makes the triplet lower than the singlet; but the optimum forms of the MO's in the singlet and triplet functions may differ considerably and the energy difference is then no longer determined solely by an exchange integral. The accurate calculation of the energies of open-shell states is thus a matter of some difficulty.

Problems

5.1. Suppose $\Psi_1^A, \Psi_2^A, \Psi_1^B, \Psi_2^B$, are wavefunctions for the ground and first excited states of two helium atoms, A and B. Write down wavefunctions to describe (with neglect of interaction):

 (i) the ground state of the whole system (two helium atoms);
 (ii) a state in which atom A is excited;
(iii) a state in which both A and B are excited;
(iv) two states in which either A or B is equally likely to be excited.
What are the energies of these states? How could you use your wavefunctions to determine an approximation to the energy with interaction included?

5.2. The helium atom is described, neglecting electron interaction, by assigning electrons to one-electron states with wavefunctions $\phi_{1s}, \phi_{2s}, \ldots$. Write down wavefunctions to describe:

 (i) both electrons in the 1s orbital;
 (ii) one electron in 1s, the other in 2s;
(iii) both electrons in 2s;
and proceed as in Problem 5.1.

5.3. The electron spin is described by one of two 'spin functions', α or β ('spin-up' or 'spin-down'). How would you describe, neglecting the interaction between spin and orbital motion, a hydrogen atom with an 'up-spin' electron in the 1s orbital? (The resultant wavefunction is a 'spin-orbital', see §3.10.) Turn back to Problem 5.2 and propose some helium atom wavefunctions with spin included.

5.4. Use the *aufbau* approach to write down two product-type wavefunctions (spin included) for the degenerate lithium atom ground state (doublet state). (Hint: add a third electron to the two already present in Li^+.)

Why is a single product physically unsatisfactory as a wavefunction? Can you set up a combination of alternative products, for the state with a resultant up-spin, which changes sign under interchange of the labels (e.g. 1, 2) for every pair of electrons? (Hint: write down products representing all allowed assignments of the electrons to the chosen spin-orbitals, and combine them together with appropriate coefficients ± 1.)

5.5. Derive the normalizing factors in equations (5.15a, b). (Hint: note that each integral 'separates' into two independent factors.)

5.6. Think of the lithium hydride molecule as consisting of two valence electrons moving around two point charges, the Li^+ 'core' and a bare proton. Propose wavefunctions of both MO and VB type to describe the electron-pair bond. How would your wavefunctions differ from those for a homopolar bond (e.g. H_2), and how might they recognize the polar character of the bond?

5.7. Make a comparison of the MO and VB functions for LiH (Problem 5.6) by describing both in terms of 'structures'. What is the significance of the structures and their weights? (Hint: expand the MO function. If in difficulty consult p. 125.)

5.8. How would you expect the coefficient of covalent and ionic structures in the VB function for LiH (Problem 5.7) to change on pulling the atoms apart? How would they change in a parallel treatment of HF?

5.9. Verify the statements leading to equation (5.31).

5.10. Find an expression for the charge density, using the wavefunction (5.29). Then derive equation (5.39). (Hint: use the same argument that leads to (5.35).)

5.11. Normalize the one-determinant wavefunction derived (see p. 135) from the spin-orbital product $\psi_1(1)\psi_2(2)\cdots\psi_N(N)$, assuming the spin-orbitals orthonormal. (Hint: note that Ψ^2 contains $N! \times N!$ terms, arising from all permutations of variables, and ask which products give a non-zero result on integrating.)

5.12. Use the normalized wavefunction of Problem 5.11, and the procedure used on p. 128, to show that the probability per unit volume of finding electron 1 in $d\tau_1$ is

$$N^{-1}\left[|\psi_1(1)|^2 + |\psi_2(1)|^2 + \cdots + |\psi_N(1)|^2\right]$$

and hence that the corresponding probability for *any* electron in $d\tau_1$ is just the sum of spin-orbital terms within the square brackets. Then show, by eliminating spin, that the electron density is a sum of *orbital* densities. (Hint: use the approach of Problem 5.11, asking what terms give non-zero results after $N-1$ integrations and how many times each term will appear; then put in the spin factors explicitly, writing $\phi_1\alpha, \phi_1\beta, \phi_2\alpha, \phi_2\beta,$ ...instead of $\psi_1, \psi_2, \psi_3, \psi_4, \ldots,$ and integrate over spin. If you have difficulty consider first the cases $N = 2$ and $N = 3$.)

6

Bonding in diatomic molecules

6.1. Some numerical calculations

Various homonuclear diatomic molecules were discussed in a qualitative way, using the MO approach, in Chapter 4. The more quantitative discussion of the covalent bond in H_2 in the last chapter has shown us that a simple MO wavefunction does not compare particularly well with other types of wavefunction (notably the VB function of Heitler and London). The simple MO function is inferior in terms of numerical accuracy and flexibility, and does not in general yield the correct dissociation limit. Nevertheless, the MO description is not bad near the equilibrium geometry and gives a charge distribution not very different from that obtained by VB theory, the best results lying somewhere between. The important fact is that, starting from the same basic set of AO's, the alternative approaches are more in agreement than in disagreement, and that as both are refined they converge towards the same end point. The electron-pair bond, whether its wavefunction be constructed by one method or the other, is characterized by a charge cloud of the form (5.36), namely

$$P = q_1\phi_1^2 + q_2\phi_2^2 + q_{12}(\phi_1\phi_2/S_{12}) \tag{6.1}$$

where ϕ_1 and ϕ_2 are strongly overlapping orbitals, one from each constituent atom, and q_{12} is positive, corresponding to a certain 'accummulation' of charge in the overlap region. The alternative theories agree that the wavefunction describing the bond has a singlet spin factor ($S = 0$) corresponding to paired spins. The main advantages of the MO approach are (i) its conceptual simplicity and (ii) its mathematical simplicity. The second advantage is reflected in the fact that probably 95 per cent of all present-day *ab initio* calculations (i.e. calculations which seriously attempt an approximate solution of the Schrödinger equation without making use of any kind of experimental data) on molecules containing more than a few electrons are performed by the MO method, with or without the refinement of CI (p. 125). Such calculations are now well documented and we shall not describe them in great detail (the standard references to a vast and rapidly growing literature are Richards *et al.* 1971, 1974, 1978) but it is useful to present some of the results in a pictorial way since they provide quantitative support for many of the concepts and arguments developed so far.

Such calculations are conveniently divided into two categories: (i) those which use an LCAO-type approximation based on a *basis set* limited to the AO's which feature in the ground-state description of the constituent atoms; (ii) those which use an augmented basis sufficient (ideally) to give an *exact* solution of the Hartree–Fock SCF equations. The best obtainable limiting values of the calculated ground-state energy (for a given molecular geometry) are then referred to, correspondingly, as the 'minimal basis limit' and the 'Hartree–Fock limit'; the latter represents the best result obtainable using a simple MO wavefunction (i.e. a single antisymmetric wavefunction corresponding to a unique assignment of electrons to orbitals), while the former suffers from the additional limitation due to representation of the MO's in terms of a very restricted set of AO's. The two levels of approximation are illustrated by the sixth and seventh functions in Table 5.1. The advantage of minimal basis calculations is that they give reasonably good wavefunctions (particularly when the AO's contain parameters such as screening constants which may be variationally optimized) which permit a very simple pictorial interpretation; for example, the charge cloud may be broken down as in equation (6.1) into a relatively small number of terms, referring to atomic and bond (overlap) regions, each with its own electron population. Calculations at the Hartree–Fock limit provide a somewhat better electron density, but there is no simple way of describing it other than by drawing contour maps (cf. Fig. 4.4) or difference-density maps (Fig. 4.7). When we are concerned only with the qualitative discussion of valence, we shall very often find minimal basis calculations adequate. Both types of calculation may be performed using an LCAO-type adaptation of the Hartree–Fock SCF method, first formulated independently by Hall (1951) and Roothaan (1951) and later developed and extended by many others.

Let us now look at the results of *ab initio* MO calculations on some of the simple first-row diatomics considered in Chapter 4.

(i) *Lithium* Li_2. The electron configuration, as discussed in §4.3, is

$$Li_2[(1\sigma_g)^2(1\sigma_u)^2(2\sigma_g)^2].$$

Calculations using a minimal basis set yield MO's which give the individual charge density contributions shown in Fig. 6.1 (*a, b, c*). The total density of the charge cloud is then a sum of MO contributions and appears in Fig. 6.1 (*d*). We could simplify the the representation of the orbitals by leaving out all but the outermost contour (of constant ψ^2 ($= 6.1 \times 10^{-5}$ electrons bohr^{-3}) and hence of constant ψ) and adding signs to indicate where ψ is positive or negative. The resultant pictures would clearly resemble the schematic representations of the MO's used in Figs. 4.8–4.10. It should be noted in particular that an electron in $1\sigma_g$ (or in $1\sigma_u$) is almost exactly equivalent to half an electron in a 1s-type density on each nucleus. The four electrons in these

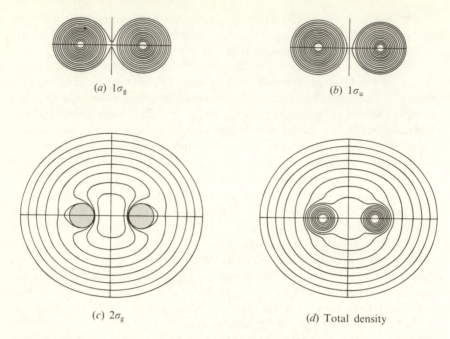

(a) $1\sigma_g$ (b) $1\sigma_u$

(c) $2\sigma_g$ (d) Total density

FIG. 6.1. Orbital contributions to the electron density in Li_2. The $1\sigma_g$, $1\sigma_u$, and $2\sigma_g$ MO's are all doubly occupied; superposition of the corresponding densities gives the total density shown. (Contours near the nuclei are too close to be shown.) (From A. C. Wahl, *Science* **151**, 961 (1966), with permission.)

two MO's in fact give almost exactly the same density as two independent K shells; the 'cancellation' of equally occupied bonding and antibonding MO's (p. 98) is virtually exact when overlap is small.

The calculated total electronic energy is† -14.8715 compared with the 'experimental' value‡ of -14.9944. The error of 0.82 per cent may seem quite small. Unfortunately, however, the dissociation energy itself is a small part of the total energy; the value computed by subtracting the total Hartree–Fock energy of the separate atoms is 0.17 and this compares very poorly with the experimental value of 1.05. The situation is quite favourable in Li_2, where there are only four high-energy inner-shell electrons, and the poor result therefore emphasizes the extreme difficulty of computing accurate dissociation energies. Even with a minimal basis the total electronic energy is only in error by 1.03 per cent, and it is therefore clear that the major defect of the computations lies in their use of the single-configuration MO wavefunction, not in the precise choice of the AO basis.

† In quoting energies we shall assume atomic units (1 hartree $= 27.2108$ eV) unless otherwise indicated.

‡ That is, inferred from spectroscopic data.

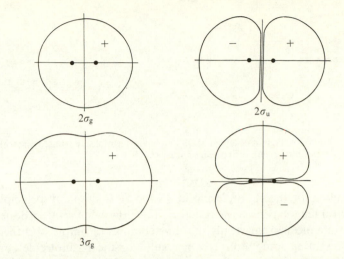

FIG. 6.2. Forms of the valence MO's in the nitrogen molecule. The boundary contour shown corresponds to an electron density of 6.1×10^{-5} electrons bohr^{-3}. (Based on results of A. C. Wahl, *Science* **151**, 961 (1966), with permission.)

(ii) *Nitrogen* N_2. The electron configuration is (p. 101)

$$N_2[KK(2\sigma_g)^2(2\sigma_u)^2(1\pi_u)^4(3\sigma_g)^2].$$

The calculated MO's are depicted in Fig. 6.2 using, as suggested above, the bounding surfaces corresponding to an electron density (ψ^2) of 6.1×10^{-5} electron bohr^{-3}. Again, the schematic diagrams in Figs. 4.8–4.10 are seen to give a reasonable qualitative picture of the MO forms. The computed total energy is -108.993 compared with the experimental value of -109.586, an error of only 0·54 per cent, and yet the computed dissociation energy of 5·31 eV falls substantially short of the experimental value of 9·90 eV.

(iii) *Fluorine* F_2. The electron configuration obtained (p. 101) by adding four more electrons in the antibonding pair of $1\pi_g$ MO's is

$$F_2[KK(2\sigma_g)^2(2\sigma_u)^2(1\pi_u)^4(3\sigma_g)^2(1\pi_g)^4]$$

and some orbital forms are shown in Fig. 6.3. Again the one new MO ($1\pi_g$) has the expected form (cf. Fig. 4.10). In this case, however, although the total energy is wrong by only 0·44 per cent the dissociation energy is computed to be -1.37 eV compared with the experimental 1·68 eV. Theory thus predicts instability against dissociation into two fluorine atoms!

The above examples indicate some of the limitations of *ab initio* SCF calculations, even when they are performed using the most sophisticated computational techniques available and effectively reach the Hartree–Fock limit. To go beyond that limit, by using for example VB, GVB, or CI methods

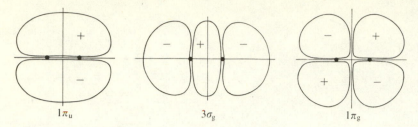

$1\pi_u$ $3\sigma_g$ $1\pi_g$

FIG. 6.3. Forms of the highest occupied MO's in the fluorine molecule. (Based on results of A. C. Wahl, *Science* **151**, 961 (1966), with permission.)

(as explained for the H_2 molecule in §5·4–5·5), is simple in principle but is computationally expensive in practice. To show what can be done and to demonstrate more dramatically the shortcomings of simple MO theory, it is worth presenting some more recent results for the F_2 molecule considered above. These results (Fig. 6.4) were obtained by Wahl and Das (1970), whose 'optimized valence configuration' method admits sufficient CI to ensure that the wavefunction dissociates to give the correct electron configurations of the separate atoms and also allows variational optimization of the forms of all the valence-shell orbitals used. Although approached from the MO point of view,

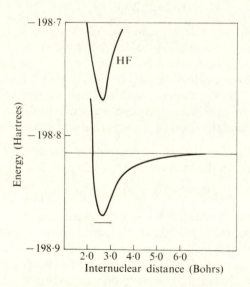

FIG. 6.4. Calculated potential energy curves for the fluorine molecule. The horizontal asymptote indicates the Hartree–Fock energy of the separate atoms; the dissociation energy is calculated relative to this limit, the experimental value being indicated by the horizontal line. The Hartree–Fock calculation for the molecule gives the curve marked HF. (Taken from A. C. Wahl and G. Das, *Adv. quantum Chem.* **5**, 261 (1970), with permission.)

the simpler forms of such a wavefunction are mathematically equivalent to a GVB function (§5.5) in which the overlapping (deformed) 'atomic' orbitals are optimized. It is therefore expected to yield energy values which will closely follow the true potential energy curve as the atoms are separated. Fig. 6.4 bears out this expectation, the theoretical estimate of the dissociation energy (1·57 eV) being quite close to the experimental value. The computational effort required to reach this level of accuracy, even for a molecule as small as F_2, is very large, but from a conceptual point of view progress beyond the Hartree–Fock limit involves little more than was encountered in our discussions of the hydrogen molecule in Chapter 5.

Ab initio calculations also provide us with numerical values of the energies of the occupied MO's at the one-configuration level of approximation. Henceforth we shall adopt the usual practice of designating individual *orbital* energies by the Greek letter ε (epsilon) to avoid confusion with total electronic energies (E). These may be correlated with ionization potentials, and are thus useful in the assignment and interpretation of photoelectron spectra; they also provide a quantitative basis for the construction of the correlation diagrams discussed in §§4.6, 4.7. A correlation diagram for N_2, obtained by Mulliken (1972), is shown in Fig. 6.5. The detailed behaviour of the various levels could not have been predicted without a rather extensive calculation, and the fact that some curves (in particular the $3\sigma_g$ and the $1\pi_u$) may cross and lie close together over a considerable range of internuclear distances explains the considerable uncertainty during past years as to the relative positions of some levels. This is emphasized in Table 6.1 which shows the MO energies in N_2, and F_2 (at the equilibrium distance) computed with (i) minimal basis and (ii) near-Hartree–Fock accuracy. Even experimentally (e.g. by

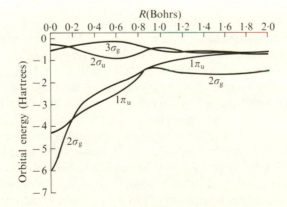

FIG. 6.5. Correlation diagram for the nitrogen molecule valence orbitals in the range $R = 0$ (united atom) to $R = R_e = 2\cdot013a_0$. (From R. S. Mulliken, *Chem. phys. Lett.* **14**, 137 (1972), with permission.)

TABLE 6.1
Orbital energies in the molecules N_2 and F_2

	N_2		F_2	
$1\sigma_g$	$-15\cdot6471$†	$-15\cdot6820$‡	$-26\cdot3593$	$-26\cdot4227$
$1\sigma_u$	$-15\cdot6442$	$-15\cdot6783$	$-26\cdot3593$	$-26\cdot4224$
$2\sigma_g$	$-1\cdot4211$	$-1\cdot4736$	$-1\cdot6259$	$-1\cdot7565$
$2\sigma_u$	$-0\cdot7137$	$-0\cdot7780$	$-1\cdot3613$	$-1\cdot4950$
$3\sigma_g$	$-0\cdot5555$	$-0\cdot6350$	$-0\cdot5461$	$-0\cdot7460$
$1\pi_u$	$-0\cdot5454$	$-0\cdot6154$	$-0\cdot6079$	$-0\cdot8052$

† The results in the first column are from minimal basis SCF calculations by Ransil (1960).

‡ The results in the second column are from extended basis SCF calculations of near-Hartree–Fock accuracy, by Cade *et al.* (1966) and by Wahl (1964).

photoelectron spectroscopy) it is not possible to resolve such uncertainties for (cf. p. 90) the MO's *do not exist*! They are artefacts of a particular *theory*, based on an independent particle model, and they possess experimental significance only at the level of approximation appropriate to that model. The measurable ionization energy, a difference of *total* electronic energies of the neutral molecule and its ion, is mathematically identifiable with an orbital energy only at the Hartree–Fock level; with *exact* wavefunctions, as we remarked at the end of §5.2, there would be *no* orbitals and no 'orbital energy' concept.

When we turn to heteronuclear molecules, the behaviour of the MO's and the energy as a function of internuclear distance is of particular interest, for such molecules sometimes dissociate not into free atoms but into ions. As the atoms are pulled apart, a point must be reached at which the electrons in a molecular orbital, shared over both centres, must 'decide' whether or not they will both go to the same atom; in other words the MO which holds them may suddenly become much more like an AO. Dramatic pictures of what actually happens have been produced by Wahl and co-workers (1968). Figure 6.6 shows three stages in the formation of the highly polar molecule LiF from *neutral* atoms. At large distances each atom has essentially a free-atom charge cloud (Fig. 6.6 (*a*)), the contours showing clearly the rather diffuse lithium 2s distribution. However, at a distance of 13·9 bohr the lithium suddenly loses its 2s electron, which leaves behind a Li^+ core, and completes the octet of the fluorine valence shell to give F^-. As the equilibrium distance is approached, the Li^+ core apparently sinks into the somewhat expanded charge cloud of the F^- ion, giving a final electron density in which the lithium nucleus retains only its K shell.

Of course most polar molecules do not dissociate into ions. The simplest description (that of GVB theory §5.5) is that the highest doubly occupied MO

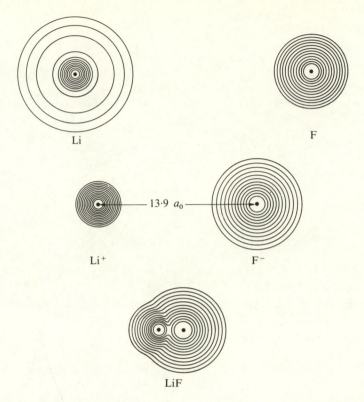

FIG. 6.6. Approach of lithium and fluorine atoms to form the lithium fluoride molecule. At R = $13.9a_0$ the ion pair becomes more stable than a pair of neutral atoms. (After A. C. Wahl *et al. Int. J. quantum Chem.* **Symp. 3** (Pt. II) 499 (1970), with permission).

then 'splits' into two singly occupied overlapping orbitals which become more and more atomic in character as separation proceeds, finally yielding the neutral atoms. A typical case is lithium hydride with configuration LiH[$(1\sigma)^2(2\sigma)^2$]. The GVB orbitals are indicated in Fig. 6.7; both are 'molecular' in character (and overlap strongly) near the equilibrium distance, but at 10 bohr have become essentially $1s_H$ and $2s_{Li}$ orbitals.

Calculations of MO energies give quantitative support to the principles advanced in Chapter 4. Thus, in calculations by Ransil on lithium hydride, the Li 1s orbital is energetically out of range of the valence AO's, but *both* 2s and 2p on the lithium atom are expected to participate, along with the hydrogen 1s, in the 2σ bonding MO, and this is confirmed by the results, as will be seen presently.

Tabulations of orbital energies, orbital forms, and electron density maps of the kind we have shown are readily available (Wahl *et al.* 1968, 1970; Bader

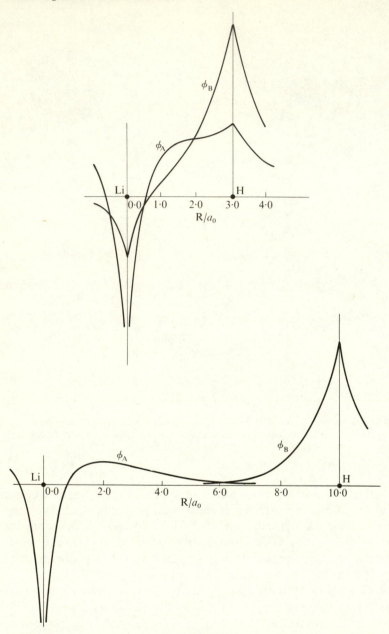

FIG. 6.7. Valence-bond description of bond breaking in LiH. (*a*) Near the equilibrium distance the MO splits into singly occupied, and strongly overlapping, 'distorted' AO's (values of ϕ_A and ϕ_B plotted along the internuclear axis). (*b*) As the separation increases, these turn into lithium 2s and hydrogen 1s orbitals. The cusps at the lithium nucleus arise from the requirement that the valence orbitals are orthogonal to the lithium K shell. (From L. Newbould, Ph.D. thesis, University of Sheffield (1977), with permission.)

1970), as also are the difference density maps such as Fig. 4.7; the latter have been used, particularly by Bader *et al.* (1967, 1968), in discussions of the origin of the electrostatic forces responsible for the bonds.

6.2. Hybridization

So far we have been content to accept in pictorial form the results obtained from large computations; these have shown us the forms of the MO's involved and, at internuclear distances where the simple MO approximation is invalid, the forms of the constituent 'atomic' orbitals on the weakly overlapping, almost free, atoms. However, we have not considered how these various orbitals are constructed from the AO's of the free atoms; by doing so we shall be able to introduce, on a firm foundation, certain ideas which were formulated more intuitively in the early days of valence theory and are still of great value at a qualitative and descriptive level.

Let us consider more fully the case of lithium hydride, looking at the LCAO forms of the MO's. *Ab initio* SCF calculations (minimal basis) by Ransil (1960), using a basis of lithium 1s, 2s, 2p,† and hydrogen 1s AO's which we denote by ϕ_{1s}, ϕ_{2s}, ϕ_{2p}, and ϕ_H, respectively, lead to the following MO's (neglecting slight mixing between ϕ_{1s} and the other AO's)

$$1\sigma \simeq \phi_{1s}$$
$$2\sigma \simeq 0{\cdot}323\phi_{2s}+0{\cdot}231\phi_{2p}+0{\cdot}685\phi_H. \tag{6.2}$$

The electron configuration

$$\text{LiH}[1\sigma^2 2\sigma^2]$$

then indicates a lithium inner shell $1\sigma^2$ virtually identical with that in the free atom, and a σ bond with two electrons in the bonding MO, 2σ. This is in agreement with what would be expected on the basis of Chapter 4, except that the 2σ MO contains an appreciable amount of 2p character instead of being simply a mixture of the singly occupied valence orbitals, ϕ_{2s} and ϕ_H, of the separate atoms. The lithium 2p AO was in fact admitted because, lying only a little higher than 2s, there is no real reason to exclude it according to the energy criterion (p. 81); in the full implementation of the LCAO method we should allow free mixing of all AO's of not too different energy and see what comes out when we solve the secular equations (§3.8), which is just what was done (by methods which will be described later) in the calculation leading to (6.2).

At first sight this result appears to invalidate our qualitative conclusion (§4.3) that reasonable MO's for diatomic molecules can be built up from overlapping *pairs* of orbitals, one on each atom. We should like to retain this

† Pointing along the bond axis, to be consistent with σ symmetry.

very simple principle, if at all possible; we can do so by rewriting (6.2) in the form

$$2\sigma = 0{\cdot}397(0{\cdot}813\phi_{2s}+0{\cdot}582\phi_{2p})+0{\cdot}685\phi_H \tag{6.3}$$

which is a linear combination of two orbitals, one on each atom, *the lithium orbital being a mixture of* ϕ_{2s} *and* ϕ_{2p}. The mixture so defined is called a *hybrid* orbital and the process of mixing is called *hybridization*. The AO coefficients in the hybrid

$$\phi_{hybrid} = 0{\cdot}813\phi_{2s}+0{\cdot}582\phi_{2p} \tag{6.4}$$

are in the ratio $0{\cdot}323:0{\cdot}231$ but have been multiplied by a common factor in order to normalize the hybrid. For this to be so we require in general

$$\int (a\phi_{2s}+b\phi_{2p})^2\,d\tau = a^2+b^2 = 1$$

assuming ϕ_{2s} and ϕ_{2p} are normalized and orthogonal (p. 70); the multiplying factor must therefore be chosen so that the sum of the squares of the coefficients in (6.4) is equal to unity. In terms of the hybrid, (6.3) becomes

$$2\sigma = 0{\cdot}397\phi_{hy}+0{\cdot}685\phi_H \tag{6.5}$$

and the bonding MO is thus formed by overlapping the hydrogen 1s AO with a lithium *hybrid*.

Hybridization, by which in general we mean mixing of AO's *on the same atom*, is the rule rather than the exception whenever we perform a full MO calculation. However, the fact that the accurate results can often be reinterpreted in terms of simple pairwise overlaps, though perhaps of hybrids instead of pure AO's, allows us to retain and improve the qualitative picture used in Chapter 4. The pictorial form of a hybrid such as (6.4) is indicated in Fig. 6.8† and gives added strength to the principle of maximum overlap (p. 82) which was at the basis of our qualitative discussions, for the 'contamination' of a pure 2s AO with a little 2p evidently gives a hybrid strongly directed towards the hydrogen 1s with a greatly enhanced overlap; the energy is lowered (it must be, because the coefficients are determined by the variation method), the wavefunction is improved, and the strength of the predicted bond is increased. Even without performing an actual calculation, we might have inferred on the basis of a maximum overlap postulate that hybridization of the lithium 2s AO (to make it 'lean' towards the hydrogen) would give an improved description of the bond; the calculation merely serves to give accurate values of the mixing coefficients.

Since hybridization can always be invoked as a means of improving overlap between AO's, it is natural to ask why the effect is large in some cases and small

---

† The hybrid shown in Fig. 6.8 has a somewhat higher 2p content than that defined in (6.4); it is an 'sp³' hybrid (p. 200).

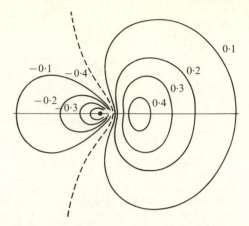

FIG. 6.8. Contour map for a hybrid orbital formed by s–p mixing.

in others. In other words, whilst recognizing that hybridization decreases the energy by putting more electron density in the overlap region (thereby strengthening the *bond*), what effects, if any, *oppose* hybridization and thus control the balance of the s and p mixing coefficients?

To find the answer, we need only recall that the 2p AO was admitted in the calculation because the lithium 2p AO is not much higher in energy than the 2s. Had it been very much higher, we should in fact have found a much smaller 2p participation. The reason is simply that for an electron in a hybrid orbital

$$\phi_{hy} = a\phi_{2s} + b\phi_{2p} \tag{6.10}$$

the expectation value of the energy is

$$\varepsilon_{hy} = \int \phi_{hy} \hat{h} \phi_{hy} \, d\tau$$

$$= a^2 \int \phi_{2s} \hat{h} \phi_{2s} \, d\tau + b^2 \int \phi_{2p} \hat{h} \phi_{2p} \, d\tau.$$

The other integrals vanish for symmetry reasons (cf. p. 83); those given are simply the 2s and 2p orbital energies. Hence

$$\varepsilon_{hy} = a^2 \varepsilon_{2s} + b^2 \varepsilon_{2p}. \tag{6.11}$$

Since a^2 and b^2 are positive fractions, adding up to unity, the hybrid orbital energy is a weighted mean lying somewhere in the interval between ε_{2s} and ε_{2p}. The energy *drop* due to bonding, favoured by increasing b (which is zero for a pure 2s AO), is thus offset by a *rise* in the 'atomic' component of the energy due to electrons in the hybrid. If ε_{2p} is much higher than ε_{2s} it takes a lot of energy to mix in some 2p, and hybridization is strongly resisted.

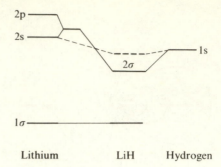

FIG. 6.9. Correlation diagram for formation of lithium hydride (schematic). Incorporation of some of the lithium $2p_\sigma$ AO in the 2σ MO lowers the energy considerably, corresponding to stronger bonding. The broken line indicates the result obtained using only the 2s AO on the lithium.

We now begin to understand at a deeper level the energetics of bond formation. To relate this understanding to the simpler qualitative picture used throughout Chapter 4, let us consider again the correlation diagram (§4.7). We may admit the hybridization in LiH in two stages: first mixing the lithium 2s and 2p to form a hybrid, and then overlapping the hybrid with the hydrogen 1s to form the bond. The result is shown schematically in Fig. 6.9. It must be stressed that the diagram is drawn in this way only to present energy relationships; hybridization is not a physical effect but merely a feature of our theoretical description, and bonding does not actually occur as a two-stage process. Nevertheless, the diagram indicates how the energy spent on hybridization has its reward in the improved overlap which results in much stronger bonding; the broken line indicates the result without hybridization. The hypothetical 'state' of the lithium atom with electron configuration $\text{Li}[1s^2\phi_{hy}]$ is known as a 'valence state'. It is not a real spectroscopically observable state at all, but rather the result of an imaginary dissociation of the molecule in which the lithium orbitals are 'frozen' in the forms employed in describing the molecule.

The valence state concept is discussed more fully in Chapter 7, but we note at this point that the rise in energy when some 2p character is mixed into a 2s orbital may be described as 'promotion', in which a fraction of a 2s-type electron density is removed and replaced by a fraction of a 2p type. This is because the density for an electron in ϕ_{hy} is

$$P_{hy} = \phi_{hy}^2 = a^2\phi_{2s}^2 + b^2\phi_{2p}^2 + 2ab\phi_{2s}\phi_{2p}. \tag{6.12}$$

If this is integrated, it gives unity (1 electron) with a^2 coming from the 2s density, b^2 from the 2p density, and nothing from the third term since 2s and 2p are orthogonal. $\phi_{2s}\phi_{2p}$ has the same general shape as a p orbital, with positive and negative lobes on opposite sides of a nodal plane; it contributes zero to the integral but is the term which 'polarizes' the density $P_{h s}$, making it 'lean' in the direction of the positive lobe. In the spirit of the discussion in §4.5 we may say the 2s and 2p AO's have populations a^2 and b^2, respectively. Thus the electron configuration Li[$1s^2\phi_{hs}$] could also be indicated formally as Li[$1s^2 2s^{a^2} 2p^{b^2}$]; the lithium valence state corresponding to the LiH calculation above would thus be represented as

$$\text{Li}[1s^2 2s^{0.661} 2p^{0.339}] \qquad (6.13)$$

as if 0·339 of an electron had been 'promoted' from a 2s state into a 2p state.

The use of populations to indicate the amount of s or p character in a hybrid has an important application when two or more hybrids are formed by mixing two or more AO's. Thus from one 2s and one 2p AO we may form *two* independent hybrids (i.e. linear combinations) with the same orthogonality property as the original AO's; if one has 30 per cent s character and 70 per cent p character (i.e. $a^2 = 0·3$, $b^2 = 0·7$), the other will have 70 per cent s character and 30 per cent p character. (The proof is simple. If ϕ_1 and ϕ_2 are orthogonal and normalized, then so are $\psi_1 = a\phi_1 + b\phi_2$ and $\psi_2 = b\phi_1 - a\phi_2$ for $\int \psi_1 \psi_2 \, d\tau = ab - ba = 0$. Also normalization requires $a^2 + b^2 = 1$ and the result follows at once. It may be generalized to any number of orbitals.) The second of the two hybrids with which the 2s and 2p AO's are replaced is thus determined once we have decided on the form of the first. The interrelationship of different hybrids is discussed in the next chapter.

Since hybridization is energetically less costly when $\varepsilon_p - \varepsilon_s$ is small, which is the case for a small number of electrons outside a compact core (i.e. resembling a point charge), the importance of hybridization is expected to be greater for elements on the left of the periodic table. In carbon monoxide, for example, the effect would be considerably more marked for carbon than for oxygen. For heteronuclear diatomics such as CO, hybridization has a dramatic effect on the correlation diagram and cannot be ignored. CO itself is an excellent example. Without admission of hybridization, we should draw the correlation diagram (valence electrons only) along the lines of Fig. 6.10: pairing those

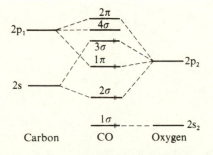

FIG. 6.10. Correlation diagram for the CO molecule, without hybridization. The highest occupied MO (3σ) is antibonding.

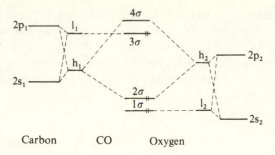

FIG. 6.11. Correlation diagram for the CO molecule, hybridization admitted. With enhanced separation of the bonding and antibonding σ MO's, the highest occupied MO (3σ) becomes a carbon lone-pair orbital, in agreement with experiment. (For clarity the π levels have been omitted.)

AO's which are closest in energy, it would appear that the lowest σ MO would be mainly oxygen 2s, holding an oxygen lone pair; the next six valence electrons would be described by $2\sigma^2 1\pi^4$ (i.e. a σ bond and two π bonds); finally $3\sigma^2$ would apparently be a σ *anti*bonding pair. This description is in conflict with experimental evidence, which points to a strong σ bond and a carbon lone pair. The effect of some slight 2s–2p mixing (the 2p AO's of σ symmetry) is shown in Fig. 6.11; the closeness of h_1 (mainly carbon 2s) and h_2 (mainly oxygen 2p), together with their improved overlap, gives a greatly enhanced splitting into bonding and antibonding levels; the 3σ MO is then essentially the carbon orbital labelled l_1, which is a lone pair hybrid consisting mainly of the 2p AO but with some admixture of 2s. The antibonding MO is then unoccupied, the top two electrons being very nearly l_1^2; moreover, just as h_1 points *into* the bond, the second hybrid l_1 points *out of* the bond (Fig. 6.12). (As above, if $h_1 = a2s + b2p$, its orthogonal partner is $l_1 = b2s - a2p$.) The electron density associated with $1\sigma^2 2\sigma^2 3\sigma^2$ (i.e. the σ electron density) is thus expected

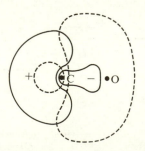

FIG. 6.12. Carbon lone-pair hybrid in carbon monoxide. This is the orthogonal partner of the bonding hybrid (shown by a broken line) and points in the opposite direction, i.e. *away from* the oxygen.

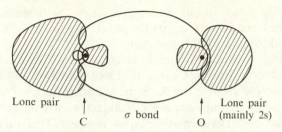

Lone pair ↑   ↑ Lone pair
C σ bond   O   (mainly 2s)

FIG. 6.13. Schematic representation of the expected electron density in CO (π-type contributions not shown). The carbon lone pair, with more 2p character, should project somewhat further than the oxygen lone pair, which is more heavily 2s.

to have the form shown schematically in Fig. 6.13. This picture is in complete accord both with accurate calculations† on CO and with the observed properties of the molecule. In particular, the CO group binds very easily to transition-metal ions, attaching itself by the electron-rich carbon end which is attracted by the positive ion. The groups CN^- and NO^+ are isoelectronic with CO and also occur frequently in transition-metal complexes, attaching themselves in a similar way.

To summarize, the concept of hybridization is of vital importance in qualitative valence theory, allowing us to anticipate the main results of complete wave-mechanical calculations whilst yet retaining the simple pictorial principles (e.g. pairing of orbitals with similar energies and good overlap) introduced in Chapter 4. In subsequent chapters we find the concept no less valuable in dealing with polyatomic molecules.

6.3. Polar bonds

In heteronuclear diatomics there is frequently a considerable flow of charge from one atom to the other, depending on the relative 'electron-attracting powers' of the two atoms concerned. We say the bond becomes 'polar'. The polarity of the bond is revealed experimentally in the electric dipole moment of the molecule; in MO theory it is reflected in a disparity of the coefficients of the two AO's in the bonding MO. Thus, if we consider a bond described by two electrons (one from atom A, one from B) in the normalized MO of (4.3), namely

$$\psi = N(\phi_A + \lambda\phi_B) \tag{6.14}$$

the limits $\lambda \to 0$ and $\lambda \to \infty$ would concentrate the two electrons entirely on atom A or entirely on atom B, respectively ($\psi \to \phi_A$ or $\psi \to \phi_B$). We should obtain a purely *ionic* bond, represented by A^-B^+ or A^+B^- as the case may be. For $\lambda \simeq 1$, however, there would be no appreciable polarity, the electrons would be equally shared between the two atoms, and we should speak of a

† Contour maps of the MO's are given by Huo (1965); some of these are reproduced in a later section (p. 179).

purely *covalent* bond. In intermediate situations, with limited migration of electrons from one atom towards the other, the bond would be polar and might be said to show some 'ionic character'.

The polarity of a bond indicates an 'electronegativity' difference between the atoms it connects, Pauling's original (1960) definition of electronegativity being 'the power of an atom in a molecule to attract electrons to itself'. We now look for a quantitative measure of ionic character and consider its relationship to observable properties such as the dipole moment. This will prepare us for a discussion of electronegativity in the following section (§6.5).

First, we proceed using simple MO theory. To describe the charge shifts, we develop the 'population analysis' introduced in §§4.5 and 5.4.† The electron density associated with two electrons in the MO of (6.14) is clearly

$$P = 2\psi^2 = 2N^2[\phi_A^2 + 2\lambda\phi_A\phi_B + \lambda^2\phi_B^2]. \tag{6.15}$$

The normalizing factor N may be fixed by requiring $\int P \, d\tau = 1$, since the amount of charge in the distribution is two electrons. Thus, since $\int \phi_A^2 \, d\tau = \int \phi_B^2 \, d\tau = 1$ and $\int \phi_A \phi_B \, d\tau = S$ (the overlap integral), we obtain on integrating (6.15) $2 = 2N^2[1 + 2\lambda S + \lambda^2]$. This result agrees, of course, with (4.4):

$$N^2 = \frac{1}{1 + \lambda S + \lambda^2}. \tag{6.16}$$

It is convenient to write (6.15) in the standard form

$$P = P_{AA}\phi_A^2 + P_{BB}\phi_B^2 + 2P_{AB}\phi_A\phi_B \tag{6.17}$$

where the coefficients are fixed by the numerical values of λ and S. The electron populations of the associated orbital and overlap regions (cf. p. 129) are now determined by integrating the electron density (6.17) over all space. The three regions (three terms) give contributions $q_A = P_{AA}$, $q_B = P_{BB}$, $q_{AB} = 2P_{AB}S$, respectively, and $q_A + q_B + q_{AB} = 2$ (the number of electrons in the distribution). Reference to (6.15) then shows that the orbital and overlap populations are

$$q_A = 2N^2, \qquad q_B = 2\lambda^2 N^2, \qquad q_{AB} = 4\lambda N^2 S \tag{6.18}$$

where N^2 is given in (6.16). In terms of populations, (6.17) may be written alternatively as (cf. §5.4)

$$P = q_A d_A + q_B d_B + q_{AB} d_{AB} \tag{6.19}$$

where

$$d_A = \phi_A^2, \qquad d_B = \phi_B^2, \qquad d_{AB} = \phi_A\phi_B/S \tag{6.20}$$

† For a discussion of the generality of the concept see McWeeny (1954; 1955). For further elaboration in LCAO MO theory see Mulliken (1955).

are all *normalized* density functions—two orbital densities and an 'overlap density'.

When $\lambda \neq 1$, $q_A \neq q_B$ and the resultant shift of charge gives the bond ionic character. We adopt the convention that in A—B the electron flow is from A to B (giving A^+B^-) and note that this corresponds to $\lambda > 1$. If q_A and q_B were initially equal (covalent bond, no ionic character) and q electrons moved from A to B, the population difference would become $q_B - q_A = 2q$. The bond would become completely ionic for $q_B = 2$, $q_A = 0$, corresponding to transfer of one electron ($q = 1$), and it is therefore convenient to adopt $\frac{1}{2}(q_B - q_A)$ as a measure of the 'fractional ionic character' of the bond:

$$\text{FIC} = \tfrac{1}{2}(q_B - q_A). \tag{6.21}$$

This index was introduced as a convenient 'polarity parameter' by Klessinger and McWeeny (1965), in calculations on polyatomic molecules. According to MO theory, the fractional ionic character of the bond described by the MO of (6.14) follows from (6.16) and (6.18) as

$$\text{FIC} = \frac{\lambda^2 - 1}{1 + 2\lambda S + \lambda^2} \tag{6.22}$$

with limits 0 for a covalent bond ($\lambda = 1$) and 1 for a completely ionic bond ($\lambda \to \infty$). We expect this quantity to be related, at least approximately, to the dipole moment associated with the bond.

To obtain the dipole moment we evaluate the electric moment of the charge cloud (6.19), indicated schematically in Fig. 6.14 (*a*) where distances (z) are measured from the mid-point of A — B and no restrictive assumptions have been made about the forms of the orbitals. Since $-eP$ is the amount of charge per unit volume the electric moment is

$$-e \int zP \, d\tau = -e[q_A z_A + q_B z_B + q_{AB} z_{AB}]$$

where $\bar{z}_A$, for example, is the z co-ordinate of the centroid of the orbital

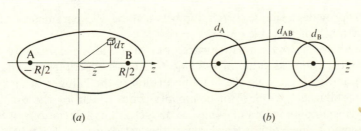

(a) (b)

FIG. 6.14. Evaluation of bond dipole moment. (*a*) Amount of charge in $d\tau$ is $-eP \, d\tau$, giving electric moment contribution $-ezP \, d\tau$ in the z direction. (*b*) Densities d_A, d_B, d_{AB}, of which P is a weighted sum. The centroids of these densities are at the nuclei and (approximately) the bond mid-point, respectively.

distribution $d_A = \phi_A^2$, i.e.

$$\bar{z}_A = \int z\phi_A^2 \, d\tau .$$

The dipole moment arises as the sum of the electronic part (from the bonding electrons) and the electric moment of the remaining positive ions (each having lost one electron to the bond). If we assume the latter contribution to arise from point charges $(+e)$ at $z = R/2$ and $z = -R/2$ (Fig. 6.14 (a)), the resultant moment about the mid-point will be zero and hence the total dipole moment will be the electronic term deduced above, namely,

$$\mu = -e[q_A \bar{z}_A + q_B \bar{z}_B + q_{AB} \bar{z}_{AB}]. \tag{6.23}$$

This is exactly what we should expect for three point charges $-eq_A$, $-eq_B$, $-eq_{AB}$, located at points $\bar{z}_A$, $\bar{z}_B$, $\bar{z}_{AB}$, respectively. There is clearly no direct relationship between μ and the FIC unless further simplifying assumptions are made; these are normally that (Fig. 6.14 (b)) (i) ϕ_A and ϕ_B are ordinary AO's (i.e. not hybrids) centred on their respective nuclei, giving $\bar{z}_A = -\frac{1}{2}R$, $\bar{z}_B = +\frac{1}{2}R$, and (ii) the centroid of the overlap density $\phi_A \phi_B$ is more or less at the mid-point of A—B, giving $\bar{z}_{AB} = 0$. On inserting these approximations in (6.23), we obtain at once

$$\mu = -\tfrac{1}{2}eR(q_B - q_A) = -eR \times (\text{FIC}) \tag{6.24}$$

which is a simple proportionality between dipole moment and fractional ionic character. The minus sign simply indicates that the left-hand atom is the positive end of the dipole, corresponding to our assumption $q_B > q_A$ (electron flow from left to right). In MO approximation, substitution from (6.22) yields a dipole moment of magnitude

$$|\mu| = \frac{eR(\lambda^2 - 1)}{1 + 2\lambda S + \lambda^2} \tag{6.25}$$

and thus provides a simple correlation between the MO parameter λ and the experimentally measurable quantity μ.

Figure 6.15 shows the relative variations of μ and λ according to (6.25), giving S the rounded value of $\frac{1}{3}$. Table 6.2 shows experimental data for a series of molecules (taking the bonding orbital of the halogen atom to be a pure p orbital, but see §6.5 for comments) from which, by the use of Fig. 6.15, corresponding values of λ may be deduced. Small deviations of S from the assumed value would scarcely affect the last column in the table, or the appearance of the curve. The values of λ are such that in each case $\lambda > 1$ for the halogen atom orbital. It will be seen from this table that except for very polar molecules, such as HF or KCl, λ does not differ greatly from unity.

TABLE 6.2

Fractional ionic character and MO parameter λ from observed dipole moments

Molecule	μ/ea_0† (expt.)	R/a_0	μ/eR (=FIC)	λ
HF	0·716	1·74	0·412	1·80
HCl	0·405	2·40	0·168	1·28
HBr	0·326	2·66	0·123	1·17
HI	0·176	3·04	0·058	1·10
KCl	2·48	5·27	0·470	2·00

† μ/ea_0 measures the dipole moment in atomic units ($ea_0 = 8\cdot478\ 10^{-30}$ C m); similarly for the bond length (R/a_0). The ratio gives μ/eR (=FIC, from equation (6.24)) and the parameter λ from (6.25). The overlap integral has been given a rounded value of $\frac{1}{3}$. For an example of accurate dipole moment determination see Van Dijk and Dymanus (1970).

Before commenting on the approximations just made, we note that VB theory gives a result of exactly the same form, namely (6.23), the only difference being (cf. p. 129) that the estimates of q_A and q_B are not quite the same as those calculated by the MO method. Indeed, so long as the wavefunction is constructed using only two AO's, ϕ_A and ϕ_B, the charge density is bound to appear in the form (6.19), and the analysis leading to (6.24) is therefore valid even when both methods are refined to the point at which they agree, as

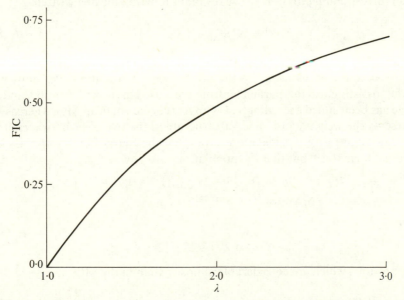

FIG. 6.15. Relationship between fractional ionic character and λ for a polar bond (MO theory). Overlap is taken to be $S = \frac{1}{3}$.

discussed in §5.3; the only differences arise from somewhat different initial estimates of q_A and q_B.

In the VB approach the charge density is not just a sum of orbital contributions, as in MO theory, but must be evaluated from the full two-electron wavefunction, exactly as we did for the molecule H_2 (p. 128). This time, however, the covalent (Heitler–London) function will not suffice and we must add the ionic terms. The best possible function would be (5.23), namely

$$\Psi = c_1 \Psi_{AB} + c_2 \Psi_{A^+B^-} + c_3 \Psi_{A^-B^+} \qquad (6.26)$$

where, omitting spin factors,

$$\Psi_{AB} = N_1 [\phi_A(1)\phi_B(2) + \phi_B(1)\phi_A(2)] \qquad (6.27a)$$

$$\Psi_{A^+B^-} = N_2 [\phi_B(1)\phi_B(2)] \qquad (6.27b)$$

$$\Psi_{A^-B^+} = N_3 [\phi_A(1)\phi_A(2)] \qquad (6.27c)$$

are the covalent and the two ionic structures, respectively. Again (see p. 134) omission of the singlet spin factor has no effect on the calculated electron density. In a homonuclear molecule such as H_2 the coefficients c_2 and c_3 are equal by symmetry, but in the heteronuclear case the structure which puts both electrons on the more electronegative atom will be strongly favoured on energetic grounds. With our convention that B is the more electronegative atom, this suggests that $c_2 \gg c_3$ and that we might reasonably drop the term in (6.26) corresponding to A^-B^+. The resultant function may be written

$$\Psi = N[\Psi_{AB} + \lambda' \Psi_{A^+B^-}] \qquad (6.28)$$

and again the amount of ionic character will depend on a single parameter λ', with N determined as usual by the normalization condition. We must be careful to distinguish this parameter from the λ used in the MO method, and a prime has been added accordingly; λ is an AO coefficient in an MO, whereas λ' represents the weight of an ionic VB structure in the two-electron wavefunction. A straightforward calculation shows that the charge density appears in the usual form (6.19) but that the populations are now

$$q_A = D^{-1}, \qquad q_B = D^{-1}[1 + 2S\sqrt{\{2(1+S^2)\}}\lambda' + 2(1+S^2)\lambda'^2]$$
$$q_{AB} = 2D^{-1}S[S + \sqrt{\{2(1+S^2)\}}\lambda']$$

where

$$D = (1+S^2)(1+\lambda'^2) + 2S\sqrt{\{2(1+S^2)\}}\lambda'.$$

These results, which are considerably clumsier than in the MO case, yield

$$(\text{FIC})_{VB} = \tfrac{1}{2}(q_B - q_A) = \frac{(1+S^2)\lambda'^2 + S\sqrt{\{2(1+S^2)\}}\lambda'}{(1+S^2)(1+\lambda'^2) + 2S\sqrt{\{2(1+S^2)\}}\lambda'} \qquad (6.29)$$

which may be compared with the MO approximation (6.22). The only way to simplify this expression is to make a somewhat unsatisfactory approximation, commonly made in VB theory, by neglecting overlap. On putting $S = 0$ we obtain

$$(\text{FIC})_{\text{VB}} \simeq \frac{\lambda'^2}{1+\lambda'^2}. \tag{6.30}$$

The corresponding dipole moment, with the same assumptions as those leading to (6.25), is

$$|\mu| = eR(\text{FIC})_{\text{VB}} = \frac{eR\lambda'^2}{1+\lambda'^2}. \tag{6.31}$$

The curve relating the FIC to λ', through (6.30), is shown in Fig. 6.16 (broken curve); the more accurate result given by (6.29) is indicated by the full curve, drawn for the representative overlap $S = \frac{1}{3}$. The observed values of μ for some hydrogen halides yield the values of FIC and λ' shown in Table 6.3.

The values of λ' inferred from dipole-moment measurements have a certain pictorial appeal, since (with the assumption that Ψ_{AB} and $\Psi_{\text{A}^+\text{B}^-}$ are approximately orthogonal) the relative weights of the two structures, A—B and A^+B^-, in (6.28) are $1/(1+\lambda'^2)$ and $\lambda'^2/(1+\lambda'^2)$; the covalent structure is *non*-polar and the FIC given in (6.30) is thus, as expected, simply the weight

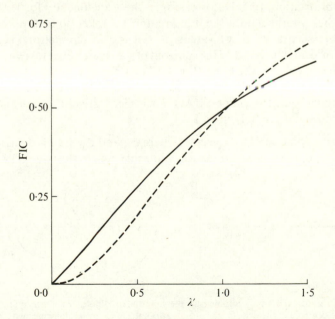

FIG. 6.16. Relationship between fractional ionic character and λ' for a polar bond (VB theory). The broken curve corresponds to neglect of overlap; the full curve corresponds to $S = \frac{1}{3}$.

TABLE 6.3

Fractional ionic character and VB parameter λ'

Molecule	HF	HCl	HBr	HI
FIC	0·41	0·17	0·12	0·06
λ'	0·84†	0·45	0·37	0·25

† Values estimated from the approximate VB relationship (6.30).

with which A^+B^- appears. In fact, however, the values so calculated probably have little quantitative significance and at best indicate only general trends.

Finally, we must comment on the approximations (p. 156) on which the dipole moment formula (6.24) rests. Let us take them one by one.

(i) When ϕ_A and ϕ_B, the overlapping orbitals used in constructing the electron-pair wavefunction, are ordinary AO's, then $\bar{z}_A = -\frac{1}{2}R$ and $\bar{z}_B = \frac{1}{2}R$. This is the case for either the MO function (6.14) or the VB (Heitler–London) function (6.26) provided hybridization is neglected, but actual calculations, as outlined in §6.2, should warn us that the simple description of a bond in terms of an overlapping *pair* of AO's may frequently be unrealistic unless the AO's are allowed to distort by hybridization. In LiH, for example, the wavefunction for the bond is rather poor if ϕ_A (on the lithium atom) is taken to be a pure 2s AO; more accurately, as we saw in §6.2, it is a 2s–2p mixture, i.e. an s–p hybrid, and in that case the centroid of the ϕ_A^2 density moves away from the nucleus into the bond (Fig. 6.17 (*a*)). We may say that inclusion of hybridization results in an *atomic dipole* owing to deformation of the originally centrosymmetric AO. Such effects are often large and will be discussed more fully in §6.5.

(ii) For a homonuclear molecule, where ϕ_A and ϕ_B are of identical form, $\bar{z}_{AB} = 0$, but in a heteronuclear molecule, where the two AO's differ in

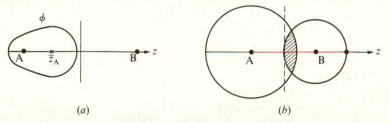

(a)   (b)

FIG. 6.17. Origin of the atomic dipole and the homopolar dipole. (*a*) when hybridization is admitted the centroid of the density $d_A = |\phi_A|^2$, at point $\bar{z}_A$, no longer coincides with nucleus A, and (*b*) when ϕ_A and ϕ_B are dissimilar the centroid of the overlap density is no longer at the origin (bond mid-point).

form, the centroid of the overlap density $\phi_A\phi_B$ will be displaced from the bond mid-point (see Fig. 6.17 (b)) giving $\dot{z}_{AB} \neq 0$. This means that the overlap density contributes to the dipole moment; there is a 'bond dipole'. This effect is present even when ϕ_A and ϕ_B are simple AO's, provided they differ in size, and even in the homopolar situation where there is no overall shift of charge from A to B (i.e. when $\lambda = 1$ in the MO (6.14) or $\lambda' = 0$ in the VB function (6.26), corresponding to a purely covalent bond with no ionic character); the resultant dipole is for this reason sometimes called a 'homopolar dipole', and it may be quite large, as we shall see when we discuss a particular example (§6.5).

(iii) We have considered only the charge distribution in a single electron-pair bond, assuming that the moment due to the remaining nuclei and electrons could be calculated as if it arose from two positive ions, regarded as point charges situated at $\pm R/2$. This is clearly a crude approximation. In the first place there may be other bonds (e.g. the π bonds in N_2), in which case we should have to treat each in a similar way and sum the results. Secondly, the positive cores remaining after dealing with the bonding electrons (and these may now be singly, doubly, or triply ionized for single, double, and triple bonds) can certainly not be regarded as positive point charges; the nuclei are point charges but they may be surrounded by inner-shell and lone-pair electrons whose orbitals may be strongly distorted, or polarized, by comparison with the centrosymmetric AO's of the free atoms. A considerable contribution to the total dipole moment may thus arise from polarization of the non-bonding electrons. Again, we shall see later (p. 178) that such effects may be very important.

It is now clear why the values of λ and λ' indicated in Tables 6.2 and 6.3, estimated from observed dipole moments, can only be notional and can at best indicate general trends.

6.4. Electronegativity

In introducing the subject of polar bonds, we quoted Pauling's definition of the electronegativity of an atom; a flow of electron density from one atom to another arises as a result of a difference in their 'electron attracting powers', i.e. from an *electronegativity* difference. We must now try to give this idea more quantitative substance. How can we assign quantities $x_A, x_B, \ldots$ to atoms A, B,... such that when $x_B > x_A$ the bond A—B will be polarized in the direction corresponding to electrons flowing from A to B? A set of x values, for all elements, constitutes an *electronegativity scale*, but it is clear that there is nothing unique about such a scale for if any constant is added to all the x's their definitive property will not be affected (cf. the choice of zero on a temperature scale), nor will it be affected if all the x's are *multiplied* by the same constant (cf. the size of the degree on a temperature scale). Consequently several scales are commonly used and may be brought roughly into coincidence by shifting the zero and changing the units. Unlike a temperature scale, however, an electronegativity scale with a precise experimentally verifiable significance cannot be set up; the electron-attracting power of an atom in a molecule is essentially a *theoretical* concept and electronegativity values are useful only in indicating *trends* and establishing *rough correlations* between measurable quantities.

At least three electronegativity scales are currently in widespread use; we shall consider them one by one, indicating at the same time some of the purposes to which they have been put.

The Pauling scale. Pauling's aim was to assign electronegativity values to the elements in such a way that $x_B - x_A$ could be used to predict the ionic character of the bond A—B in a VB description. He noted that the strength of a polar bond A—B, as measured by dissociation energy $D(A—B)$, could be greater than that of either of the non-polar bonds A—A and B—B and introduced a quantity Δ_{AB}, which he called the 'ionic-covalent resonance energy', to measure the stabilization due to increasing ionic character. Pauling assumed a VB wavefunction (6.28), and defined Δ_{AB} as the difference between the *actual* bond energy $D(A—B)$ and the quantity $D_{cov}(A—B)$ calculated by assuming the molecular wavefunction to be a *single covalent structure* (corresponding to $\lambda = 0$ in (6.28)). Whilst $D_{cov}(A—B)$ could in principle be calculated according to (5.26), it is customary to adopt an empirical procedure by taking $D_{cov}(A—B)$ as the geometric mean of $D(A—A)$ and $D(B—B)$. The difference Δ_{AB}, namely $D(A—B) - D_{cov}(A—B)$, then represents the stabilization (energy lowering) associated with the admission of ionic character into the wavefunction; this quantity must be positive because (§3.6) improvement of the wavefunction can only *lower* the calculated energy. Originally, Pauling proposed the arithmetic mean in computing Δ values; this was discarded in favour of the geometric mean when it was occasionally found to give negative Δ values. Usually, however, the discrepancies are quite small. Table 6.4 shows

TABLE 6.4
Ionic-covalent resonance energies

Bond	HF	HCl	ClF	LiH	NaH
Bond energy (kJ mol^{-1})	565	430	364	241	218
D_{cov} (geometric mean)† (kJ mol^{-1})	237	322	177	213	176
Δ (kJ mol^{-1})	328	108	185	28	43

† Based on the following bond energies (kJ mol^{-1}): H_2 (432), Li_2 (105), Na_2 (72), F_2 (130), Cl_2 (240) (see Table 7.7).

TABLE 6.5

Pauling electronegativity values (x)†

H						
2·2						
Li	Be	B	C	N	O	F
0·98	1·57	2.04	2·55	3·04	3·41	3·98
Na	Mg	Al	Si	P	S	Cl
0·93	1·31	1·61	1·90	2·19	2·58	3·16
K	Ca		Ge	As	Se	Br
0·82	1·00		2·01	2·18	2·55	2·96
Rb						I
0·82						2·66
Cs						
0·79						

† The numbers are chosen so that $x_A - x_B$ gives the ionic–covalent resonance energy directly in eV according to (6.32).

some Δ values estimated in this way from experimental dissociation energies; the bonds usually described as 'more polar' do indeed correspond to larger values of Δ.

Again on purely empirical grounds, Pauling adopted $\sqrt{\Delta}$ as a measure of polar or ionic character and then tried to set up a scale by postulating that

$$\sqrt{\Delta_{AB}} \simeq |x_A - x_B| \qquad (6.32)$$

a proportionality constant being set equal to unity when all quantities are measured in eV. The choice of the square root rests on the experimental observation that for any three bonds A—B, B—C, A—C (all assumed polar from left to right)

$$\sqrt{\Delta_{AB}} + \sqrt{\Delta_{BC}} \simeq \sqrt{\Delta_{AC}}. \qquad (6.33)$$

Equation (6.32) is compatible with (6.33) because it yields

$$\sqrt{\Delta_{AB}} + \sqrt{\Delta_{BC}} = (x_B - x_A) + (x_C - x_B) = x_C - x_A$$

which by postulate is $\sqrt{\Delta_{AC}}$, in accordance with (6.33). Note that, by assumption, $x_C > x_B > x_A$, and the differences, as written, are all positive. The scale built up by Pauling in this way (with subsequent corrections and additions based on improved dissociation energies) is reproduced in Table 6.5. It shows the expected increase in electronegativity from left to right along any row of the periodic table and a decrease down any column (cf. p. 44). The values in Table 6.5 may be used for roughly assessing the stabilization of any given bond resulting from its ionic character; thus, for HCl, the 'ionic-covalent

TABLE 6.6
Dipole moments and electronegativity differences

Molecule	HF	HCl	HBr	HI
$x_A - x_B$	1·78	0·96	0·76	0·46
μ(debyes)	1·82	1·03	0·83	0·45

resonance energy' should be $(3·0-2·1)^2$ eV, in rough agreement with the value in Table 6.4.

Since electronegativity differences are indicators of ionic character, it is not surprising that they also show a correlation with dipole moments. A remarkable, but fortuitous, relationship is that $x_B - x_A$ is roughly equal to the dipole moment μ_{AB} (A^+B^-) expressed in the old 'Debye units' (1 debye $= 3·334 \times 10^{-30}$ Cm). Table 6.6 shows the agreement for the hydrogen halides. From (6.24) the implication is that the fractional ionic character of a bond A—B, as defined in (6.21), is roughly proportional to the electronegativity difference $x_B - x_A$:

$$(\text{FIC})_{AB} \simeq \text{constant} \times (x_B - x_A). \tag{6.34}$$

In the early literature, with fractional ionic characters estimated from experimental dipole-moment values using (6.24), efforts were made to obtain a more precise relationship by drawing a smooth curve of FIC against $x_B - x_A$, chosen to give a best fit to the experimental points. One such fit, due to Hannay and Smyth (1946), is shown in Fig. 6.18 and is represented by the equation

$$\text{FIC} = 0·16(x_B - x_A) + 0·035(x_B - x_A)^2. \tag{6.35}$$

Various 'rival' curves have been proposed (for a review see Gordy 1955), based on alternative measures of ionic character, but in view of the many approximations and theoretical uncertainties surrounding such relationships it is questionable whether anything is to be gained by proceeding beyond a simple straight-line fit. Electronegativities are useful only for the *semi-quantitative* discussion of *trends* along series of somewhat similar bonds, and for that purpose the values shown in Table 6.5 are adequate.

Mulliken-type scales. Shortly after Pauling's introduction of his electronegativity scale, Mulliken (1934; 1935) proposed an alternative definition of an 'absolute' electronegativity in terms of precisely measurable quantities. (A more sophisticated justification was given by Moffitt (1949).) To avoid confusion, we henceforth use the Greek letter χ to denote Mulliken's electronegativity which is defined by

$$\chi_A = \tfrac{1}{2}(\text{IP}_A + \text{EA}_A) \tag{6.36}$$

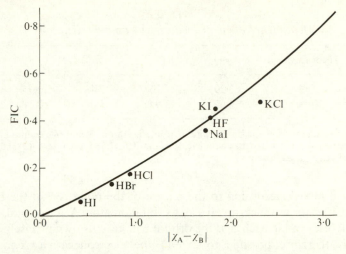

Fig. 6.18. Fractional ionic character plotted against electronegativity difference $|\chi_A - \chi_B|$, using the Hannay–Smyth formula. Experimental values are shown by dots.

where IP_A is the ionization potential of atom A and EA_A is its electron affinity. Mulliken argued that the bond A—B would be well represented by a single covalent structure if there were no particular preference for either one of the ionic structures A^+B^- or A^-B^+, i.e. if $E(A^+B^-) \simeq E(A^-B^+)$. Now the energy required to create A^+B^-, for A and B sufficiently well separated, would be $IP_A - EA_B$ by definition, while for A^-B^+ it would be $IP_B - EA_A$. The criterion for a non-polar bond, therefore, would be $IP_A - EA_B = IP_B - EA_A$, or

$$IP_A + EA_A = IP_B + EA_B$$

but this is the definitive property of electronegativity values, and $IP_A + EA_A$ is therefore an acceptable measure of electronegativity. In fact, Mulliken adopted the mean of IP_A and EA_A as indicated in (6.36).

The advantage of Mulliken's definition is that it gives a theoretically well-defined basis for the calculation or experimental measurement of electronegatives. The resultant scale may also be brought into remarkably close coincidence with Pauling's empirical scale by shifting the zero and changing the units (p. 162) according to

$$x_A = 0.336(\chi_A - 0.615). \tag{6.37}$$

Values of x calculated in this way from Mulliken electronegativities do not in general differ greatly from those shown in Table 6.5. The disadvantage of Mulliken's definition is that it does not strictly give an electronegativity at all, in Pauling's sense of the term, since it refers to *free atoms* and not to atoms 'in a molecule'! However, the theory is flexible enough to allow a choice of

TABLE 6.7
Dependence of electronegativity† on hybridization

Hybridization	sp	sp^2	sp^3	p
C	3·29	2·75	2·48	1·75
N	4·67	3·94	3·68	2·28

† Computed for a valence orbital (s-p mixture) from Mulliken's definition (6.36) and converted to the Pauling scale by use of (6.37). (See J. Hinze (1963) and references cited therein.)

theoretical model, according to the nature of the bond A—B; the bond in question may, for instance, involve an s orbital on atom A, a p orbital, or even an s–p hybrid, and in each case the orbital will have its own precisely defined IP; it is therefore possible to extend the electronegativity concept to distinguish between different valence states of an atom by introducing 'orbital electronegativities'. Extensive tabulations of valence state IP's have been made by Pritchard and Skinner (1955). The dependence of electronegativity on the degree of hybridization of the orbital concerned is in fact quite marked. Thus, for carbon and nitrogen the variation of χ values with hybridization (converted to the Pauling scale by use of (6.37)) is shown in Table 6.7. Since a 2s electron is more tightly bound than a 2p electron, the increase of electronegativity with increasing s content is as expected; the commonly accepted value in Table 6.5 ($x \simeq 2.5$) corresponds to a hybrid with only 25 per cent s character which, as will be seen in Chapter 7, is typical of saturated carbon compounds. There is some experimental evidence for such a variation of electronegativity, for the strength of binding of a proton (and hence the acidity) is related to the electronegativity of the atom to which it is attached.

In recent years the Mulliken definition has been extended still further, notably by Jaffe, Hinze, Whitehead, and co-workers† to allow for the *variability* of electronegativity, according to the amount of electronic charge associated with an atom *in its molecular environment*—more in the spirit of Pauling's original definition. The values in Table 6.5 refer to neutral atoms and do not allow for the flow of electrons towards, or away from, the atom.

The essence of the method may be indicated using, once again, a single bond A—B, but similar considerations may be applied to multiply bonded atoms in polyatomic molecules. We have seen that the number of electrons formally associated with orbital ϕ_A, *in the molecule*, is q_A (the 'orbital population' used in §6.4). For a free atom with one valence electron in ϕ_A, we should have $q_A = 1$, while the anion and cation would correspond, respectively, to $q_A = 2$ and $q_A = 0$. We are now going to suppose that the energy of the electrons

† See for example Hinze *et al.* (1963); Baird and Whitehead (1964); Whitehead *et al.* (1965).

associated with ϕ_A in the field of all the remaining electrons (whose energy is assumed constant) is a smooth function of the number of electrons ϕ_A possesses. However, here we meet a difficulty, which we must dispose of before continuing.

In a free atom the number of electrons 'in' ϕ_A is, of course, an integer, but in a molecule, *even in the absence of any polarization of the bond*, q_A is reduced from its integral value by the flow of electrons into the bond region. However, if we are interested in the gross features of any charge transfer *from one atom to another* we shall not wish to make the refinement of partitioning the density into *three* parts, containing q_A, q_B, and q_{AB} electrons, respectively; it is more convenient to regard the overlap density as belonging half to atom A and half to atom B, taking

$$Q_A = q_A + \tfrac{1}{2}q_{AB}, \qquad Q_B = q_B + \tfrac{1}{2}q_{AB} \qquad (6.38)$$

to be the numbers of electrons formally associated with the orbitals ϕ_A and ϕ_B, respectively. Q_A and Q_B are usually called‡ the '*gross* orbital populations' of ϕ_A and ϕ_B, while q_A and q_B represent the *net* values which remain when some of the charge has 'gone into the bond'. We now see at once from (6.38) that

$$q_B - q_A = Q_B - Q_A$$

and that the FIC defined in (6.21), along with all our subsequent discussion, is therefore unaffected if we use the Q's instead of the q's; the Q's serve equally well in describing the gross features of the charge transfer. If we separate the atoms, of course, $q_{AB} \to 0$ and $Q_A \to q_A$, $Q_B \to q_B$, and we therefore still notionally associate $Q_A = 1, 2$, and 0 with the monovalent atom, its anion, and its cation, respectively.

We may now continue the argument. The energy of the electrons associated with ϕ_A is assumed to be a function of Q_A, with three known experimental points on the curve corresponding to $Q_A = 0, 1, 2$. Let us assume a simple quadratic variation:

$$E_A = a_A + b_A Q_A + c_A Q_A^2. \qquad (6.39)$$

Clearly, we should put $a = 0$, since there is no energy term when there are no electrons in ϕ_A ($Q_A = 0$); the energy of the ϕ_A electrons in the positive ion is thus $E_A^+ = 0$. For the neutral atom and the anion, we have

$$E_A^0 = b_A + c_A \qquad (Q_A = 1)$$
$$E_A^- = 2b_A + 4c_A \qquad (Q_A = 2).$$

‡ The terminology is that of Mulliken (1955). For other discussions of the gross populations and their applications see McWeeny (1951; 1952; 1954).

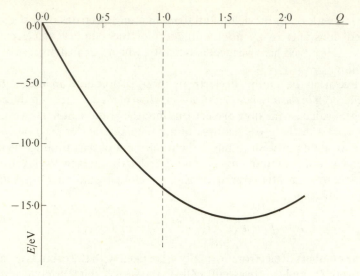

FIG. 6.19. Variation of energy with orbital population Q.

However, the orbital ionization potential and electron affinity are defined by

$$IP_A = E_A^+ - E_A^0 = -(b_A + c_A)$$
$$EA_A = E_A^0 - E_A^- = -(b_A + 3c_A)$$

(6.40)

and we may therefore determine the constants b_A and c_A by solving two simultaneous equations; the result is

$$b_A = -\tfrac{1}{2}(3IP_A - EA_A), \qquad c_A = \tfrac{1}{2}(IP_A - EA_A).$$

(6.41)

It is therefore possible to draw a simple curve showing how the energy associated specifically with an AO ϕ_A changes as electrons flow towards or away from ϕ_A, changing its population accordingly. Such a curve is shown in Fig. 6.19, for the 2p orbital in an oxygen atom. The part of the curve to the left of the vertical line at $Q = 1$ corresponds to removal of electrons to give a (fractionally) positive ion, the part to the right $(Q > 1)$ corresponds to acceptance of electrons, i.e. negative ionization. The electronegativity associated with orbital ϕ_A of the atom in the molecule may now be defined† as a measure of its 'thirst' for electrons, i.e. as the rate at which the corresponding energy term decreases towards the minimum as electrons are added. In symbols,

† This definition was first proposed, in essence, by R. P. Iczkowski and J. A. Margrave (1961). Note the signs in (6.42) and elsewhere which differs from that in some of the literature cited; this results from our use of population (Q) rather than the corresponding electric charge ($-eQ$) or the net charge on the ion.

$$\chi_A = -\frac{\partial E_A}{\partial Q_A}.$$ (6.42)

Thus the negative slope of the curve in Fig. 6.19, at any point, may be taken as a *Q-dependent orbital electronegativity*. By differentiating (6.39) we obtain, explicitly,

$$\chi_A = -(b + 2cQ_A) = \tfrac{1}{2}(3IP_A - EA_A) - (IP_A - EA_A)Q_A.$$ (6.43)

At the point $Q_A = 1$ we find at once the neutral atom value

$$\chi_A^0 = -(b_A + 2c_A) = \tfrac{1}{2}(IP_A + EA_A)$$ (6.44)

and thus retrieve Mulliken's definition, which is apparently appropriate for an approximately neutral atom. As the atom attracts electrons towards it the electronegativity *decreases* (decreasing magnitude of the slope in Fig. 6.19) until, beyond the minimum, the partially negative ion would eventually *repel* further electrons, the slope becoming positive and χ becoming negative. Data for hydrogen and the halogens are collected in Table 6.8.

TABLE 6.8

Representative data for orbital electronegativity calculations

	b	c	χ^0
H	−20·02†	6·42	7·18
F	−29·54‡	8·68	12·18
Cl	−20·75	5·67	9·41
Br	−17·80	4·70	8·40
I	−17·25	4·57	8·11

† All values in electron volts.
‡ Values for the halogens assume a pure 2p orbital (no hybridization).

The inclusion of a dependence of electronegativity on the type of *orbital* involved in bond formation (including any hybridization) and on its *population* represents an interesting and revealing refinement of the original Mulliken definition with many potential applications. In particular, the Q dependence of χ permits a simple discussion of charge shifts in bonds, in which the empirical relationship (6.34) is replaced by a simple physical principle—the 'principle of equalization of orbital electronegativities'. This principle† simply asserts that electrons will flow from A to B (assuming B more electronegative

† Apparently first formulated by Sanderson (1945)—for further developments see Iczkowski and Margrave (1961), Hinze *et al.* (1963), Baird and Whitehead (1964).

TABLE 6.9

Fractional ionic characters based on orbital electronegativity

	HF	HCl	HBr	HI
FIC from (6.46)	0·166	0·089	0·082	0·069
FIC 'observed'†	0·412	0·168	0·123	0·058

† That is, based on experimental dipole moments using (6.24), see Table 6.2.

than A) until, with χ_B falling and χ_A rising, the two electronegativities become equal: $\chi_A = \chi_B$. This implies from (6.43) that

$$b_A + 2c_A Q_A = b_B + 2c_B Q_B$$

but since there are two electrons in the bond

$$Q_A + Q_B = q_A + q_B + q_{AB} = 2$$

and solution of the two simultaneous equations gives

$$Q_A = \frac{b_B - b_A + 4c_B}{2(c_A + c_B)}, \qquad Q_B = \frac{b_A - b_B + 4c_A}{2(c_A + c_B)}. \tag{6.45}$$

The fractional ionic character is then found to be, using (6.21) and (6.38),

$$\text{FIC} = \tfrac{1}{2}(Q_B - Q_A) = \frac{\chi_B^0 - \chi_A^0}{2(c_A + c_B)}. \tag{6.46}$$

The principle thus leads to a relationship exactly parallel to (6.34), but the result now has a more fundamental basis and the proportionality constant is fixed in terms of the 'charge coefficients' c_A and c_B.

Some results based on (6.46), for the elements listed in Table 6.8, are shown in Table 6.9. The fact that a better correlation with experiment is obtained from the formula of Hannay and Smyth, particularly for HF and HCl, is not unexpected; in the first place their fit is purely empirical and uses the Pauling empirical electronegativities, and secondly Table 6.8 determines orbital electronegativities for halogen valence orbitals of pure p type, when in fact a small but appreciable amount of hybridization is certainly present. As hybridization becomes less important the agreement improves.

Now that more information is available from *ab initio* calculations, further efforts might profitably be devoted to the rationalization of orbital electronegativities. It is none the less reassuring to find that widely different definitions and procedures all lead to such qualitatively similar results, and even to numerical electronegativity values which do not differ greatly (when the scales are shifted into approximate coincidence) except for atoms in highly ionic situations or in unusual valence states.

TABLE 6.10

Comparison of electronegativity values based on different definitions

	Li	Be	B	C	N	O	F
Pauling	0·98	1·57	2·04	2·55	3·04	3·41	3·98
Mulliken-Jaffé[†]	0·84	1·40 (sp)	1·93 (sp^2)	2·48 (sp^3)	3·68 (sp^3)	3·04 (p)	3·90
Allred-Rochow	0·97	1·47	2·01	2·50	3·07	3·50	4·10

[†] Converted to the Pauling scale. The hybridization assumed is shown in parentheses: for C, N, and O the calculated orbital electronegativity depends strongly on the amount of s character, about 10–15 per cent giving values closer to those of Pauling.

The Allred-Rochow scale. Another widely accepted scale, with a very direct physical interpretation, was introduced by Allred and Rochow (1958). They proposed that the force exerted by an atom on a 'test' electron at a distance equal to the empirical covalent radius (§7.9) would form a suitable measure of its power to attract electrons. The simplest way of estimating the force is to use Slater's rules to define an effective nuclear charge Z_e; in estimating this effective charge Allred and Rochow counted *all* the electrons in the atom (not $N-1$, as in §2.6); but relative values are not strongly dependent on this choice. The attractive force is then simply $Z_e e^2/\kappa_0 r_{cov}^2$. To bring the scale into line with Pauling's scale, Allred and Rochow defined the electronegativity as

$$x = 1·282 \frac{Z_e}{r_{cov}^2} + 0·744 \tag{6.47}$$

where r_{cov} is measured in bohrs. Again, it is reassuring to find that electronegativity values computed in this way are not, in general, widely different from those based on definitions of the Pauling or Mulliken type. The main advantage of this scale is its simplicity and the ease with which it may be extended to the heavier atoms in the periodic system, for which other values have not been obtained; the scale is therefore widely used in transition-metal chemistry. Some Allred – Rochow values are shown in Table 6.10, where the various scales are compared; more complete tables are available elsewhere (see, for example, Huheey 1972).

6.5. Polar bonds: experimental implications

The dipole moment of a molecule is perhaps the most obvious indicator of bond polarity but, for reasons already mentioned (p. 160), not always the most reliable one for quantitative purposes. The distribution of charge in a bond may be investigated experimentally by many other methods, but the interpretation of the results in terms of simple concepts such as electronegativity difference or degree of hybridization is never easy.

Let us consider a few examples to illustrate some of the difficulties.

(i) *Dipole moments*. These involve merely the calculation of the electric moment of the charge cloud. The simple picture which leads to (6.24) is inadequate because it ignores the presence of atomic dipoles and bond (or homopolar) dipoles and recognizes only a shift of charge between the two centres; it also ignores the asymmetrical distribution of the non-bonding electrons, treating them as part of a compact positively charged core.

It is not difficult to estimate the size of such effects. Fig. 6.17 (*a*) shows very clearly that the centre of mean position in a hybrid may be at some distance from the nucleus. Thus, for a hybrid $(s + \lambda p_z)/(1 + \lambda^2)^{1/2}$, the required centroid is at a distance $\bar{z}$ from the nucleus, where

$$\bar{z} = \int z \frac{(s + \lambda p_z)^2}{1 + \lambda^2} \, d\tau = (1 + \lambda^2)^{-1} (z_s + \lambda^2 z_p + 2\lambda z_{sp}).$$

Now z_s and z_p, which give the centroids of the s and p_z AO's, are zero, and so

$$z = \frac{2\lambda}{1 + \lambda^2} z_{sp} = \frac{2\lambda}{1 + \lambda^2} \int z s p_z \, d\tau. \tag{6.48}$$

For pure s ($\lambda = 0$) and pure p ($\lambda = \infty$) $\bar{z}$ vanishes; but for hybrids of s and p, $\bar{z}$ may be quite large, with a maximum value of 1 in the digonal hybrids. If s and p_z are the AO's appropriate to carbon, $\bar{z}_{sp}$ has the numerical value $0.89a_0$. This asymmetry of charge means that in the process of preparing a carbon atom for bond formation, we introduce very substantial atomic dipoles. The magnitude of such a dipole is $\mu = e\bar{z}$ per electron, or in the present example $0.89ea_0$. As in Table 6.2 we use the atomic unit of electric moment, ea_0, the SI equivalent being $ea_0 = 8.478 \ 10^{-30}$ C m. Such dipoles are quite large, as Fig. 6.20 shows for the case of carbon where each electron may contribute as much as $0.89ea_0$. This is almost equivalent to moving one electron between centres 1 bohr apart! In considering the polarization of non-bonding electrons, some of which may occupy hybrid orbitals, even larger dipoles might be expected, a lone *pair* giving *twice* the moment just calculated.

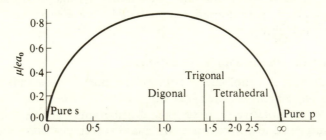

FIG. 6.20. The atomic dipole for a carbon hybrid. (Horizontal scale indicates weight of the p orbital, $\lambda^2/(1 + \lambda^2) \to 1$ for $\lambda \to \infty$. Values of λ are shown.)

The other effect we must consider arises from the asymmetry of the overlap density d_{AB} in (6.20), which as Fig. 6.17 (*b*) suggests may be considerable even for two s orbitals. If we take two 1s AO's where (using atomic units)

$$\phi_A = \pi^{-1/2}\exp(-r_A), \qquad \phi_B = \left(\frac{k^3}{\pi}\right)^{1/2}\exp(-kr_B)$$

then k determines the ratio of the sizes. It follows that

$$z_{AB} = \frac{1}{S_{AB}}\frac{k^{3/2}}{\pi}\int z\exp(-r_A - kr_B)\,d\tau. \qquad (6.49)$$

The values of the moment integral and the overlap integral (S_{AB}) may be obtained from formulae first listed by Coulson (1942). This gives the homopolar dipole moment associated with the overlap density as $\mu = e\bar{z}_{AB}$ per electron (d_{AB} being normalized to unity); in the expression (6.23) for the bond dipole this is weighted with the value of the overlap population q_{AB}. The dependence of μ on k is shown in Fig. 6.21. When $k = 1$, as in a homonuclear molecule, μ vanishes, but even a small asymmetry can easily lead to a moment of $0.1ea_0$. To estimate such effects with any confidence it is clearly necessary to have quite accurate wavefunctions.

(ii) *Nuclear shielding constants.* An electron is a 'spin $\frac{1}{2}$' particle with an angular momentum quantum number $s = \frac{1}{2}$. Nuclei, however, may have various values of the spin angular momentum, characterized by a nuclear spin quantum number I (e.g. ^1H, ^{13}C have $I = \frac{1}{2}$; ^{12}C, ^{16}O have $I = 0$; ^{14}N has $I = 1$). A nucleus with non-zero spin has an associated *magnetic dipole* $\mu_{n..}$ (analogous to that of the electron, (3.60)) and a corresponding coupling energy with an applied magnetic field. The coupling energy depends on the nuclear

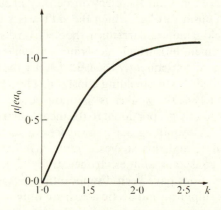

FIG. 6.21. The homopolar dipole for two unlike 1s orbitals. (The effective nuclear charges are taken as e and ke, and the internuclear distance as $2a_0$).

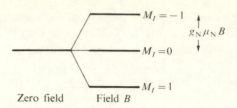

FIG. 6.22. Splitting of nuclear spin energy levels (for $I = 1$) by a magnetic field. Transitions between such levels occur in NMR spectroscopy.

spin state, characterized by a quantum number M_I, and is (cf. (3.61))

$$\Delta E_{\text{nuc}} = -g_N \mu_N M_I B \qquad (M_I = I, I-1, \ldots, -I). \qquad (6.50)$$

The sign difference arises because the nucleus carries a positive charge. Here g_N is the 'nuclear g value' and μ_N is the 'nuclear magneton' given by $\mu_N = eh/2M_p$ where M_p is the proton mass.

The splitting of nuclear spin energy levels by the field (Fig. 6.22) is the nuclear Zeeman effect (cf. p. 73) and for a given field the spacing of the levels is $g_N \mu_N B$. Transitions between such levels may be induced by an oscillating radiofrequency field and occur for corresponding frequencies, such that $h\nu = g_N \mu_N B$; they give a nuclear magnetic resonance (NMR) signal characteristic of the nucleus.

NMR signals as a means of 'fingerprinting' the various nuclei present in a compound, and of investigating many aspects of molecular structure and dynamics, have acquired a tremendous importance in chemistry. Their main interest for valence theory arises from the fact that they depend not only on the nucleus but also on its electronic environment. The 'position' of the signals (i.e. the absorption frequency) in general experiences a 'chemical shift' which depends largely on the electron density around the nucleus in question. The mechanism through which the shift arises is indicated in Fig. 6.23. The external field B induces currents in the electron distribution (i.e. some orbital angular momentum) which generate an *opposing* field (directly proportional to B), and consequently the field 'felt' by the nucleus is $B(1 - \sigma)$ where σ is the so-called 'NMR shielding constant'. The 'thinner' the electron cloud around the nucleus, the weaker is the induced current, and we might thus expect σ to be roughly proportional to the mean electron density near the nucleus. In other words, by pulling charge away from a nucleus we shall reduce the chemical shielding, and this suggests, for a given bond, a correlation between shielding constants and electronegativity differences. Efforts to establish such correlations have been disappointing, however. It is true that acid protons (e.g. in sulphuric and acetic acids) are only weakly shielded, and that shifts in organic molecules can sometimes be correlated with the electron withdrawing power of various substituent groups, but the induced electronic

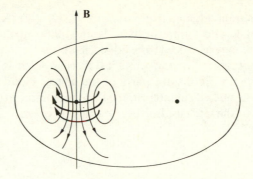

FIG. 6.23. Origin of the 'chemical shift' of NMR signals. The induced currents in the electron distribution set up a secondary field which opposes the applied field B, producing a 'shielding' effect.

currents extend over the whole molecule and the shifts thus represent the net effect of a large number of contributions, not lending themselves to simple quantitative representation. Detailed discussions are available elsewhere (e.g. Carrington and McLachlan (1967)).

(iii) *Electric field gradients.* When the nuclear spin is non-zero the *electric* field it produces resembles that of a non-spherical charge. The electric *dipole* moment of the nucleus is always zero, but there may be a very small, but non-vanishing, *quadrupole* moment Q. When the field at the nucleus, due both to the electron cloud and any other nuclei, is non-uniform there is again a small coupling energy which depends on the nuclear spin state. This time, instead of (6.50), we find (note that for $I = 0, \frac{1}{2}$ there is no quadrupole moment)

$$\Delta E_{\text{quad}} = \frac{eQ}{4} q_{zz} \left\{ \frac{3M_I^2 - I(I+1)}{I(2I-1)} \right\} \tag{6.51}$$

where the 'electric field gradient' $q_{zz} = (\partial^2 \phi / \partial z^2)$ (ϕ being the electrostatic potential) is evaluated at the nucleus and the axis of quantization is taken as z axis. Transitions between levels with different values of M_I are observed in nuclear quadrupole resonance (NQR) and, provided Q is known, give us information about the field gradients.

To calculate q_{zz} we need only apply classical electrostatics to the charge density function P. The result is

$$q_{zz} = \frac{-e}{\kappa_0} \int \frac{(3 \cos^2 \theta - 1)}{r^3} P(\mathbf{r}) \, d\tau \tag{6.52}$$

where $P(\mathbf{r})$ is the charge density at the point whose position vector relative to the nucleus is $\mathbf{r}$ and θ is the inclination of this vector to the z axis.

To obtain the field gradient at nucleus A due to an A—B bond pair we insert (6.19) and obtain contributions from d_A, d_{AB}, and d_B, to which we must add a nuclear contribution, arising from nucleus B, $(2Z_B e/\kappa_0 R_{AB}^3)$. When it is remembered that ϕ_A and ϕ_B may be hybrids, and that contributions from non-bonding electrons may also need to be added, it is clear that the result will arise as a sum of many terms. In semi-empirical discussions it is usual to neglect contributions from more distant parts of the charge cloud since, when r is measured from the nucleus of interest, the r^3 denominator in (6.52) will favour the region close to the nucleus. Further analysis (Problem 6.9) then leads to a result of the form

$$q_{zz} = \frac{e}{\kappa_0}(2 - q_A)I_{aa}^A. \tag{6.53}$$

In other words the field gradient measures the extent to which the halogen *fails* to complete its p shell, vanishing for $q_A \to 2$. The effect of s–p hybridization is the subject of Problem 6.10.

Although in the 1950's there was much controversy about the relative importance of polar character, hybridization, nature of the lone pairs, etc. (see for example the original papers by Townes and Dailey (1949) and other references contained in (1955), or other reviews: Orville–Thomas (1957) and Lucken (1963)), it is now realized that such concepts once again have only a 'notional' significance (though they are still widely used in discussing general trends) and that only highly accurate *ab initio* calculations can yield reliable values of field gradients.

(iv) *Chemical shifts in ESCA.* In §1.5 reference was made to a relatively new experimental technique ESCA (electron spectroscopy for chemical analysis), in which the ionization energies necessary to remove atomic inner-shell electrons were inferred from measurement of the kinetic energies of the ejected photoelectrons (Fig. 1.3); these ionization energies depend on molecular environment, being larger if an atom is deficient in electrons and smaller if it is rich in electrons, and must therefore be related to bond polarities and electronegativities. In recent years ESCA studies (apart from their obvious use in chemical analysis as a means of showing what atoms are present) have been of great value as a means of investigating electronic structure. The basic reference is Siegbahn *et al.* (1967) but many brief accounts are available (e.g. Hollander and Jolly (1970)). Here we simply indicate the connection between the 'chemical shifts' in ESCA signals and the electron populations of the regions surrounding the atom concerned.

At the Hartree–Fock level of approximation each orbital energy ε_i may be identified with the negative of an ionization energy $(-I_i)$, so the ESCA peaks allow us to measure (within the framework of this approximation) orbital energies. We therefore ask how a typical inner-shell orbital energy depends on its molecular environment. Let us take for illustration a carbon 1s electron; we can write

$$I_{1s} \simeq -\varepsilon_{1s} = -\{\varepsilon_{1s}(C^{4+}) + V_{val} + V_{ext}\} \qquad (6.54)$$

where $\varepsilon_{1s}(C^{4+})$ is the energy the electron would have if all the valence electrons were removed (the 1s orbital remaining 'frozen'); V_{val} is the potential energy of the electron in the field of the valence electrons associated with the carbon atom and V_{ext} is that arising from the electrons and nuclei of the atoms to which the carbon is bonded. The first term is characteristic of the carbon atom and may be regarded as a constant, but the others clearly depend on electronic environment and will therefore vary from molecule to molecule, thus determining the chemical shifts. As a first approximation we might assume surrounding atoms to be electrically neutral and neglect V_{ext}. The shift will then depend on V_{val} and this will be directly proportional to the electron population P_C of the valence shell of the carbon atom. This will be a *gross* population (see p. 167) in which the electrons in the overlap regions have been divided between the bonded atoms so as to make the charge density appear as a sum of 'atomic' contributions.

Plots of the shift $\Delta\varepsilon_{1s}$ against populations (or net charges) often do give a reasonably good straight-line correlation (Fig. 6.24); the shifts may

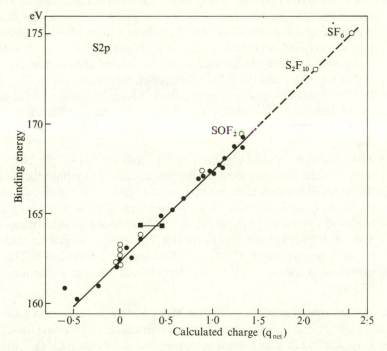

Fig. 6.24. Binding energy plotted against net charge on the sulphur atom in various compounds. (After Hamrin *et al.* (1968) with permission; the net charge is estimated semi-empirically, using electronegativity values, cf. §6.7.)

consequently be used to infer the valence states or formal oxidation numbers of atoms in large molecules or crystals for which computations may not be feasible. Clearly, the environmental effects contained in V_{ext} will not be negligible when strongly polar bonds are involved, but when V_{ext} is taken into account the good correlation is often restored (see for example Carver *et al.* (1974)). Much work remains to be done in this area (particularly in going beyond the approximation (6.54), to allow for the 'relaxation effects' which result from the ionization), but ESCA studies will undoubtedly continue to provide interesting insights into heteropolar bonding and valence electronic structure in general. Interesting reviews of recent applications may be found in Orchard *et al.* (1972–74).

To end this section let us consider a specific molecule, carbon monoxide. The occupied orbitals in CO may be described as follows† (Figs. 6.25):

(i)   There are the two inner shells $C(1s)^2$ and $O(1s)^2$.

(ii) There are three occupied σ-type MO's 1σ, 2σ, 3σ and two occupied π-type MO's $1\pi_x$, $1\pi_y$; the σ-type orbitals are compounded out of $C(2s)$, $C(2p_z)$, $O(2s)$, $O(2p_z)$, and the π-type orbitals from $C(2p_x)$, $O(2p_x)$ and $C(2p_y)$, $O(2p_y)$.

(iii) The 1σ orbital is very tightly bonding (ionization potential about 43 eV) and is mainly composed of $O(2s)$ with a smaller amount of a nearly digonal carbon orbital $C(2s) + C(2p_z)$ chosen to overlap positively in the bond region between the nuclei. Thus the $(1\sigma)^2$ charge density is concentrated between the nuclei, on the side of the oxygen nucleus rather more than on the side of the carbon nucleus.

(iv) The 2σ orbital is chiefly a combination $O(2s)$–$O(2p_z)$ concentrated near the O nucleus, but with its centre a little on the side farther away from C; these $(2\sigma)^2$ electrons contribute relatively little to the binding energy.

(v)   The 3σ orbital is almost entirely a digonal hybrid $C(2s)$–$C(2p_z)$ which is strongly directed away from the carbon nucleus. Since both the 2σ and 3σ electrons are concentrated in distinct regions of the molecule, there is relatively little interaction between them. The 3σ electrons, with ionization potential 13 eV, are the easiest to remove from the molecule.

(vi) The $1\pi_x$ orbital is $0.42C(2p_x) + 0.81O(2p_x)$, giving about four times as high a π population on O as on C. This is quite a polar bond. The $1\pi_y$ orbital is, of course, just the same, but rotated through 90° around the CO axis.

The dipole moment arises from all four MO's, but the contribution from the lone-pair electrons (v) is sufficiently large to compensate that arising from (iii), (iv), and (vi), leading to a calculated total moment in direction C^-O^+. Ransil

† The minimal basis set results were given by Ransil (1959; 1960); more accurate results have been given by Huo 1965.

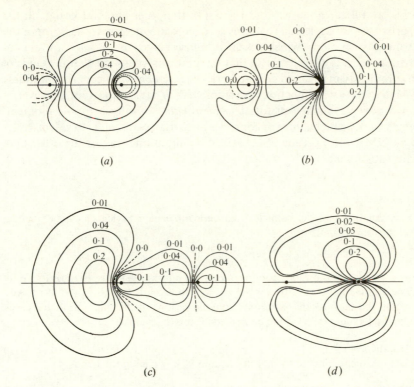

FIG. 6.25. Forms of the occupied valence MO's in the CO molecule (carbon on the left). (a) 1σ, (b) 2σ, (c) 3σ, (d) 1π. (Adapted from W. M. Huo, *J. Chem. Phys.* **43**, 624 (1965), with permission.)

shows that if we add together the total amount of each of the various atomic orbitals involved we can write for the final AO populations

$$\text{Carbon:} \quad 1s^{2\cdot0}, \quad 2s^{1\cdot68}, \quad 2p_z^{0\cdot96}, \quad 2p_x^{0\cdot62}, \quad 2p_y^{0\cdot62}$$
$$\text{Oxygen:} \quad 1s^{2\cdot0}, \quad 2s^{1\cdot85}, \quad 2p_z^{1\cdot51}, \quad 2p_x^{1\cdot37}, \quad 2p_y^{1\cdot37}.$$

Again (cf. p. 167) these are *gross* populations, all bond contributions being equally divided between the two atoms; summation of the gross populations then accounts for all 14 electrons.

According to this description the bond is properly described as a triple bond, since there are six bonding electrons $(1\sigma)^2(1\pi_x)^2(1\pi_y)^2$, but owing to the considerable asymmetry of the π orbitals the strength of the bond is lower than with a normal triple bond, as in N_2. However, there is a small additional bond order coming from the relatively non-bonding $(2\sigma)^2(3\sigma)^2$.

The actual forms of the MO's, shown in the contour maps of Fig. 6.25, are rather more accurate than Ransil's calculations would have given, but it is clear that their general nature can be understood in terms of appropriate hybrids, and less readily without them. On the other hand, although all the properties we have referred to in this section relate directly to the form of the electron distribution, which itself is represented as a sum of orbital contributions, it is not surprising that integrations involving charge clouds of such complexity may fail to give numerical agreement with experiment unless very high quality wavefunctions are employed. What *is* important, however, is that we now have some pictorial *understanding* of why a dipole moment, arising as a delicate balance of contributions, may be difficult to calculate, or why an NMR (or NQR or ESCA) peak is shifted one way or the other by a certain substituent.

Problems

6.1. A calculation on a heteronuclear diatomic molecule AB yields a σ-type bonding MO of the (normalized) form

$$\psi = 0\cdot5s_A + 0\cdot3p_A + 0\cdot7s_B - 0\cdot2p_B$$

where p_A and p_B are p orbitals pointing along the internuclear axis. Express the result as a linear combination of two normalized hybrid orbitals h_A and h_B. What can you infer about the nature of the two atoms from the various coefficients? (Hint: first check that you understand the working behind equations (6.3) and (6.4).)

6.2. Use the argument on p. 151 to set up lone-pair hybrids, for the molecule AB in Problem 6.1, orthogonal to the hybrids h_A and h_B involved in the bond.

6.3. The molecule AB in Problems 1 and 2 has a σ bond and one σ lone pair on each atom. How could you describe numerically the occupation of the s and p_σ AO's in the valence state of each atom? (Hint: express the electron density for $A[1_A^2 h_A]$ and $B[1_B^2 h_B]$ in terms of s and p AO's, as in equation (6.12).)

6.4. Given that the MO of Problem 1.1 is normalized, infer the value of the overlap integral between the hybrids h_A and h_B. Hence obtain the orbital populations and the fractional ionic character of the σ bond. (Hint: $\psi = c_A\phi_A + c_B\phi_B$ is normalized if $c_A^2 + c_B^2 + 2Sc_Ac_B = 1$. Write the charge density in (6.15) in terms of c_A and c_B, and use (6.21).)

6.5. Take a wavefunction of the form (6.28) to describe the bond A–B in Problem 6.4, and use the approximation (6.30) to obtain a VB function which gives the same FIC as the MO function. How would the VB estimate of FIC be changed if overlap were not neglected? (Hint: use (6.29) with the value of λ already found and with S from Problem 6.4, to recalculate FIC.)

6.6. Use the Hanney–Smyth formula, with neglect of the quadratic term, to infer the electronegativity difference between the atoms A and B of the molecule AB in Problem 6.5. Comment on the value of this interpretation in the light of Table 6.7.

6.7. The ionization potentials and electron affinities of hydrogen, oxygen, and fluorine are

	H	O	F
IP	13·60	13·62	17·42 eV
EA	0·76	1·47	3·40 eV

Calculate the Mulliken electronegativities and their Pauling equivalents (using equation (6.37)). Compare your results with the values in Table 6.5.

6.8. From the data of Problem 6.7 obtain the population-dependent electronegativities, defined in equation (6.43), for H, O, and F. Plot the results against Q and devise a graphical method for estimating FIC's using the equalization principle. Check your results for OH and HF against those given by equation (6.44). (Hint: remember $Q_A + Q_B = 2$ for an electron-pair bond.)

6.9. For a filled p shell on a given atom A the electron density $P(r)$ is spherically symmetrical (Problem 2.9) and the integral (6.52), which gives the field gradient q_{zz}^A in a molecule AB, vanishes. Use this fact to derive (6.53), taking account only of the populations of AO's on the nucleus in question, and identify the integral $I_{\sigma\sigma}^A$. Why is it a fairly good approximation to neglect the contribution from nucleus B?

6.10. Show that the method used in Problem 6.9 still applies, even when the σ bond pair employs a hybrid $\phi_A = N(s + \lambda p_\sigma)$ and there is a corresponding polarized σ lone pair. Hence derive a formula analogous to (6.53) but containing the hybridization parameter. Compare your treatment with that of Townes and Dailey (1949). (Hint: express the valence-shell charge density as a sum of a spherical term and a p_σ-type contribution.)

7

Polyatomic molecules: Structure and shape

7.1. Localized bonds in polyatomic molecules: bond properties

We have now reached the stage when we can discuss polyatomic molecules. The next three chapters are concerned with different types of polyatomic molecule: first, the quantum-mechanically simplest, then the whole field of carbon chemistry, and finally molecular complexes of the various types encountered in transition-metal chemistry.

First we use the MO approach. When we attempt to apply MO methods to polyatomics, we are met by an immediate difficulty. This can best be illustrated by means of an example. Let us consider the methane molecule CH_4. According to the principles of Chapter 4, which proved so effective for diatomic molecules, we shall argue that the 10 electrons of methane are distributed so that two of them complete the K-shell of carbon and the remaining eight occupy MO's of a polycentric character embracing all five nuclei. Such a description presents more problems than it solves, for it is well known that the C—H bond has characteristic properties, such as its length, force constant, and polarity, which, while not exactly constant, vary only relatively little from molecule to molecule. For example, the existence of a characteristic infrared CH vibration frequency in the region of $3000 \, \text{cm}^{-1}$ is used not only to verify the presence of CH bonds in an unknown molecule but also to estimate how many such bonds there are. Since all the bond properties which we have listed depend upon the details of the electron distribution, it is hard to see why the CH bond should be so reproducible in character and so relatively insensitive to the nature of the other substituents around the carbon atom when the MO's themselves extend over all of them. In simple terms bond properties seem to imply localized distributions of charge, but the MO method seems to require delocalized orbitals. The only satisfactory way out of the dilemma is that a possible alternative description can be found in which, despite the polyatomic nature of the molecule, the MO's are effectively bicentric. As long ago as 1931, Hund (1931; 1932) remarked that chemical intuition and experience force us to seek a replacement of the anticipated non-localized orbitals by localized ones. In this way we expect the bonds of a polyatomic molecule each separately to resemble a bond such as those discussed in Chapters 4 and 6.

Energy considerations support this view. The fact that the heat of formation of H_2O ($916 \, kJ \, mol^{-1}$) differs by only about 10 per cent from twice the heat of formation of the OH radical ($2 \times 416 \, kJ \, mol^{-1} = 832 \, kJ \, mol^{-1}$) leads us immediately to the conclusion that the two OH bonds in H_2O closely resemble the one bond in the OH radical. Precisely the same situation is found in the paraffin series, where the assumptions that the C—C and C—H bond energies are 346 and $411 \, kJ \, mol^{-1}$ respectively lead to molecular heats of formation in error by only 1 or 2 per cent. The assumption of somewhat different values can give equally good agreement. Indeed, the very existence of tables of bond energies (for numerical values see Cottrell (1958) and Dasent (1970)) forces us to the conclusion that usually bonding electrons are localized in the region of one particular bond—hence the tremendous value of the electron-pair concept, to which the last two chapters were devoted.

It should be added at this point that completely delocalized MO's *can* be set up and their details calculated. We shall return to this matter later (§7.3). However, it is important that we should emphasize at once the distinction which we have made between localized and non-localized MO's. Of course, if localized MO's can be used, it is better to do so, for they are vastly easier to imagine and handle than are non-localized MO's and they preserve the idea of a bond between two of the atoms in a polyatomic molecule, represented in conventional chemical diagrams by the symbol A—B, as in the molecule A—B—C. If we do not use localized MO's for those simple cases where they can be used, we do violence to the long chemical tradition dating from G. N. Lewis's famous paper of 1916 on electron-pair bonds. Only for certain types of molecule (notably the aromatic and conjugated compounds in Chapter 8) does the language of localized MO's break down completely, and except in such cases (and in dealing with excited states, which fall largely outside the scope of this book) the electron-pair concept survives almost intact. The localized MO's used in the MO description of such pairs are conveniently called 'bond orbitals'.

It is possible to give a plausible justification for the use of bond orbitals if we invoke the criterion of maximum overlapping of §4.3. Let us illustrate it in terms of the water molecule H_2O. In Fig. 7.1 the plane of the molecule coincides with the plane of the paper, which is the xy plane, and the oxygen atom is at the origin. Such figures are purely schematic. The p-type AO's, for example, are usually somewhat elongated in order to simplify the drawing. Now the available AO's for our LCAO wavefunctions are the two 1s orbitals of the hydrogens (H_1 and H_2, say) and the $2p_x$ and $2p_y$ orbitals of oxygen (p_x and p_y, say). This is because the $2p_z$ electrons are not able to combine with the hydrogen 1s AO's on account of their antisymmetry in the xy plane and the oxygen 2s electrons are too tightly bound (estimated ionization potential 32 eV, as compared with 13 eV for the $2p_z$ oxygen electrons) to be of much use by themselves for molecule formation. Let us place the H atoms as shown on

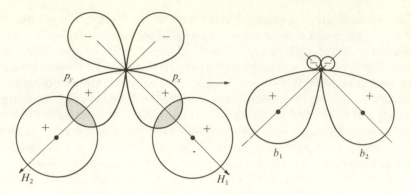

FIG. 7.1. Formation of localized bond orbitals in the molecule H_2O.

the left of Fig. 7.1, directly along the directions of the $2p_x$ and $2p_y$ orbitals of O. Then there is strong overlap between H_1 and p_x and between H_2 and p_y, but practically none at all between any other pair. Thus, instead of a linear combination of all four AO's we shall take just two at a time. Indeed we use H_1 and p_x to form a bond orbital

$$b_1 = N(p_x + \lambda H_1) \qquad (7.1a)$$

and in a similar way we form

$$b_2 = N(p_y + \lambda H_2) \qquad (7.1b)$$

the same normalizing constants (N) and polarity parameters (λ) being used to ensure that both bonds are alike. These two bond orbitals are shown—a little schematically—on the right-hand side of the figure. To a first approximation they are quite independent of one another, so that if we replace one hydrogen atom H_2 by some other group, e.g. a methyl group CH_3, we shall expect to change b_2 but make little alteration in b_1. In other words the electrons in an O—H bond have characteristic wavefunctions. This is the basis of the approximate constancy of the energy, length, and other characteristic properties of this bond.

It should by now be evident that what is important is the *localization* of the wavefunction for each bond, not the precise way in which it is constructed. We could just as well have adopted a VB approach in which, instead of assigning two electrons to b_1 to describe the left-hand OH bond, we constructed a Heitler–London-type wavefunction from the two overlapping AO's H_1 and p_x. The electron density in the bond would have a similar form in each case, being a superposition of orbital and overlap terms just as in (6.17), though the two approximations might give somewhat different values for the populations of H_1, p_x, and their region of overlap.

7.2. Separability of localized bonds

The discussion just given suggests that the approximate independence of the two O—H bonds in H_2O is connected with the non-overlapping character of the AO's employed in describing them. Mathematically, this is expressed by the vanishing of the corresponding overlap integrals; the overlap between the orbitals H_1 and H_2 is usually neglected because of their separation (much greater than that in the hydrogen molecule) as, similarly, is that between H_1 and p_y (and H_2 and p_x). That between p_x and p_y is zero because they are AO's on the same atom and are thus rigorously orthogonal (p. 70), as is obvious also from their symmetry:

$$\int p_x p_y \, d\tau = 0. \tag{7.2}$$

There is, however, no real reason at this stage for building the bond orbitals from p functions directed *perpendicularly* to each other as in Fig. 7.1, for a p function has vector properties (Fig. 7.2) and a function with the axis in the xy plane but directed at an angle θ to p_x is thus

$$p = \cos \theta \, p_x + \sin \theta \, p_y. \tag{7.3}$$

We might, therefore, have started with any two p-type AO's, p_1 and p_2, say, perhaps inclined at 100° instead of 90°. That being so there seems to be no reason for H_2O to be a 'bent' molecule at all; we could have constructed bond orbitals by pointing p_1 and p_2 in opposite directions along the same axis and bringing up H_1 and H_2 to form a *linear* molecule H—O—H. What is so special about the non-overlapping nature (i.e. orthogonality) of the bond orbitals?

To answer this question, which is clearly important for stereochemistry, we

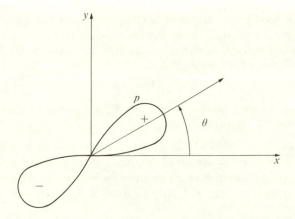

FIG. 7.2. Vector property of a p orbital. A p orbital pointing in the direction θ is related to p_x and p_y exactly as if all three were unit vectors (equation (7.3)).

must re-examine our motives in looking for a *localized* description of each bond, when we know perfectly well (and shall verify in the next section) that a direct application of the LCAO method would lead to *non*-localized MO's. We want to use localized MO's, if we can, in order to present the wavefunction, and the electron density to which it leads, in a form which can readily be visualized in terms of 'separable' parts. When this is done we can discuss the energy of the molecule and the factors affecting its shape in terms of the energies of its separate parts (e.g. bonds, inner shells, lone pairs) and the interactions between them. The important thing about requiring ortho-gonality of different bond orbitals is that it ensures (as can be shown mathematically) a 'separability' of the bonds in the sense that each will then make its own additive contribution to the total charge cloud of the molecule; to obtain the electron density we simply superimpose the densities associated with the separate parts. We could perfectly well construct the wavefunction from non-orthogonal orbitals, but the calculation of the electron density would then be exceedingly difficult and we should be unable to visualize its origin easily in terms of the densities of individual bonds, nor should we be able to argue in physical terms about factors affecting the energy as a whole—for the physics and chemistry would be lost in the mathematics.

The use of orthogonal bond functions is thus forced upon us by the need for simplicity in the way we resolve the charge cloud into its constituent parts. The general result† can be put quite simply: we may represent each bond by an electron-pair function (of either MO or VB type) and, provided the orbitals used in constructing each pair function are orthogonal to those used in constructing every other pair, the charge density will appear as a sum of parts—one part for each bond, just as if the others were absent, with no 'interference' effects.

To put this mathematically, we construct a wavefunction Ψ_A for bond A, another function Ψ_B for bond B, etc., exactly as in Chapter 5, and set up a product function $\Psi_A(1, 2)\Psi_B(3, 4)\ldots$ (cf. p. 112). We then enforce antisymmetry on the wavefunction by permuting the variables and summing, exactly as for an orbital product (p. 135). The theorem quoted above, namely that

$$P = P_A + P_B + P_C + \ldots$$

then follows provided the orbitals from which Ψ_A is constructed are orthogonal to those from which Ψ_B is constructed, etc.

The orthogonality of the functions p_x and p_y, expressed in (6.2), therefore fits them perfectly for the description of different electron pairs. Consequently, it is legitimate to picture the electron density as the superposition of two O—H bond densities, as in §7.1, and to discuss the energy of the molecule

---

† First derived, in essence, for electron pairs by Hurley, Lennard-Jones and Pople (1953) (see also Parks and Parr (1958)), and generalized to include separable parts containing any number of electrons (e.g. inner shells) by McWeeny (1959).

accordingly in terms of the energies of the separate bonds and their interactions.

We are now in a position to continue the discussion in §7.1. In the case of H_2O, where (7.1a) and (7.1b) are the bond orbitals, we once more expect greatest binding if the overlap of H_1 and p_x is as great as possible. This has already been achieved by our decision to place the H atoms along the x and y axes. It shows us, however, why the valence angle is expected to be about 90°. It will always be about 90° if the central atom has two unpaired p electrons with which to form bonds. Thus we should expect SH_2 to be approximately right angled, and indeed it is (92°).

Arguments of this kind provide the basis of a theory of directed valence, and so lead us to the very heart of the theory of stereochemistry. The theory rests upon two main ideas: (1) the possibility of using *localized* MO's, i.e. bond orbitals; (2) the criterion of maximum overlapping.

We may generalize the argument used above for H_2O. If a central atom has three unpaired p electrons with which it may form bonds, then the bond directions are expected to be at right angles. In fact, all p bonds of this kind are roughly at right angles to each other. Nitrogen, with an atomic configuration

$$N[(1s)^2(2s)^2(2p_x)(2p_y)(2p_z)]$$

provides an example. We satisfy the orthogonality condition at the N atom itself if we use the $2p_x$, $2p_y$, and $2p_z$ AO's, and we obtain maximum overlapping with the three attached hydrogens in a molecule such as ammonia if these are brought up along the x, y, z directions. The ammonia molecule should therefore be a pyramid whose apical angle is in the region of 90°. The same is true of the trivalent compounds of phosphorus and arsenic.

Each of the p-type AO's we have just been using has symmetry around its axis. This means that the bond which is formed, with an orbital such as (7.1a) or (7.1b), is itself symmetrical around the line of nuclei. It may therefore be called a σ bond, by analogy with the situation (§4.7) in diatomic molecules. Of course, if some of the adjoining bonds are highly polar their resulting electric field may succeed in disturbing the σ character of the first bond. There is, however, no evidence to suggest that such distortions are large. Nevertheless, although our argument has suggested 90° as the characteristic valence angle, the observed angles are somewhat greater (104° 31' for H_2O,† 107° for NH_3, 93·5° for PH_3, 92° for AsH_3, 93° for H_2S).

The main reasons for these discrepancies are that (a) we have disregarded the effect of electrostatic interactions between different electron pairs, and (b) we have ignored the possible consequences of hybridization, already known from §6.2 to be far reaching. Before discussing these factors, however, we first turn to the alternative treatment of H_2O using non-localized MO's.

---

† These and many other geometrical angles and distances in molecules, are taken from the excellent compilation of Bowen *et al.* (1958) and its supplement Sutton *et al.* (1965).

7.3. Non-localized orbitals

So far, we have stressed the desirability of using localized MO's wherever possible. It is not necessary to do this, however, for the whole theory can be developed in terms of non-localized orbitals. When this is done, it turns out that there are some interesting relations between the two types of MO. It will be sufficient if we illustrate this by a single example. Let us therefore consider non-localized MO's for H_2O,[†] and suppose that the AO's involved in the binding of the molecule are the two oxygen 2p AO's, p_x and p_y say, and the two hydrogens, H_1 and H_2. For the moment let us suppose that the HOH angle (Fig. 7.3) is 2α; one of our objects will be to see for what value of α the total energy is a minimum. If the valence angle is $90°$, then $\alpha = 45°$.

As usual, we discuss the valence electrons only, regarding the oxygen 'core' (nucleus plus K shell) as a point charge, and we try to set up MO's by linear combination of p_x, p_y, H_1, and H_2, assuming as in §7.1 that the doubly occupied 2s and $2p_z$ AO's describe oxygen lone pairs and are not involved in the bonding. Now a glance at Fig. 7.3 shows that if we introduce axes a, b bisecting H_1OH_2 internally and externally, then the molecule is symmetrical across Oa. More precisely, it is symmetrical with respect to reflection in the plane zOa. This implies, entirely analogously to §4.4, that the allowed MO's must be either symmetrical or antisymmetrical across this plane. Therefore H_1 and H_2 must always appear in the combinations $H_1 + H_2$ (in a symmetric MO) or $H_1 - H_2$ (in an antisymmetric MO); such combinations of AO's, with simple symmetry properties, are often called 'symmetry orbitals'. Each MO of given symmetry will be constructed by linear combination of symmetry orbitals of the same type. Evidently p_x and p_y are neither symmetric nor antisymmetric for this reflection and we therefore seek to replace them by a more suitable pair. The vector character of p orbitals shows that instead of a linear combination of p_x and p_y, we could equally well have chosen a linear combination of p_a and p_b. This follows because

$$p_a = \frac{1}{\sqrt{2}}(p_x + p_y) \qquad p_b = \frac{1}{\sqrt{2}}(p_x - p_y) \qquad (7.4a)$$

$$p_x = \frac{1}{\sqrt{2}}(p_a + p_b) \qquad p_y = \frac{1}{\sqrt{2}}(p_a - p_b). \qquad (7.4b)$$

Now p_b is antisymmetric in the plane zOa and p_b is symmetric. This means that the allowed MO's are combinations of p_a with $H_1 + H_2$, or of p_b with $H_1 - H_2$. If we call them ψ_1 and ψ_2 then

$$\psi_1 = N_1[p_a + \mu_1(H_1 + H_2)] \qquad (7.5a)$$

[†] The first MO discussion was by Hund (1931) and (1932). Since then there have been many detailed calculations: see, for example, Ellison and Shull (1955) (revised by Pitzer and Merrifield (1970) and Arrighini and Guidotti (1970), who made essentially a 'best-possible' one-configuration MO calculation).

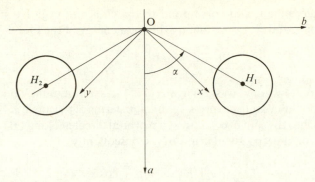

FIG. 7.3. Choice of axes in constructing non-localized MO's for H_2O.

$$\psi_2 = N_2[p_b + \mu_2(H_1 - H_2)] \tag{7.5b}$$

where μ_1 and μ_2 are two constants depending on the angle and the electronegativities of H and O.

We are now ready to write down the secular equations (3.41). If we do so, and then eliminate the constants μ_1 and μ_2 (effectively the ratios $c_1 : c_2$ in the standard form (3.3b)), they give us the secular determinants (continuing to use ε, instead of E, to denote *one*-electron (orbital) energies)

$$\begin{vmatrix} \alpha_p - \varepsilon & \beta_a \\ \beta_a & 2(\alpha_H - \varepsilon) \end{vmatrix} = 0 \tag{7.6a}$$

$$\begin{vmatrix} \alpha_p - \varepsilon & \beta_b \\ \beta_b & 2(\alpha_H - \varepsilon) \end{vmatrix} = 0. \tag{7.6b}$$

The first equation gives the symmetric solutions, of type (7.5a); the second gives the antisymmetric solutions of type (7.5b). α_p denotes roughly the energy of an oxygen 2p electron (either $2p_x$, $2p_y$, $2p_a$, or $2p_b$); α_H is roughly the energy of a hydrogen electron. β_a and β_b are defined in terms of the Hamiltonian $\hat{H}$ by the relations, analogous to (3.10),

$$\begin{aligned} \beta_a &= \int p_a \hat{H}(H_1 + H_2) \, d\tau \\ \beta_b &= \int p_b \hat{H}(H_1 - H_2) \, d\tau \end{aligned} \tag{7.6}$$

and, for simplicity, we have neglected all overlap integrals. The roots of these two secular determinants are

$$\varepsilon = \varepsilon_a = \tfrac{1}{2}(\alpha_p + \alpha_H) \pm \tfrac{1}{2}\{(\alpha_p - \alpha_H)^2 + 2\beta_a^2\}^{1/2}$$

and similarly $\varepsilon = \varepsilon_b$ with β_b replacing β_a. Now we have four electrons to put in the lowest-energy orbitals. These energies appear as the lowest roots of each separate determinant and correspond to the choice of negative signs in the formulae for ε_a and ε_b. Consequently the total energy of the four electrons is

$$E = 2\varepsilon_a + 2\varepsilon_b = 2(\alpha_p + \alpha_H) - \{(\alpha_p - \alpha_H)^2 + 2\beta_a^2\}^{1/2} - \{(\alpha_p - \alpha_H)^2 + 2\beta_b^2\}^{1/2}.$$
(7.8)

What we are now going to show is that (7.8) has a minimum value when $2\alpha = 90°$. To do this we must express β_a and β_b in terms of the angle α. Now p_a can be resolved into the sum of $\cos\alpha$ times a p orbital directed along OH_1 plus $\sin\alpha$ times a p orbital perpendicular to OH_1. Consequently

$$\int p_a \hat{H} H_1 \, d\tau = \beta_{OH} \cos\alpha$$
(7.9)

where β_{OH} denotes the resonance integral between a hydrogen orbital and an oxygen 2p orbital directed straight towards it. The term in $\sin\alpha$ vanishes on account of symmetry. Similarly

$$\int p_a \hat{H} H_2 \, d\tau = \beta_{OH} \cos\alpha$$

and consequently (7.7a) becomes

$$\beta_a = 2\beta_{OH} \cos\alpha.$$
(7.10a)

A similar argument gives

$$\beta_b = 2\beta_{OH} \sin\alpha.$$
(7.10b)

Combining (7.10a) and (7.10b) with (7.8) we have for the total energy

$$E = 2(\alpha_p + \alpha_H) - \{(\alpha_p - \alpha_H)^2 + 8\beta_{OH}^2 \cos^2\alpha\}^{1/2} - \{(\alpha_p - \alpha_H)^2 + 8\beta_{OH}^2 \sin^2\alpha\}^{1/2}.$$
(7.11)

It is a matter of straightforward calculus to verify that this expression has its least value when $\alpha = 45°$. This shows that the right-angled model is the most stable.

Anyone who compares this theory with the earlier account in §7.1 cannot fail to recognize how most of the pictorial character has been lost in these non-localized orbitals. Even the justification for a right-angled molecule seems to depend on the introduction of specific quantities, such as α_p, α_H, and β_{OH}, which were not apparently necessary in the earlier discussion. There is also the fact that the whole analysis has proceeded on the basis of complete symmetry in the $z0a$ plane, so that if one of the hydrogens was replaced by another atom, the calculation would break down, leaving us no explanation for the experimental individuality of the O—H bond. Also, the theory is still highly approximate because we have neglected *all* overlap integrals (even between p_x and H_1, p_y and H_2), all electron interaction effects (simply adding the energies of individual electrons to obtain (7.11)), and even the nuclear repulsion energy!

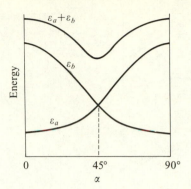

FIG. 7.4. Variation of H_2O orbital energies, ε_a and ε_b, with valence angle 2α.

These are fair comments, and in our later discussions we shall find, for example, that the interaction effects are large and of paramount importance. We shall also find (§8.9) that in the equilibrium situation there is an approximate proportionality between the total energy and the sum of the orbital energies. This being so, we may hope that our conclusions may with some justification be based on the latter. Nevertheless, we *have made a calculation*, however crude, which allows us to follow the way in which the energies of the two bonding orbitals change when the angle α varies from $0°$ to $90°$, and this gives us considerable insight into the nature of the molecule. Figure 7.4 shows the variation of the two energies ε_a and ε_b, and also of their sum $\varepsilon_a + \varepsilon_b$. At the equilibrium position $\varepsilon_a + \varepsilon_b$ is least and the angle $\alpha = 45°$; also $\varepsilon_a = \varepsilon_b$ and the two curves cross. There is, of course, no objection to this crossing since they correspond to wavefunctions with different symmetries. When $\alpha = 45°$ it follows that $\mu_1 = \mu_2$ (and $N_1 = N_2$) in (7.5a) and (7.5b). We may reasonably suppose that even when some of the refinements which we have omitted are included, this approximate equality of μ_1 and μ_2 will still obtain. However, for $\alpha < 45°$ the approximations made in setting up (7.6a) and (7.6b) begin to break down and the curves lose all meaning for $\alpha \to 0$. Nor will the crossing point be exactly at $\alpha = 45°$.

The kind of argument we have just given was developed in detail by Walsh (1953) and led to the formulation of 'Walsh's rules' for predicting the equilibrium shapes of small molecules. We develop such topics later in another context (Chapter 10).

To conclude the present section, we look for a connection between the description just given, in terms of non-localized MO's, and the more qualitative discussion of §7.1, in terms of bond orbitals. The forms of the MO's (7.5a) and (7.5b) are indicated in Fig. 7.5 (a, b) and the MO description of the electron configuration is

$$H_2O[K2s^2 2p_z^2 \psi_1^2 \psi_2^2]. \tag{7.12}$$

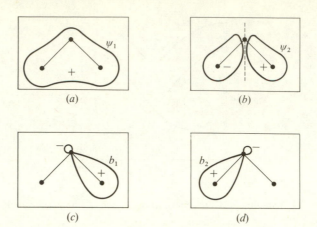

FIG. 7.5. Relationship between non-localized MO's (ψ_1 and ψ_2) and localized MO's (bond orbitals, b_1 and b_2) in H_2O.

The bonding arises because both ψ_1 and ψ_2 have forms which give a concentration of electron density in the O—H bond regions; this part of the density is in fact

$$P_{\text{bond}} = 2\psi_1^2 + 2\psi_2^2. \tag{7.13}$$

It is possible, however, to recast this result in a form corresponding to the use of *localized* MO's, for if we take the sum and difference of ψ_1 and ψ_2, defining normalized orbitals

$$b_1 = (\psi_1 + \psi_2)/\sqrt{2}, \quad b_2 = (\psi_1 - \psi_2)/\sqrt{2} \tag{7.14}$$

we achieve strong localization. Thus, b_1 is large in the region of the right-hand bond, where ψ_1 and ψ_2 (Fig. 7.5 (a, b)) have the same sign and reinforce each other, and very small in the left-hand bond, where they 'cancel'. Similarly, by superimposing ψ_1 and $-\psi_2$, we obtain a b_2 which is the mirror image of b_1, concentrated mainly in the left-hand bond. In fact, b_1 and b_2, shown in Fig. 7.5 (c, d), are just the bond orbitals introduced in the qualitative discussion of §7.1; to verify this we need only substitute $(7.5a, b)$ in (14), remembering that $\mu_1 = \mu_2$ for $\alpha = 45°$, obtaining

$$b_1 = \frac{1}{\sqrt{2}} [N_1(p_a + p_b) + 2N_1\mu_1 H_1].$$

On using $(7.4b)$ this becomes

$$b_1 = N_1[p_x + \sqrt{2}\mu_1 H_1] \tag{7.15a}$$

which is exactly equivalent to $(7.1a)$. Similarly, we find

$$b_2 = N_1[p_y + \sqrt{2}\,\mu_1 H_2] \tag{7.15b}$$

exactly equivalent to (7.1b).

The remarkable thing about the introduction of the localized orbitals is that we can put two electrons in b_1 and two in b_2 and obtain exactly the same description of the molecule; *the results are mathematically equivalent*! In terms of bond orbitals, the electron configuration would be described as

$$H_2O[K2s^2 2p_z^2 b_1^2 b_2^2] \tag{7.16}$$

instead of (7.12), and the bonding part of the charge density would be

$$\begin{aligned} P_{\text{bond}} &= 2b_1^2 + 2b_2^2 \\ &= (\psi_1 + \psi_2)^2 + (\psi_1 - \psi_2)^2 \quad \text{(from (7.14))} \\ &= 2\psi_1^2 + 2\psi_2^2 \end{aligned}$$

exactly as in (7.13). Whether we use localized or non-localized orbitals is thus, in a sense, quite immaterial; they simply offer alternative ways of describing the molecule and of resolving its electron density into parts. The advantage of the bond orbitals is that the parts have a more immediate chemical interpretation. Of course, if μ_1 and μ_2 were not exactly equal the cancellation of the H_2 terms which led to (7.15a) would not be perfect and b_1 would contain a little bit of H_2; the bonds would not be quite so well represented by overlapping *pairs* of AO's, and the results of using (7.15a) and (7.15b) would not *exactly* reproduce those of the more complete MO treatment. Broadly speaking, however, there is often little to choose between the two nearly equivalent ways of looking at the molecule; we can always pass from a description in terms of completely delocalized MO's to one in which the orbitals are much more strongly localized, by a transformation such as (7.14). In a one-configuration approximation, not only the charge density but also the whole many-electron wavefunction can be shown to be unchanged (mathematically 'invariant')[†] against such a change of description (see, for example, Coulson (1949)).

7.4. The perfect-pairing approximation. Interaction between bonds

We have seen that even in polyatomic molecules the bonds may often be regarded as localized, each with its own characteristic properties and its own 'personal' wavefunction. Exactly as for a single electron-pair bond, it does not matter much whether each pair is described by a localized MO function or by a VB function of the Heitler–London type.[‡] The localized bond description,

---

[†] The implications of this invariance have been discussed extensively by Lennard–Jones and co-workers: see, in particular, Lennard–Jones (1949) and for a useful review Pople (1957).

[‡] See, for example, detailed numerical comparisons for the molecule H_2O by McWeeny and Ohno (1960).

with its strong chemical appeal, was in fact first developed within the framework of VB theory; the wavefunction then corresponded to a single 'structure', in the sense of §5.3, in which the spins of the electrons in each bond were 'paired' (i.e. antiparallel coupled) to give a singlet wavefunction. The resultant 'perfect-pairing approximation', like the localized MO approximation, depended for its validity on the possibility of arranging the singly occupied AO's of the constituent atoms into strongly overlapping *pairs*, so as to obtain an obvious and unique pairing scheme to describe the whole set of bonds. Within each pair the overlap should be large and between different pairs it should be negligible; these are the features which characterize the 'perfect-pairing approximation' with our extended interpretation developed in §7.2.

We shall make no use of the energy expressions derived by VB theory for this situation, which are now of mainly historical interest, but present instead an energy formula which depends on the above approximations alone (in particular on the orthogonality of the orbitals used in constructing different pair functions) and which is valid even with individual pair functions of considerable complexity (e.g. for the 'VB+ionic' or 'MO+CI' functions used in §5.4). The formula needed (given in general form in McWeeny 1959, 1960) applies whenever we can distinguish a number of weakly interacting groups of electrons (e.g. inner shells, bond pairs, lone pairs), A, B, C,... say, each described by its own wavefunction ($\Psi_A, \Psi_B, \ldots$), assuming the orbitals for any two different groups are orthogonal; it reads

$$E = E_A + E_B + E_C + \ldots + G_{AB} + G_{AC} + G_{BC} + \ldots \qquad (7.17)$$

where, for example, E_A is the energy of the electrons of Group A in the field of all the nuclei, with all other electrons taken away, and G_{AB} is the energy of interaction between the electrons of Group A and those of Group B. Obviously the molecular conformation of lowest energy, corresponding to the equilibrium geometry, will result when the negative terms $E_A, E_B, \ldots$ reach their lowest values (most stable bonds) compatible with not too large values of the interaction terms $G_{AB}, G_{AC}, \ldots$ which are positive, corresponding to mutual repulsion between the electrons of different groups. Formula (7.17) thus provides a secure, and physically transparent, basis for discussion of the factors affecting molecular geometry. We are already familiar with the conditions for strong bonding (i.e. large negative E values for individual bonds); the new feature, which we must now consider in dealing with polyatomic molecules, is the strength of the repulsions between different 'non-bonded' electron pairs. The G's have a simple electrostatic interpretation, for

$$G_{AB} = J_{AB} - K_{AB} \qquad (7.18)$$

where the 'exchange term' K_{AB} is small provided groups A and B overlap only slightly, and when this is the case

$$G_{AB} \simeq J_{AB} = \frac{e^2}{\kappa_0} \int \frac{P_A(1)P_B(2)}{r_{12}} d\tau_1 d\tau_2. \tag{7.19}$$

This is simply the Coulomb repulsion between two charge clouds, one of density P_A (for the electrons of Group A) and the other of density P_B (for the electrons of Group B).[†] If, for example, A and B are bonds, each charge density may be calculated from an electron-pair wavefunction, exactly as in Chapter 5, just as if the other bonds were absent.

The equilibrium geometry of any molecule to which the perfect pairing approximation applies is now seen to be determined by a competition; each bond tries to become as strong as possible (e.g. by adjusting its orbitals to secure maximum overlap), in competition with all other bonds, and in such a way as to keep the bonds as 'separate' as possible so that their mutual repulsions do not become too large. Formula (7.17) provides a basis for a much used 'model' first proposed by Sidgwick and Powell (1940) and later developed very extensively by Nyholm and Gillespie (1957) (see also Gillespie (1972)), to which we turn presently. First, however, we must consider how the energies of the individual bonds may be lowered by maximizing the overlap within each electron pair, for we know from §6.2 that the inclusion of hybridization (which we ignored in §7.1 and §7.3) can lead to a dramatic increase in overlap.

7.5. Inclusion of hybridization: H_2O

It was noted in §6.2 that the description of the bond in LiH was much improved by using, instead of a pure 2s AO on the lithium atom, a *hybrid* of the form

$$h = N(s + \lambda p). \tag{7.20}$$

Here the constant λ determines the s–p mixing and normalization requires the squares of the s and p coefficients to add up to unity, giving

$$N^2 = 1/(1 + \lambda^2).$$

The overlap with the hydrogen 1s AO was thereby increased and the energy lowered. The superposition of s- and p-type AO's to give a hybrid was indicated in Fig. 6.8, and clearly any such mixing will produce a strongly directed orbital.

If we allow free mixing of s, p_x, p_y, and p_z orbitals in the valence shell of an atom, it is possible to form up to four hybrids pointing in different directions, and these will permit a considerable variety of multiple-bonding situations. Let us turn back to the water molecule, by way of illustration. Instead of using p_x and p_y (Fig. 7.1), overlapping with H_1 and H_2 respectively, we might

---

† $-eP_A(1)d\tau_1$ is the amount of charge in $d\tau_1$ at point 1, while $-eP_B(2)d\tau_2$ is that in $d\tau_2$ at point 2; their energy of repulsion is obtained on dividing by $\kappa_0 r_{12}$ and the integral represents the sum of such contributions for all pairs of volume elements in both clouds.

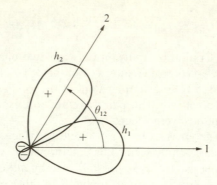

FIG. 7.6. Evaluation of overlap between two hybrids h_1 and h_2 formed from p orbitals (p_1 and p_2) pointing along axes 1 and 2.

introduce hybrids of the form

$$s + \lambda p_x, \quad s + \lambda p_y$$

which would lean directly towards the hydrogens and therefore give larger overlap and lower energy. However, a moment's reflection shows that these hybrids have a non-zero overlap and would therefore (p. 186) be unacceptable as components of two different bond orbitals. We are thus forced to look for the condition that two hybrids must satisfy in order to be orthogonal.

The condition is a simple one, with far-reaching consequences. Let us consider any two hybrids (Fig. 7.6), one pointing along the direction of a p-type AO p_1 and the other along p_2. They may therefore be written

$$h_1 = N_1(s + \lambda_1 p_1), \quad h_2 = N_2(s + \lambda_2 p_2). \tag{7.21}$$

The overlap integral is thus, assuming p, p_1, and p_2 normalized,

$$\int h_1 h_2 \, d\tau = N_1 N_2 \left[1 + \lambda_1 \int p_1 s \, d\tau + \lambda_2 \int s p_2 \, d\tau + \lambda_1 \lambda_2 \int p_1 p_2 \, d\tau \right]. \tag{7.22}$$

Now the two middle terms vanish by symmetry and we therefore need only the final integral, the overlap of two arbitrarily directed p orbitals. In Fig. 7.6 we have used p_1 to define the x axis ($p_1 = p_x$) and, remembering the vector character of p functions (p. 185), we can resolve p_2 into its components:

$$p_2 = \cos \theta_{12} p_x + \sin \theta_{12} p_y.$$

The required overlap integral is thus

$$\int p_1 p_2 \, d\tau = \cos \theta_{12} \int p_x^2 \, d\tau + \sin \theta_{12} \int p_x p_y \, d\tau = \cos \theta_{12}.$$

On inserting this value in (7.22) and requiring the result to vanish, we obtain the required orthogonality condition:

$$\lambda_1\lambda_2 = -1/\cos\theta_{12}. \tag{7.23}$$

The angle between the orthogonal hybrids is thus related to the relative amounts of s–p mixing they exhibit. Since λ_1 and λ_2 are positive (to give hybrids pointing along the positive directions of p_1 and p_2), $\cos\theta_{12}$ is negative and the hybrids must be inclined at an angle greater than 90°.

If now we return to H_2O and allow the orbitals to be constructed using hybrids (with $\lambda_1 = \lambda_2$ to ensure symmetry), then the hybrids must be inclined at an angle such that $\cos\theta_{12} = -1/\lambda^2$. Since this angle is greater than 90°, the effect of admitting hybridization is to open out the H_2O valence angle. If we assume the hybrids point directly along the bonds and put $\theta_{12} = 104\cdot5°$ (the observed angle), we obtain $\lambda = 1\cdot998$ and since the amounts of s and p character are N^2 and $N^2\lambda^2$ respectively this indicates about 80 per cent p character and 20 per cent s character. In terms of the energy analysis of the last section, hybridization allows the energies of the two bond pairs (E_A and E_B, say) to fall, owing to the enhanced overlap between h_1, H_1 and between h_2, H_2, and at the same time allows the bonds to move apart, thus reducing the repulsion energy J_{AB}. The expected stabilization is thus considerable and the opening of the H—O—H angle, above 90°, is certainly in the direction of the experimentally observed geometry.

The simple picture of §7.1 is modified in one other important way. In the original treatment we assumed the 2s AO held a lone pair of electrons and was *not* mixed with the other orbitals, but now we have allowed mixing of *three* AO's (s, p_x, p_y) leaving only a $2p_z$ lone pair. We have used two of the mixtures, h_1 and h_2, in forming the bond orbitals, but this leaves a third linear combination (i.e. another hybrid) $s + \lambda_3 p_3$, say, which will be free to hold two electrons and thus to accommodate a lone pair. The direction of p_3 is fixed (Fig. 7.7) by the need for orthogonality to both h_1 ($= s + \lambda p_1$) and

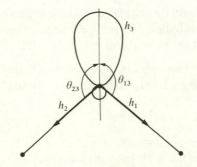

FIG. 7.7. Form of the lone pair hybrid in H_2O.

h_2 $(=s+\lambda p_2)$, for (7.23) gives $\cos\theta_{13} = -1/\lambda\lambda_3$ and $\cos\theta_{23} = -1/\lambda\lambda_3$. Thus $\theta_{13} = \theta_{23}$ and the lone-pair hybrid sticks out to the rear of the oxygen, bisecting the H—O—H angle. With $\theta_{12} = 104 \cdot 5°$, we must have $\theta_{13} = \theta_{23} = 127 \cdot 75°$, and this leads to an s content of 60 per cent and a p content of 40 per cent. The total s content of all the hybrids is $2(0 \cdot 20) + (0 \cdot 60) = 1$ (we have used *one* s orbital) and the total p content is $2(0 \cdot 80) + (0 \cdot 40) = 2$ (we have used *two* p orbitals); this gives a useful check on the arithmetic.

Our picture of the electronic structure of H_2O is now rather satisfactory. Not only have we understood the nature of the energy balance which results in the H—O—H angle opening above 90°, as observed, we have also discovered that the remaining part of the electron distribution is at the same time 'polarized', the 2s oxygen lone pair being forced out to the 'rear' of the oxygen atom. As in the case of CO (p. 153), this redistribution of lone-pair electron density has important chemical implications (see, for example, Pass (1973)). It is a key factor in 'hydrogen bonding' in which the electron-rich region is attracted by an electron-deficient hydrogen atom in some other chemical group; such bonds provide the mechanism for base pairing in the DNA helix, and also hold water molecules together in the ice crystal (Chapter 12).

7.6. Principal types of s–p hybridization

In molecules such as H_2O two or more hybrids may have exactly the same values of λ as a consequence of molecular symmetry, for otherwise the corresponding bonds would not be identical in form. Such hybrids are said to be *equivalent*; they are alike in all respects except orientation in space. For an atom with a valence shell containing s and p orbitals we can distinguish three principal types of hybridization.

   (i) By mixing an s with *one* p orbital (leaving the other two unchanged) we can set up *two* equivalent hybrids.

  (ii) By mixing an s and two p orbitals (leaving the third unchanged) we can set up *three* equivalent hybrids.

 (iii) By mixing an s and all three p orbitals we can set up *four* equivalent hybrids.

To determine the forms of these hybrids, and the angles between them, we need only observe that each hybrid of a set (being equivalent) will have the same s content $1/(1+\lambda^2)$ and the same p content $\lambda^2/(1+\lambda^2)$.

Case (i). With only two hybrids the condition that the total s content is 1 gives $1/(1+\lambda^2) = \frac{1}{2}$ for each hybrid; hence $\lambda^2 = 1$ and (7.20) gives

$$h = \frac{1}{\sqrt{2}}(s+p).$$

The two hybrids correspond to different choices of direction for the p orbital, and according to (7.23) the angle between the p's is given by $\cos\theta = -1$. Hence $\theta = 180°$ and the two hybrids point in opposite directions along the same axis:

$$h_1 = \frac{1}{\sqrt{2}}(s+p_1), \quad h_2 = \frac{1}{\sqrt{2}}(s+p_2) \qquad (p_1, p_2 \text{ at } 180°). \qquad (7.24)$$

If we use p_1 to define the positive z axis, $p_1 = p_z$ and $p_2 = -p_z$, but of course the choice of axes if immaterial and Fig. 7.8 (a) tells us all we need to know. We call h_1 and h_2 *digonal* or *sp hybrids*; their s content is $\frac{1}{2}$.

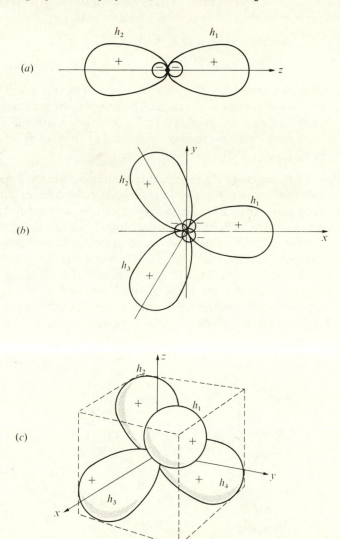

FIG. 7.8. Principal types of s–p hybridization: (a) digonal, hybrids oppositely directed along the same axis; (b) trigonal, hybrids pointing along three axes in a plane inclined at 120°; (c) tetrahedral, hybrids pointing towards the corners of a regular tetrahedron.

Case (ii). With three equivalent hybrids, the s content of each must be $\frac{1}{3}$ and a similar argument gives

$$h_1 = \frac{1}{\sqrt{3}}(s+\sqrt{2}p_1), \quad h_2 = \frac{1}{\sqrt{3}}(s+\sqrt{2}p_2), \quad h_3 = \frac{1}{\sqrt{3}}(s+\sqrt{2}p_3) \quad (7.25)$$

p_1, p_2, p_3 are coplanar and inclined at $120°$ to each other. We call h_1, h_2, h_3 *trigonal* or sp^2 *hybrids* (Fig. 7.8 (b)); their s content is $\frac{1}{3}$.

Case (iii). In exactly the same way we find

$$h_i = \frac{1}{2}(s+\sqrt{3}p_i) \qquad (i = 1, 2, 3, 4) \tag{7.26}$$

p_1, p_2, p_3, p_4 are all inclined at $109° 28'$ to each other. In other words the four hybrids point from the centre to the corners of a regular tetrahedron. We call h_1, h_2, h_3, h_4 *tetrahedral* or sp^3 *hybrids* (Fig. 7.8 (c)); their s content is $\frac{1}{4}$. Here, and in Case (ii), the hybrids p_i are easily expressed in terms of p_x, p_y, p_z, using the vector properties (p. 185).

Although the above cases are highly special and are, strictly, appropriate only for atoms forming two, three or four *identical* hybrid bonds, they often provide a rough approximation in less symmetrical situations. Thus, the carbon atom in methane, which is tetrahedral, forms four identical CH bonds, and the four identical electron pairs would be described by overlapping four tetrahedral hybrids with the four hydrogen 1s AO's (Fig. 7.9 (a)); however, even when one bond is somewhat different, as in CH_3F (Fig. 7.9 (b)), the tetrahedral set of hybrids still gives a good account of the electronic structure—the C—F electron pair is not quite the same as a C—H pair, but the geometry is still roughly tetrahedral around the carbon atom. The tetrahedral set is, in fact, generally adequate to describe the bonding of quadrivalent carbon in saturated molecules. In a similar way, carbon in a graphite layer

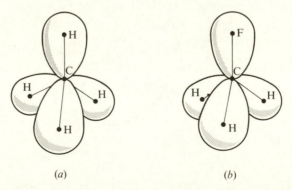

(a)   (b)

FIG. 7.9. Tetrahedral system of electron-pair bonds: (a) CH_4 (bonds formed by overlap of carbon sp^3 hybrid with hydrogen 1s); (b) CH_3F (CF bond formed from carbon sp^3 hybrid and fluorine 2p).

(Fig. 7.10 (*a*)) is apparently forming three identical localized bonds, at 120°, and the trigonal set is appropriate, but carbon in the benzene molecule (Fig. 7.10 (*b*)) is also quite well described using the trigonal set, though the C—H bond differs somewhat from the other two C—C bonds. Trigonal hybridization is characteristic of carbon in aromatic hydrocarbons. Finally, when carbon occurs in a linear molecule, such as O—C—O or H—C—N, we may anticipate that it forms bonds with digonal, or roughly digonal, hybrids, the remaining 2p electrons accounting for any remaining π bonding, just as they did in the molecule CO (p. 151).

It is clear that the recognition of three principal types of s–p hybridization has, at a stroke, opened the way to the understanding of a vast area of stereochemistry, which we shall develop systematically in due course. At this point, however, we wish to relate the discussion to that of earlier sections.

When hybridization was first encountered (§6.2) we introduced also the idea of a *valence state* as a means of describing the electron configuration of an atom *in the molecule*. Thus, if the Li—H bond was broken without allowing the Li electronic structure to 'relax', the lithium atom would be left with one electron in a slightly hybridized orbital. Its valence state would be Li[$1s^2h$] with $h = as + bp$, and in terms of s and p AO's this could be written as

$$\text{Li}[1s^2 2s^{a^2} 2p^{b^2}] \qquad (7.27)$$

where a^2 and b^2 indicate, respectively, the s and p contents of the hybrid. In the case of lithium (p. 151) the 'deformation' of the atom corresponded to changing about 34 per cent of the 2s electron density of the normal ground state into a 2p electron density; in more picturesque language (not to be taken literally) we spoke of 'promoting' 0·34 of an electron from the 2s AO to the 2p AO. The promotion required energy, but this was more than paid for by the greatly increased bonding power of the hybrid. Similar considerations apply to a multiply bonded atom; we therefore need to examine the valence state in typical bonding situations, and also the effect of hybridization on bond strength.

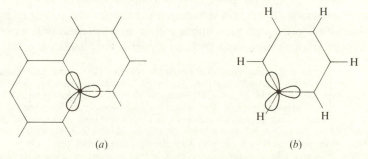

(*a*)   (*b*)

FIG. 7.10. Trigonal hybrids: (*a*) carbon in a graphite layer showing appropriate sp² hybrids (each C—C bond formed from one overlapping pair); (*b*) carbon sp² hybrids in benzene.

First we note that in the case of H_2O the electron configuration of oxygen in the ground state, namely $O[K\ 2s^2 2p^4]$ or

$$O[K\ 2s^2 2p_z^2 2p_x 2p_y] \tag{7.28}$$

correctly predicts a divalent atom, forming a bent (90°) molecule.† Thus, hybridization is essentially a *refinement* which allows us to understand why the H—O—H angle is greater than 90°. The electron configuration in the hybridized valence state may be written as

$$O[K\ 2p_z^2 l^2 h_1 h_2] \tag{7.29a}$$

where we use l to distinguish the lone-pair hybrid previously called h_3 in §7.5. On adding up the s and p contents of l, h_1 and h_2 (doubling the first for two electrons), we obtain the equivalent description

$$O[K\ 2p_z^2 2s^{1\cdot6} \bar{2p}^{2\cdot4}] \tag{7.29b}$$

where the $2p^{2\cdot4}$ refers to the total electron population of the 2p AO's in the molecular plane. The $2s^2$ closed shell of the simple description (7.28) has thus been slightly 'opened' in the valence state by promotion of about 0·4 electrons to the 2p shell.

Similar considerations apply to the nitrogen atom, whose ground-state configuration

$$N[K\ 2s^2 2p^3]$$

obviously indicates trivalence and the formation of molecules with pyramidal shape.

In the case of carbon, however, with configuration

$$C[K\ 2s^2 2p^2]$$

no sensible interpretation of the observed (variable) valence can be obtained without invoking hybridization! Apparently, the atom should be *di*valent, forming bent molecules like H_2O, but nothing could be further from the observed facts. The carbon atom is astonishingly versatile in its bonding; promotion to a rich variety of valence states occurs easily, owing to a small 2s–2p energy difference, and the number and strengths of the bonds it can form when hybridization is permitted leads to a particularly wide range of stable

---

† Even here, without any actual promotion, we should note that the valence state is somewhat higher in energy than the spectroscopic ground state (^3P). In the latter there is both orbital and spin angular momentum, coupled to give a low-energy resultant (Hund's rules, p. 40), while in the valence state such coupling is broken; the orbital angular momentum is 'quenched' by the approach of other atoms and the spins are uncoupled in readiness to recouple within the electron-pair bonds. We do not need to worry about the details, but should be aware that the electron *configuration* does not specify either the valence state or the ground state completely; one configuration may give rise to many spectroscopic states, of which a given valence state is generally a mixture. For most purposes, however, we shall find a statement of the valence *configuration* sufficient.

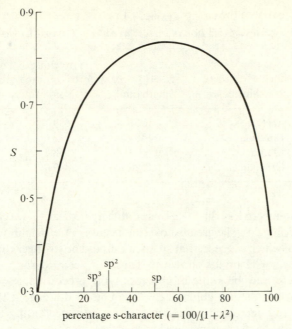

percentage s-character $(= 100/(1 + \lambda^2))$

FIG. 7.11. Dependence of overlap on hybridization parameter λ for two hybrids in a C—C bond at fixed internuclear distance.

compounds. The three principal types of hybridization all lead to valence states in which one electron has been promoted from the 2s AO to a 2p AO. For example, the tetrahedral valence configuration

$$C[K\, h_1 h_2 h_3 h_4] \qquad\qquad (7.30a)$$

is equivalent, since the s content of each hybrid is $\frac{1}{4}$ (and p content is $\frac{3}{4}$), to

$$C[K\, 2s2p^3]. \qquad\qquad (7.30b)$$

In this case, then, promotion is a much more substantial effect than in oxygen, but it occurs readily because (i) the s–p interval is only about half that in oxygen and (ii) the number of bonds which can be formed is increased from two to four,† and the return in bond energy is therefore great.

Finally, we need quantitative confirmation that hybridization can 'pay for itself' by enhancing the overlap and thereby strengthening the bond. A glance at Fig. 7.11 shows that this is indeed the case. The curve shows the overlap integral S for two hybrids of the form $s + \lambda p$ directed towards each other, at a fixed internuclear distance, as a function of the percentage s character (i.e. $100/(1 + \lambda^2)$). The astonishing thing about this curve is that, although pure s or

† Cf. oxygen, where a promotion 2s → 2p would still leave only *two* singly occupied orbitals available for bonding.

TABLE 7.1
Properties of CH bonds involving different hybridization

Hybridization	Molecule	C—H bond length (nm)	Stretching force constant ($N m^{-1}$)	Approximate bond energy ($kJ mol^{-1}$)
sp	Acetylene	0·1061	639·7	500
sp^2	Ethylene	0·1086	612·6	440
sp^3	Methane	0·1093	538·7	411
(p)	CH radical	0·1120	449·4	330

p overlap amounts to less than 0·5, by suitable hybridization we can obtain an overlap exceeding 0·8. The greatest overlap occurs in the neighbourhood of sp hybridization, which suggests that sp bonds should be stronger than either sp^2 or sp^3. Data for CH bonds shown in Table 7.1 seem to support this view. Although there may be some doubt about the precise values of the bond energies shown in the last column, there can be no doubt that the increased bond length and decreased force constant reveal a decreasing bond energy associated with the decreasing overlap of atomic orbitals as we go down the table.

7.7. Hybridization involving d orbitals

Although in principle any number of AO's may be used in a full MO treatment of a molecule, those which participate most strongly in the MO's that describe the valence electrons are those which belong to the valence shells of the constituent atoms. Thus in H_2O we built up the MO's from hydrogen 1s AO's and oxygen 2s and 2p AO's, but did not include, say, hydrogen 2p or oxygen 3d AO's. Had we done so in an actual calculation, we should have found them to occur with very small coefficients, partly because they correspond to too high an energy (and a large energy difference inhibits mixing, as we know from §4.8) and partly because they are too diffuse to be effective as a means of describing the electron density in the relatively compact regions of the bonds. However, as we move to heavier elements the participation of other types of AO becomes important. Thus, sulphur—a group VI element, like oxygen, with a similar configuration $S[K L 3s^2 3p^4]$— has 3d AO's with an energy not so far above that of 3p as to preclude some mixing. This gives sulphur the capacity to form a richer variety of compounds than oxygen, for example both a tetrafluoride SF_4 and a hexafluoride SF_6. Similarly, in the nickel atom, the 3d, 4s, and 4p levels (cf. Fig. 2.11) all lie within about 4 eV of each other and there is thus no reason why we should not expect hybridization involving d, s, and p orbitals. Pauling (1931) was the first to show that, by suitable combination of such orbitals, very strongly directed hybrids could be formed, giving co-ordination numbers and valence angles quite different from those expected using s-, p-, or d-type AO's separately.

We must now study the consequences of admitting the participation of d orbitals. Again, it is possible to define principal types, i.e. sets of two, three, four, five, or six hybrids appropriate to specific highly symmetrical molecules, and again, with slight distortion, these can be used to describe less symmetrical molecules. The number of possibilities is now much increased and a full discussion is more complicated.

Table 7.2 summarizes the more important types of hybridization involving s, p, and d orbitals, irrespective of principal quantum number provided they lie fairly close in energy.

TABLE 7.2
Important types of hybridization

Co-ordination number of hybrids	Atomic orbitals used	Resulting hybrids
2	sp	Linear
	dp	Linear
	sd	Bent
3	sp^2	Trigonal plane
	dp^2	Trigonal plane
	d^{2}s	Trigonal plane
	d^{2}p	Trigonal pyramid
4	sp^3	Tetrahedral
	d^{3}s	Tetrahedral
	dsp^2	Tetragonal plane
5	dsp^3	Trigonal bipyramid
	d^{3}sp	Trigonal bipyramid
	d^{4}s	Tetragonal pyramid
6	d^{2}sp^3	Octahedral
	d^{4}sp	Trigonal prism

Instead of deriving some of the hybrids mathematically we shall merely indicate their formation pictorially in a few of the most important cases.

(i) *Octahedral hybrids* (d^2sp^3). The terminology d^{2}sp^3 means that we mix *two* of the d orbitals with the s and *three* p orbitals. Sometimes the order of the letters is made to correspond to ascending values of the principal quantum number of the AO's that (lying close together in energy) are being mixed; in that case we might distinguish between sp^{3}d (or 3s3p^{3}3d) for sulphur, and dsp^3

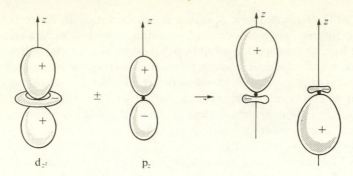

FIG. 7.12. Combination of d_{z^2} and p_z AO's to form two dp hybrids.

(or $3d4s4p^3$) for nickel, but here we shall not make any distinction between 'outer' and 'inner' hybrids, respectively, as the two types are essentially similar in appearance and properties.

An octahedral set of hybrids comprises six strongly directed mixtures pointing along the positive and negative x, y, and z axes; it is obtained by mixing $d_{x^2-y^2}$ and d_{z^2} AO's (see Fig. 2.9) with s, p_x, p_y, and p_z AO's. It is apparent (Fig. 7.12) that taking the sum and difference of p_z and d_{z^2} would yield two hybrids pointing in opposite directions along the z axis; in themselves they provide an example of linear ds hybridization (Table 7.2). Similarly, super-position of s, p_x, and $d_{x^2-y^2}$ (Fig. 7.13) would give an spd hybrid directed principally along the positive x axis, while reversal of the sign of p_x would give a similar hybrid pointing along the negative x axis. Two more similar hybrids could be set up using s, p_y and $d_{x^2-y^2}$ by suitable choice of signs of the mixing coefficients. We should then have four equivalent 'in-plane' (or

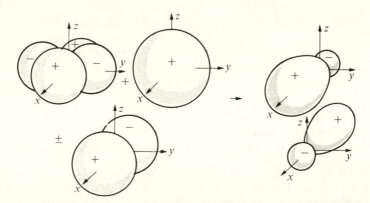

FIG. 7.13. Combination of $d_{x^2-y^2}$, s, and p_x AO's to form two dsp hybrids. A similar pair, pointing along the positive and negative y axes, is obtained by reversing the sign of $d_{x^2-y^2}$ and using p_y in place of p_x.

'equatorial') hybrids and two equivalent 'axial' hybrids, the two sets differing slightly in form because we had constructed them from different selections of AO's. In fact, however, if we permit a little inter-set mixing we can achieve six hybrids of absolutely *identical* form. These six d^2sp^3 hybrids are (assuming all AO's normalized)

$$h_{\pm z} = \frac{1}{\sqrt{6}}(s \pm \sqrt{3}\,p_z + \sqrt{2}\,d_{z^2}) \tag{7.31}$$

which point up and down the z axis, together with two more pairs obtained by replacing z by x and by y, in turn. The function d_{x^2}, for example, looks exactly like d_{z^2} in Fig. 7.12 but lies along the x axis instead; it can be written† as a linear combination of $d_{x^2-y^2}$ and d_{z^2},

$$d_{x^2} = \tfrac{1}{2}\sqrt{3}\,d_{x^2-y^2} - \tfrac{1}{2}d_{z^2}. \tag{7.32}$$

The fact that the three pairs of functions, of which (7.31) gives one, are the same except for choice of axis (x, y, or z) then shows the complete equivalence of the six hybrids.

(ii) *Bipyramidal hybrids* (dsp^3). By mixing d_{z^2} with s, p_x, p_y, and p_z it is obviously possible to obtain two axial hybrids (namely $d_{z^2} \pm p_z$) and three equatorial hybrids of essentially sp^2 form (i.e. the trigonal combinations of s, p_x, and p_y given in (7.25)). In this case, although some s character may be added to the axial hybrids and some d_{z^2} character to the equatorial hybrids, without changing their directions and general shape, it is not possible to obtain five exactly equivalent forms (as it was in the octahedral case). The trigonal bipyramid clearly has one unique axis of symmetry and the axial hybrids reflect this uniqueness by being somewhat different from those in the equatorial plane.

(iii) *Tetragonal planar hybrids* (dsp^2). If we mix only four orbitals, $d_{x^2-y^2}$, s, p_x, and p_y, we obtain the set of four square planar hybrids already referred to in (i) in deriving the octahedral set. The pair lying on the x axis (Fig. 7.13) is given by

$$h_{\pm x} = \tfrac{1}{2}(s \pm \sqrt{2}\,p_x + d_{x^2-y^2}) \tag{7.33}$$

and the second pair follows on interchanging x and y (noting that $d_{y^2-x^2} = -d_{x^2-y^2}$). All four hybrids are identical except for orientation.

The sets of hybrids defined above, which exhibit the highest symmetries compatible with co-ordination numbers of 6, 5, and 4, respectively, will clearly be appropriate for molecules which exhibit a similar high symmetry. At this point we give just one example of each type, since more detailed discussions will be given later.

---

† Cf. equation (2.20) used in discussing relationships among d-type AO's.

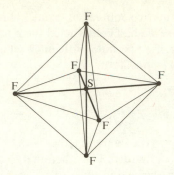

FIG. 7.14. Octahedral molecule SF_6. The broken lines indicate the edges of a regular octahedron; when this is inscribed in a cube the fluorines lie at the centres of the six-faces. All S—F bonds are equivalent.

Sulphur hexafluoride (Fig. 7.14) has the symmetry of a regular octahedron, and d^2sp^3 hybridization at the sulphur atom would appear to be appropriate. If we indicate the six hybrids by $h_1, h_2, \ldots, h_6$ the change of electron configuration in setting up the sulphur valence state would be indicated by

$$S[K \ L \ 3s^2 3p^4] \rightarrow S[K \ L \ h_1 h_2 h_3 h_4 h_5 h_6] \tag{7.34}$$

in which every hybrid is singly occupied and available to form an electron-pair bond with a 'ligand' atom, namely a fluorine with a singly occupied 2p AO. If we express the hybrids in the form (7.31) and work out the AO populations as we did for s–p hybrids (e.g. in going from (7.29a) to (7.29b)), we obtain an equivalent description of the valence state electron configuration

$$S[K \ L \ 3s^{1\cdot5} 3p^{3\cdot0} 3d^{1\cdot5}] \tag{7.35}$$

corresponding to promotion of 0·5 electrons from the 3s charge cloud and 1·0 from the 3p shell into the previously 'empty' 3d shell. As in the case of carbon (p. 203) the energy increase associated with this (imaginary!) process is likely to be more than compensated for by the formation of *six* strong hybrid bonds, where the ground-state configuration $(3s^2 3p^4)$ would have suggested only two comparatively weak bonds involving the singly occupied 3p AO's. However, oxygen, with a similar valence electron configuration $(2s^2 2p^4)$, would be quite unable to engage in six co-ordination because the lowest-lying d-type AO (3d) is energetically quite out of reach of the 2s and 2p electrons.

Although the promotion energy needed is a vitally important factor in assessing the different valence possibilities of a given atom, a word of caution is necessary. If d_{z^2} in (7.31) really were a 3d orbital obtained from, say, a Hartree–Fock calculation on the sulphur atom ground state, it would be much more diffuse (i.e. 'larger') than either 3s or 3p, and as Maccoll first pointed out

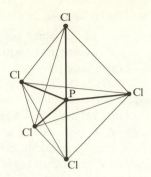

FIG. 7.15. Trigonal bipyramidal molecule PCl_5. The broken lines indicate the trigonal bipyramid. The two axial bonds (vertical) differ somewhat from those in the equatorial (horizontal) plane.

(1950; see also 1954 for numerical confirmation) it is then hard to see how it would effectively combine with the latter to give strongly directed hybrids. We must remember, however, that the optimum orbitals in the molecule should be determined by the variation method and that the 3d-type AO may thus be strongly contracted if that contraction leads to a lower energy and hence a better wavefunction. Such 'orbital contraction', first proposed by Craig and Magnusson (1956; see also Craig and Zauli 1962), is now known to occur.† The energy of a free-atom uncontracted d orbital must therefore be viewed with some caution in assessing the likelihood of its effective participation.

As an example of bipyramidal hybridization, we may take phosphorus pentachloride (Fig. 7.15). There are potentially five valence electrons on the phosphorus atom and the formation of a valence state would be indicated by

$$P[K \, L \, 3s^a 3p^a] \rightarrow P[K \, L \, h_1 h_2 h_3 h_4 h_5]. \qquad (7.36)$$

The five singly occupied hybrids would then overlap with singly occupied chlorine 3p orbitals to give five electron-pair bonds, the axial bonds differing slightly in length from the equatorial bonds.

Finally, for an example of a square planar compound we must go to the transition metals. Nickel, for example, forms the complex ion $Ni(CN)_4^{2-}$, with CN groups equidistant from the Ni along the x and y axes; the bonding may be described using the hybrids of type (7.33). However, the transition metals comprise a whole field by themselves, to which we turn in Chapter 9; for atoms earlier in the periodic table (e.g. silicon) the preferred four-co-ordinated shape is tetrahedral rather than square planar and this can be achieved without invoking d orbitals.

† For *ab initio* calculations see, for example, Hillier and Saunders (1970), and Gianturco *et al.* (1971). Even for a free atom, a similar contraction occurs in the valence state (see, for example, Coulson and Gianturco (1968).

7.8. Factors determining molecular shape

We have now almost reached the point at which we can turn from principles and methods to a discussion of the valence propensities of particular atoms as we work through the periodic table group by group. In this chapter we are considering only those molecules in which the bonds can be well described by a unique set of strongly localized electron pairs, i.e. where the perfect-pairing approximation of §7.4 offers a fairly accurate picture of the structure. We thus rule out certain special classes of molecule (e.g. conjugated molecules, transition-metal compounds) in which a considerable degree of non-localized bonding occurs; such molecules are considered separately in later chapters.

Before attempting a general survey, however, we must make a few general observations. The use of formula (7.17) as a basis for discussion of the factors which determine molecular shape has been outlined in §7.4. The total energy is a sum of (i) electron-pair energies, each pair in the field of all the nuclei, (ii) repulsion energies between different electron pairs, and (iii) repulsion energies between the nuclei. The binding energy comes from (i) alone, this being the only negative term, and the magnitude of this term is increased by maximizing the overlap of the AO's (or hybrids) participating in each bond; in this connection we must bear in mind that increasing the overlap (and lowering the energy) in some pairs may lead to less favourable hybridization (and a raising of the energy) in other pairs. We also know, from §7.5, how the directions of different hybrids are related by orthogonality conditions such as (7.23). The energy differences associated with small variations in geometry are usually much smaller than the binding energy, and are governed largely by the repulsion terms (ii) and (iii); we must therefore pay special attention to these terms in discussing molecular shape.

The validity of the electrostatic interpretation we are using depends on the condition that different pairs be described not only by mutually *orthogonal* ørbitals but by *differently localized* orbitals. The distinction is made in Fig. 7.16. In Fig. 7.16 (*a*) there is one s-type (A) and one p-type (B) electron pair; the corresponding term G_{AB}, defined in (7.18), contains a very substantial 'exchange' contribution (K_{AB}) because although the orbitals are orthogonal (by symmetry) they still overlap in so far as they share the same region of space.

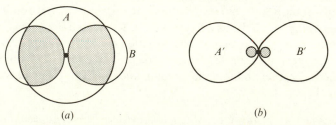

Fig. 7.16. Overlap of electron pairs: (*a*) orbitals *A* and *B* of s and p type respectively, large region of overlap; (*b*) *A'*, *B'* defined by new s–p combinations, small region of overlap.

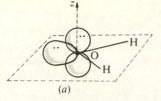
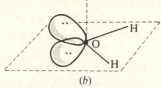

FIG. 7.17. Alternative choices of the lone pairs in H_2O: (*a*) one p-type lone pair, one roughly sp^2, substantial overlap region; (*b*) two equivalent lone pairs, one above the plane, one below, small overlap region. (For simplicity the bond pairs are omitted.)

In 7.16 (*b*), however, where the pairs are described by two digonal hybrids, A' and B', the exchange contribution is very much smaller owing to the greatly reduced overlap of the pairs. Now since (cf. p. 193) the corresponding four-electron wavefunctions will be mathematically equivalent, the total energies calculated using the alternative choices of orbitals will be identical; the only difference will be that the K_{AB} term (which is less easy to visualize than a coulombic interaction (7.19) between two charge clouds) may be neglected if we separate the electron pairs as in Fig. 7.16 (*b*), but may not be neglected if we use the alternative description Fig. 7.16 (*a*). *In order to obtain the most transparent interpretation of energy relations in the molecule we should therefore describe different electron pairs by orbitals which avoid each other as far as possible.* In fact this choice of orbitals is the one which recognizes in the simplest possible way a large part of the electron correlation referred to in §5.4 (that associated with the Pauli principle) by putting electrons of the same spin as far apart as possible, i.e. in differently localized pairs[†]; most of the remaining electron correlation is between the electrons of opposite spin within individual pairs and may be taken into account by refining the separate pair functions.

To put these principles into practice let us turn back to the H_2O molecule, described in §7.5 in terms of roughly trigonal hybrids (Fig. 7.17 (*a*)). Can we take this description one stage closer to the separated pair picture by assigning the two lone pairs of electrons to alternative hybrids (l_1 and l_2, say) instead of to one hybrid (h_3) and an oxygen $2p_z$ AO? A moment's reflection shows that we can. As in §7.3 there is nothing to stop us using (cf. (7.14))

$$l_1 = (h_3 + p_z)/\sqrt{2}, \qquad l_2 = (h_3 - p_z)/\sqrt{2}$$

instead of h_3 and p_z. The orthogonality property is preserved, and the total electronic wavefunction and charge density are unchanged. The only change is a formal one, corresponding to the change of description

$$H_2O[K\, h_3^2 p_z^2 b_1^2 b_2^2] \rightarrow H_2O[K\, l_1^2 l_2^2 b_1^2 b_2^2] \tag{7.37}$$

[†] Mathematical discussions of this observation have been given by, for example, Lennard–Jones (1949 and 1952).

but the lone-pair hybrids l_1 and l_2 are now exactly similar in form (Fig. 7.17 (b)) and as far apart as they can get. Because we have reduced their region of overlap (and hence a substantial exchange term), energy relations in the molecule will be more accurately inferred, using simple electrostatics, from Fig. 7.17 (b) than from Fig. 7.17 (a). In other words, we can think of the oxygen valence shell as comprising an octet in which the four electron pairs keep as far apart as possible and therefore adopt a roughly tetrahedral arrangement—a striking reconciliation of the conflicting ideas of Lewis (emphasizing the formation of electron pairs) and Langmuir (the completion of the octet).

The observations we have just made provide the quantum-mechanical basis for a well-established empirical approach to the interpretation of molecular shape, developed largely by Gillespie (1972) and often referred to as the 'valence-shell-electron-pair-repulsion' (VSEPR) theory. To see how the approach applies it is useful to consider a whole series of molecules (Fig. 7.18) in which the octet can be recognized. Clearly CH_4 (tetrahedral), NH_3 (trigonal pyramidal), and H_2O (V-shaped) are all *electronically* tetrahedral, or roughly so, and might be expected to show bond angles of about 109°. The ion NH_4^+ should also be tetrahedral and may be regarded as formed by protonation of NH_3, the proton going into the lone pair; indeed this provides a striking picture of the basicity of NH_3. Deviations from the tetrahedral angle are attributed in VSEPR theory to differences in the strength of the interactions among lone pairs and bond pairs; empirically these appear to follow the order

$$(\text{lone pair/lone pair}) > (\text{lone pair/bond pair}) > (\text{bond pair/bond pair}) \tag{7.38}$$

and this rule would suggest an NH_3 angle slightly less than the tetrahedral value (it is in fact 107°) and an H_2O angle even smaller (it is 104·5°). At a deeper level we might attribute the enhanced repulsive effect of lone pairs to (i) a lone-pair orbital being concentrated nearer the nucleus than a bond pair, thus increasing (inverse-distance) repulsion energies, and (ii) a lone pair being 'fatter', its electrons being attracted by only one nucleus instead of two, and therefore apparently 'repelling' other orbitals when the requirements of orthogonality are recognized. We might also expect (and shall find) that any

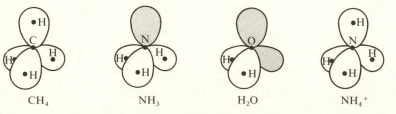

Fig. 7.18. Series of electronically similar molecules in which there are four roughly tetrahedral electron pairs. Lone pairs are indicated by shading.

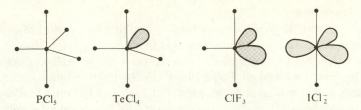

| PCl$_5$ | TeCl$_4$ | ClF$_3$ | ICl$_2^-$ |

FIG. 7.19. Series of electronically similar molecules in which there are five trigonal bipyramidal electron pairs. Lone pairs are indicated by shading. The halogens are the ligands.

factors which favour increased s content of a lone-pair hybrid will have the same effect of making it appear more repulsive, for as a result it will fall back towards the nucleus. Such factors will include (a) a large s–p energy difference and (b) high electronegativity of the atom.

A second example of a family of electronically similar molecules is shown in Fig. 7.19, where this time dsp^3 hybridization may be invoked in setting up lone-pair and bond-pair orbitals and there are five pairs around the central atom instead of an octet. Again, although the molecular shapes are bipyramidal, (irregular) tetrahedral, T-shaped, and linear, respectively, all four molecules show a common disposition of electron pairs. The rule (7.38) suggests, correctly, how any ambiguities in position of the bond pairs and lone pairs may be resolved; thus the TeCl$_4$ structure shown is expected to be more stable (with diminished repulsions) than one with an axial lone pair. A third family is shown in Fig. 7.20 and similar remarks apply.

Finally, we should remind ourselves again that hybridization is not a physical effect; it merely provides a means of constructing orbitals suitable for describing localized electron pairs and for interpreting their interrelationships. It would be quite wrong to say that, for example, CH$_4$ was tetrahedral because the carbon atom was sp^3 hybridized. The equilibrium geometry of a molecule depends on *energy* and energy only; the use of hybrids, or of orbitals of any kind, simply provides us with the means of recognizing, in a pictorial way, the origin of the main contributions to the energy and the way in which they vary with geometry.

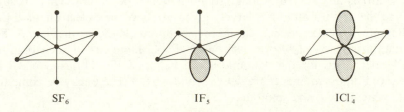

| SF$_6$ | IF$_5$ | ICl$_4^-$ |

FIG. 7.20. Series of electronically similar molecules in which there are six roughly octahedral electron pairs. Lone pairs are indicated by shading. The halogens are the ligands.

7.9. Valence rules

We are now ready to work our way through the first few rows of Mendeleev's periodic table, shown in its modern form at the end of this book, discussing the valence properties of the elements. The structure of the table and its role in chemistry are described in Puddephatt (1972). Each column of the table represents a chemical group, the atoms of each group being marked by a common valence-shell electron configuration and consequent similarity of properties, and we shall therefore proceed group by group.

Our picture of a normal single bond (σ bond) is based on two electrons, usually one from each atom, with opposed spins and approximately described by a wavefunction of VB or localized-MO type. This wavefunction is built up from two orbitals, associated with the two atoms and chosen to overlap as strongly as possible. In polyatomic molecules, and in diatomics too, these orbitals will be appropriate hybrids, and the hybrids around any one central atom must be orthogonal. Such hybrids are energetically 'profitable' only if their component s, p, and d orbitals have roughly similar energies. The hybrids for the two atoms of a bond will normally point directly towards each other, leading to a straight bond; if steric considerations prevent this, the bonds will be bent and reduced somewhat in strength. A double bond is a combination of a σ bond and a π bond; a triple bond is $\sigma\pi^2$. In all this it makes relatively little difference whether we are using the VB model or the localized MO model. The important thing is that the valence shell of an atom should contain singly occupied AO's (or hybrids), and the valence number of an atom is the number of such orbitals in either its normal ground state or some readily accessible valence state. Let us take the groups in order, beginning with the alkali atoms.

7.9.1. Group I: the alkali atoms

The characteristic of all group I atoms is that they have one valence electron outside an inner closed-shell structure. In the case of lithium this is a 2s electron, with sodium it is a 3s, and so on. With one unpaired valence electron we expect monovalence, and this is just what we find. Examples are Li_2 and LiF. The outermost orbital is very diffuse and its ability to overlap with an orbital from some other atom is small. Consequently covalent bonds involving these atoms are weak and long. Stronger bonds occur in polar molecules, where the group I atom has largely given up its valence electron, and, then being much smaller in size, can approach more closely to the ligand. This situation is revealed by the very large dipole moments of such diatomic molecules. Thus for the diatomic alkali halides CsF, KBr, and NaCl the observed dipole moments are 7·9, 9·1, and 8·5 D.† These may be compared with water (1·8 D) and ammonia (1·5 D).

---

† Although not a 'permitted' unit in the SI, the Debye is still widely used owing to its convenient magnitude. $1\,D = 3\cdot334 \times 10^{-30}\,C\,m$.

TABLE 7.3
Hybridization in alkali-metal diatomic molecules

Molecule	Li_2	Na_2	K_2	Rb_2	Cs_2
Percentage p character	14·0	6·8	5·5	5·0	5·5

Since the s-type valence electron occupies a large-size orbital with no directional character, it can overlap reasonably well with a number of neighbours simultaneously located around it. Partly for this reason these atoms have a strong tendency to form metals, in which the bonds are completely delocalized (Chapter 11).

In small molecules, such as Li_2, a group I atom tends to show a small degree of s–p hybridization. The percentage of p character in diatomic molecules of this group is indicated in Table 7.3. These estimates are due to Pauling (1949). Very few *ab initio* calculations have been made for diatomic molecules of group I elements and it is difficult to obtain satisfactory dissociation energies without considerable elaboration of the wavefunction. The bonds are thus essentially s bonds, as the considerations of Chapter 4 would suggest. The charge cloud for Li_2 has already been shown (Fig. 6.1); it comprises two 'cores' embedded in the $2\sigma_g^2$ distribution. If we remove one of these electrons, to obtain Li_2^+, the bonding power of the one remaining $2\sigma_g$ electron is sufficient to keep the molecule intact. Indeed the mutual repulsion between the $2\sigma_g$ electrons in Li_2 is about equal to the bonding power of each, so after removing one of them the dissociation energy is not much altered. In fact the value for Li_2^+ is 1·55 eV, slightly greater than that for Li_2 (1·12 eV). This indicates very clearly how weak these bonds are.

7.9.2. Group II

Atoms in this group (Be, Mg, Ca, ...) are characterized by a valence configuration $(ns)^2$, and would therefore have a 'normal' valence number of zero. The np AO, however, is not far above ns and all three np AO's may be expected to participate in the bonding. As an illustration of how *ab initio* calculations can contribute to our understanding of unusual valence situations, it is interesting to consider beryllium oxide. The best available MO calculations (McLean and Yoshimine 1967) for BeO ($n = 2$) yield an electron configuration $BeO[1\sigma^2 2\sigma^2 3\sigma^2 4\sigma^2 1\pi^4]$ which is in accord with the $^1\Sigma$ spectroscopic ground state. The 1σ and 2σ MO's serve to describe the two K shells, but the presence of two π bonds clearly suggests a beryllium valence state with configuration $2p^2$, the electrons being paired with two oxygen 2p electrons to yield $(1\pi_x)^2(1\pi_y)^2$. The 3σ and 4σ MO's provide a σ bond and a lone pair, the former containing some Be 2s and the latter mainly O 2s and $2p_z$. Since our description implies that these four electrons were provided by the oxygen atom

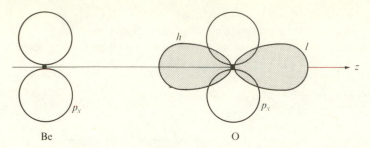

Fɪɢ. 7.21. Possible representation of valence states in BeO. The oxygen lone pairs are shaded; all 2p AO's are singly occupied. In the molecule h mixes with the (empty) Be 2s orbital to give the 3σ bonding MO. Overlap of the 2p AO's (two pairs) give $1\pi_x$ and $1\pi_y$ bonding MO's.

(originally $2s^2 2p_z^2$) the σ bond might be described as 'dative' (Fig. 7.21); of course the description is formal, referring only to the way in which we imagine the molecule to be 'assembled'. Similar considerations apply to Mg and Ca, but with Sr and heavier metals δ bonding also occurs. The bonding in diatomic molecules containing group II elements is certainly not simple.

There are, however, many linear molecules such as MgF_2, $MgCl_2$, and $HgMe_2$ in which the valence states are more familiar. Thus Hg (a group IIB element since although it has a $(6s)^2$ outer shell it also possesses a filled 5d shell) could form linear mercury dimethyl by mixing 6s and 6p AO's to give two digonal hybrids; these could overlap with the singly occupied carbon AO's of two methyl radicals to give the observed conformation. There is some support for this picture in that the dissociation of one Hg—Me bond requires $213\,kJ\,mol^{-1}$ while the second requires only $23\,kJ\,mol^{-1}$, the explanation being that after removal of one group the electronic structure around the Hg atom 'relaxes' and much of the promotion energy is recovered, the remaining bond being weakened.

An interesting difference between the group IIA and IIB elements, such as Ba and Hg, is that in the former the nearest-lying d shell is empty while in the latter it is full. Thus Hg valence states arise only from the promotion $6s \rightarrow 6p$, but in Ba the lowest energy promotion is $6s \rightarrow 5d$. Consequently the preferred valence states will differ. Whereas s–p hybridization favours linear mercury compounds, combinations of s and, say, a d_{xy} lead to hybrids inclined at $90°$; thus barium compounds such as BaF_2 are expected to be angular, not linear. These differences now seem to be well understood (Wharton, Berg, and Klemperer (1963). *Ab initio* calculations by Gole, Siu, and Hayes (1973) lead to the same conclusions).

7.9.3. *Group III*

Group III atoms have ground states with the valence electron configuration $(ns)^2(np)$, so their natural valence number is 1. Monovalence is found in BH,

BF, BBr, and BCl. More frequently, however, the atom is promoted to a valence state based on $2s2p_x2p_y$; then there will be three bonds. If all three are equivalent, as in BF_3 and BCl_3, we shall use s, p_x, and p_y to form trigonal hybrids, as in Fig. 7.8 (*b*). Then we expect a planar molecule with a threefold axis of symmetry. This is precisely what is found.

However, as soon as a fourth orbital is used the atom ceases to be trigonal. Thus, just as NH_3 can add a proton to give tetrahedral ammonium NH_4^+ (Fig. 7.18 (*d*)), so BF_3 can add a fluoride ion to give tetrahedral BF_4^-. Thus we also find BH_4^- in crystalline $NaBH_4$, KBH_4, and $RbBH_4$. In all cases the four electron pairs adopt the expected tetrahedral arrangement around the group III atom.

The tendency of a boron atom to use its fourth (empty) valence orbital and acquire a closed-shell octet of electrons is shown not only in the ions just described but also in, for example, the ammonia–borine molecule $H_3N \cdot BH_3$ in which the axial nitrogen lone pair (Fig. 7.18 (*b*)) approaches along the axis of the BH_3 group and, with participation of the boron $2p_z$ AO (which then hybridizes), becomes a B—N 'dative' bond. The resultant molecule thus establishes tetrahedral character at both ends of the axial B—N bond; this bond is polar in the sense N^+—B^-, indicating the partial donation of the nitrogen lone pair to the boron, but is not essentially different from any other covalent bond. Aluminium has analogous properties.

Group III elements do, however, also form a rich variety of compounds which we are not yet ready to deal with, many of the bonds being delocalized over more than two centres. We therefore defer further consideration of these elements to a later section (§12.3).

7.9.4. *Group IV*

Although the group IV electron configuration $(ns)^2(np)^2$ suggests a natural valence number of 2, we know already that with four electrons in the valence shell a tetrahedral valence state is sometimes strongly favoured with one electron in each hybrid. A common valence number is therefore 4. Just as carbon forms tetrahedral CH_4, silicon and germanium also form tetrahedral hydrides SiH_4 and GeH_4. We have also considered in some detail the trigonal (sp^2) and digonal (sp) valence states of carbon (§7.6), which are appropriate in planar or linear molecules where π bonding is present.

The heavier group IV elements, such as tin, are less inclined to show a valence of 4. Thus, stannous chloride $SnCl_2$ is a V-shaped molecule in which the Sn valence configuration may be described as $(l)^2(h_1)(h_2)$ where the lone-pair orbital l has a large s content ($l \sim 5s$), while h_1 and h_2 are probably not far from pure 5p AO's; this gives a deformed trigonal arrangement of two bond pairs and one lone pair, with an empty π orbital normal to the molecular plane. The increased size of such atoms (Sn ~ 0.140 nm, cf. C ~ 0.077 nm) is probably the significant factor; diminished repulsions between more distant bond pairs imply less 'need' for hybridization and an increasing tendency for the ns pair to be left undisturbed.

Another characteristic of the elements beyond carbon is their capacity to increase their valence numbers beyond 4 owing to the availability of d orbitals. Thus silicon forms the octahedral ion SiF_6^{2-} in which, with the provision of two extra electrons, the silicon (Si^{2-}) valence state may be pictured as a set of six d^2sp^3 hybrids (cf. SF_6 in Fig. 7.14).

Again, group IV atoms form compounds in which non-localized bonding is also important—in particular, the aromatic molecules of organic chemistry. Such compounds are considered in later chapters.

7.9.5. Group V

With a fifth electron in the valence shell, giving the ground state configuration $(ns)^2(np)^3$, one orbital must always be doubly occupied whatever type of s–p hybridization is invoked. The expected valence number is thus 3, and for the lighter atoms, where s–p promotion energy is low, the three bond pairs and one lone pair are expected to adopt a slightly distorted tetrahedral arrangement as in Fig. 7.18. For the larger atoms, P, As, Sb, ..., hybridization again becomes progressively less 'profitable' (as in group IV), the lone pair becoming more s like and the bond hybrids more p like. Consequently the valence angles approach 90°, as indicated in Table 7.4 which gives the experimental results.

TABLE 7.4
Valence angles in group VB hydrides

Molecule	NH_3	PH_3	AsH_3	SbH_3
Valence angle (deg)	107	93·5	92	91

We have already remarked on the readiness with which the lone pair in NH_3, projecting to the 'rear' of the nitrogen, will accept a proton to yield the tetrahedral ion NH_4^+ (see Fig. 7.18). In this process the N—H bond lengths change by less than 2 per cent, supporting our general picture of strongly localized and largely independent bonds.

The mutual repulsions among lone pairs and bond pairs, discussed in the last section, are of course not confined to the valence shell of a given atom. They are also effective in determining the shapes of larger molecules such as hydrazine H_2N—NH_2 where each NH_2 group is essentially tetrahedral but the eclipsed conformation is destabilized by the lone-pair repulsions; as a result rotation around the N—N bond occurs, so as to increase the separation of the lone pairs.

The group IV elements in the second and succeeding rows of the periodic table can show higher valence numbers as a result of d-orbital participation. Thus the penta-co-ordinated compounds PF_5, PCl_5, AsF_5, and $SbCl_5$ are all known and are all trigonal bipyramidal as in Fig. 7.15. The fact that the axial bonds in, for example, PCl_5 are weaker and longer (0.219 nm) than the equatorial bonds (0.204 nm) is not unexpected, since p–d mixing requires larger promotion energy than s–p and will, indeed, only occur to a significant degree if the 3d AO can be contracted (p. 209). An alternative picture of this type of axial bond, not involving any d orbitals at all, will be given later. Which picture is the more accurate cannot be established without very detailed calculations.

7.9.6. *Group VI*

The atoms O, S, Se, Te, and Po all have the ground-state configurations $(ns)^2(np)^4$, and of the four orbitals set up by any kind of s–p hybridization *two* must be doubly occupied. We therefore expect the valence shell to contain, in the molecule, two bond pairs and two lone pairs with mutual repulsions favouring a roughly tetrahedral arrangement. The molecule H_2O (Fig. 7.18) has already been considered from this point of view; the other extreme point of view, ignoring s–p mixing altogether, was the one with which this chapter opened (§7.1), which suggested a 90° HOH angle. The observed angles in the group VIB dihydrides are shown in Table 7.5. As in the case of group V (Table 7.4), hybridization rapidly becomes less important for the heavier atoms.

TABLE 7.5

Valence angles in group VIB hydrides

Molecule	H_2O	SH_2	SeH_2	TeH_2
Valence angle (deg)	104·5	93	91	89·5

Again, it must be emphasized that there is no *single* obvious reason for the geometry changes in the Table 7.5, which arise from a delicate balance of competing factors as discussed in §7.8, but comparison with Table 6.5 suggests the dominance of increasing electronegativity in going along a row (e.g. N to O, or P to S), which pulls the bond pairs closer to the nucleus, and the dominance of the 'size effect' (referred to under group IV) in going down a column. The tetrahedral description of the electron pairs in H_2O readily suggests the structure of the oxonium ion H_3O^+ which (like NH_4^+) may be regarded as the result of protonation of one of the lone pairs; the resultant shape will clearly be pyramidal rather than planar.

Again higher valencies exist when d orbitals are within reach. Thus $TeCl_4$ (Fig. 7.19) arises from a bipyramidal arrangement of one lone pair and four bond pairs, based on five dsp^3 hybrids, while TeF_6 (Fig. 7.20) is a regular octahedral molecule corresponding to a valence state with six singly occupied d^2sp^3 hybrids. Sulphur also forms the octahedral SF_6, although in an isolated S atom the 3d AO's lie well outside the 3s–3p shell and considerable contraction (p. 209) would thus appear to be necessary. Such contraction would certainly be encouraged by flow of electrons from the valence shell towards the highly electronegative fluorine ligands, the effective central field of the electron-deficient sulphur thus being increased, but completely definitive calculations of such effects have yet to be performed. It is even possible to put forward bonding schemes which do not involve d-orbital participation and it may be some time before we can decide which pictorial model is the more accurate.

7.9.7. Group VII

The halogens, F, Cl, Br, and I, all have one unpaired electron in an otherwise filled shell; $(ns)^2(np)^7$. The normal valence number is 1, and there is an abundance of compounds such as HF, F_2, and ClF to confirm this monovalency. Since the s–p interval for atoms near the end of any row of the periodic table is large, hybridization is inhibited and bonding usually involves an almost unmodified p orbital. Moreover, the bonds with less electronegative atoms are often fairly strongly polar. Thus in HF the observed dipole moment would suggest formal charges of $+0.4e$ on the hydrogen and $-0.4e$ on the fluorine, and Fig. 6.6 has already shown us how closely such a molecule can resemble a bare cation embedded in an F^- charge cloud. The polarity of such bonds decreases, however, with decreasing electronegativity along the series F, Cl, Br, I.

Although the halogens are less versatile than some of the earlier groups, they do form a number of remarkable molecules such as ClF_3, BrF_5, and even IF_7. The T-shaped molecule ClF_3 may be understood in terms of dsp^3 hybridization, giving three bond pairs and two lone pairs as in Fig. 7.19, and the square pyramidal BrF_5 could be described similarly (Fig. 7.20) in terms of the octahedral system. When dealing with an atom as electronegative as fluorine, however, an alternative possibility is to invoke an ionic model, drawing on the ideas of VB theory; it is useful to mention the approach here as we shall use it later in trying to understand the bonding in rare-gas compounds (§12.5).

We begin with a normal Cl—F molecule with the σ bond formed from, let us say, the singly occupied chlorine $3p_z$ and fluorine $2p_z$ AO's. Now let us take an electron from the chlorine $2p_x$ pair and give it to a second F atom to form F^- and ClF^+; the ClF^+ is then certainly able to form a normal σ bond with the *third* atom (to give a right-angled ClF_2^+) and we can imagine the whole system to be assembled as in Fig. 7.22 (a). Clearly the arrangement is lop-sided and we could not obtain two equivalent ClF bonds in this way; the only way to do so is to recognize that the arrangement in Fig. 7.22 (b) is equally satisfactory and that, regarded as VB structures (§5.3), they could be used in a 'resonance mixture' to

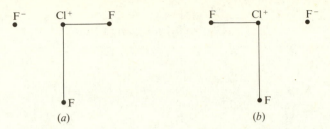

FIG. 7.22. VB description of bonding in ClF_3: (a) and (b) are alternative ionic structures.

describe a perfectly symmetrical T-shaped molecule, with two somewhat weaker, largely ionic, bonds in the linear F—CI—F region. This fits the observations (the bond length being 0·170 nm compared with the covalent bond length of 0·160 nm) and at the same time avoids hybridization involving d orbitals; the price we pay for the two extra bonds is an ionization energy, this time, instead of a promotion energy. Only detailed calculation can show which model is the more realistic. An *ab initio* discussion has been given by Guest, Hall, and Hillier (1973).

7.9.8. Group VIII

The rare-gas atoms He, Ne, Ar, Kr, Xe, and Rn all have the closed-shell configurations $(ns)^2(np)^6$ and are accordingly expected to have a valence number of zero; they are commonly called the 'inert' or 'rare' gases. During the last 15 years, however, a considerable number of compounds have been prepared and a whole field of 'rare-gas chemistry' has developed. These compounds are so exceptional as to deserve a section to themselves (§12.5).

7.10. Atomic radii, bond lengths, and bond energies

Throughout this chapter we have been concerned with molecules in which the bonds are effectively localized, each with its own individual characteristics, and it is not surprising that there have been many attempts to compile tables of bond properties and to relate such properties to those of the bonded atoms. We briefly examine two such properties—bond lengths and bond energies.

The near constancy of the length of a bond A—B suggests that we might try to determine empirical atomic radii, r_A and r_B, such that $r_{AB} = r_A + r_B$ for all the elements. Thus the bond length in the fluorine molecule is 0·142 nm and this suggests that we take $r_F = 0·071$ nm as the (covalent) radius of the fluorine atom. Of course deviations are bound to occur and we must not expect one set of empirical values to fit the bond lengths in all molecules, but the radii listed in Table 7.6 prove reasonably successful. The values all relate to σ-type single bonds of predominantly covalent character.

TABLE 7.6

Covalent radii† for atoms (in nm)

H					
0·030‡					
Li	B	C	N	O	F
0·134	0·081	0·077	0·075	0·073	0·071
Na	Al	Si	P	S	Cl
0·154	0·130	0·118	0·110	0·102	0·099
K		Ge	As	Se	Br
0·196		0·122	0·121	0·117	0·114
Rb		Sn	Sb	Te	I
0·211		0·140	0·143	0·135	0·133

† Inferred mainly from the bond length data referred to on p. 187. For a more extensive compilation see J. E. Huheey (1972).

‡ This value is applicable except for H_2 (which corresponds to a radius of 0·037 nm).

Table 7.6 may be used at once to estimate the bond lengths in most molecules, but deviations will occur (i) when double or triple bonding, or some fractional π-bond character (§8.3), is present, (ii) when the atoms differ greatly in electronegativity and the bond is strongly polar (i.e. has a large ionic character), and (iii) when the hybridization ratio varies. Such deviations are all of interest and shed light on the nature of the bonding.

The first effect is important for C, N, and O. The radii are reduced to 0·067, 0·061, and 0·057 nm, respectively, if a π bond is present, while the C and N radii fall to 0·060 and 0·055 with *two* π bonds (i.e. in a triple bond). We shall find later (§8.5) that a *fractional* π bond, arising from non-localized bonding, also produces a shortening of the 'normal' single bond.

The presence of ionic character also produces a shortening and this is still commonly represented by an approximate formula due to Schomaker and Stevenson (1941) (see also Pauling (1960), pp. 228–30)

$$r_{AB} = r_A + r_B - \beta |x_A - x_B|$$

where x_A and x_B are Pauling electronegativities and $\beta \simeq 0.009$ nm. In extreme cases, particularly in ionic crystals (§11.6) where the individual units are essentially anions and cations, this approach breaks down and it is then desirable instead to set up tables of *ionic* radii for the various anions and cations which commonly occur.

The third effect, involving change of hybridization, was studied by Coulson who used the position of the charge-density peak to give a measure of the extent to which an s–p hybrid projected from its nucleus. With this surprisingly simple model he was able to account for quite subtle variations among the lengths of C—C bonds in organic molecules, as we shall find in §8.2.

Variations of bond length also indicate variations of bond strength, but such variations are usually small (unless accompanied, for example, by major changes of valence state), and the high degree of independence of the localized electron pairs suggests that tables of *bond energies* might also be compiled in such a way that the energy of dissociation of a molecule could be obtained as a sum of the bond energies of all the bonds present. Although such additivity rules are of considerable value, we have already observed that the energy required to break the second of two similar bonds is not necessarily the same as that for the first, for the electron distribution 'relaxes' and the valence state of any atom may change appreciably with each bond broken. Thus the energy required to strip the carbon atom in methane of its four hydrogens is $425 \, kJ \, mol^{-1}$ for the first, $470 \, kJ \, mol^{-1}$ for the second, $415 \, kJ \, mol^{-1}$ for the third, and $335 \, kJ \, mol^{-1}$ for the last—altogether $1645 \, kJ \, mol^{-1}$. Since *in the methane molecule* the bonds are identical (and likely to be similar to those in other saturated molecules with C atoms in sp^3 valence states), we agree to associate a bond energy of $\sim 411 \, kJ \, mol^{-1}$ with such a C—H bond, and then we shall obtain the right dissociation energy for the whole molecule into its constituent atoms. Some bond energies arrived at by this kind of argument for covalent single bonds of the type X—X are given in Table 7.7. These bond energies may be used to estimate those of heteropolar bonds using the electronegativity considerations of §6.4.

TABLE 7.7
Bond energies† for bonds X—X (in $kJ \, mol^{-1}$)

H							He
432							0
Li	Be	B	C	N	O	F	Ne
105	208‡	347	346	167	142	130§	0
Na	Mg	Al	Si	P	S	Cl	A
72	130		222	201	268	240	0
K	Ca		Ge	As	Se	Br	Kr
49	105		188	146	172	190	0
Rb	Sr		Sn	Sb	Te	I	Xe
45	84		146	121	126	149	0
Cs							
44							

† Based mainly on up-dated values given by Darwent (1970). See also Dasent (1970) for additional values and further discussion.

‡ Huheey and Evans (1970).

§This value is still in dispute; $130 \, kJ \, mol^{-1}$ has been obtained by Dibeler, Walker, and McCulloh (1970). The previously accepted value was $155 \, kJ \, mol^{-1}$.

TABLE 7.8

Multiple-bond energies† (in kJ mol⁻¹)

	Single bond	Double bond	Triple bond
C—C	346	602	835
N—N	167	418	942
O—O	142	494	
C—N	305	615	887

† Based on values given by Darwent (1970).

As in the case of bond lengths, the above considerations must be modified when multiple bonding occurs. Some single, double, and triple bond energies are given in Table 7.8. Fortunately, extensive tables are not necessary since multiple bonding is uncommon except for a few atoms in the first two rows of the periodic table. This in itself is an interesting phenomenon, closely related to the size of the bonded atoms. For low atomic numbers the number of electrons which can engage in bonding may be a rather large proportion of the total; multiple bonds can occur and the relatively small inner shells are pulled together in spite of their mutual repulsion. For heavier atoms the number of electrons available for bonding is relatively small, while the number of non-bonding electrons and the size of the repulsive cores are both growing; the prospect of effective multiple bonding thus apparently diminishes.

Problems

7.1. Verify that p-type AO's have vector properties (equation (7.3)). (Hint: take $p = x'f(r)$ where x' is the x co-ordinate of a point referred to the *arbitrary* axis in the xy-plane; then express x' in terms of co-ordinates relative to the fixed x and y axes.)

7.2. Express the sp^2 and sp^3 hybrids, defined in (7.25) and (7.26) and shown in Fig. 7.8 (b, c), in terms of p_x, p_y, and p_z. (Hint: note the directions, relative to the given co-ordinate system, and use the vector property.)

7.3. Set up three p-type AO's pointing from the centroid towards the vertices of an equilateral triangle in the xy-plane, and show that by mixing some of p_z with each of them they may be bent out of plane, so as to point along the edges of a trigonal pyramid. How is the angle between each pair determined by the coefficient of p_z? (Hint: the angle between two unit vectors, with x, y, z components l_1, m_1, n_1 and l_2, m_2, n_2, is given by $\cos \theta = l_1 l_2 + m_1 m_2 + n_1 n_2$. Don't forget to normalize the orbitals.)

7.4. Discuss the ammonia molecule along the lines of §7.5, using hybrids constructed from the p orbitals of Problem 7.3. What value of λ in $h = s + \lambda p$ is consistent with mutually orthogonal hybrids inclined at the observed bond angle of 107°? What can you say about the form of the nitrogen lone pair?

7.5. How might the valence state of the nitrogen atom in NH_3 (Problem 7.4) be described in terms of the occupation of the s and p AO's? (Hint: use the same procedure as for the water molecule (p. 202).)

7.6. Verify equation (7.32) and express the six octahedral $d^2s\,p^3$ hybrids in terms of s, p, and the standard set of five d orbitals. (Hint: write (7.32) in terms of x^2, y^2, z^2, xy, yz, zx, and compare coefficients of corresponding terms.)

7.7. Set up hybrids for a square planar complex in the form

$$h_{\pm x} = N(s + ad_{x^2-y^2} \pm bp_x)$$

with similar expressions for $h_{\pm y}$, and show that the mixing coefficients a and b are fixed by orthogonality requirements. Hence derive the dsp^2 hybrid of equation (7.33).

7.8 Use the electronegativity values in Table 6.5 to discuss the probable character of the bonds in SF_6, assuming the six electron pairs are formed by overlap of octahedral hybrids and fluorine 2p AO's. How many electrons would be associated with the valence shell of the sulphur atom *in the molecule*, and with the s, p, d shells individually? Is the net charge predicted on the sulphur atom likely to be realistic or exaggerated? How might your simple discussion be improved? To what experimental information would you turn for support of your predictions? (Hints: you will need (6.35) (drop the quadratic term); (6.22) (neglect S); and the octahedral hybrids obtained in Problem 7.6. Try including $S = \frac{1}{2}$ in (6.22) and using the gross populations (6.38). Read again about variable electronegativity (pp. 166–170).)

7.9. Use the arguments advanced in §7.8 to discuss the probable electronic structures and resultant shapes of the following molecular species:

CH_2 (bent), CH_3 (flat), NH_3^+ (flat), H_3O^+ (pyramidal), CO_2 (linear), CF_2 (bent).

(Hints: (i) half a lone pair is not much better than no lone pair at all, (ii) the effect of π bonding is usually of secondary importance.)

8
Carbon compounds

8.1. Nature of the bonds in simple organic molecules

The carbon atom exhibits, as we have already remarked, an extraordinarily rich variety of valence capabilities, and the whole of organic chemistry derives from its properties. Carbon compounds therefore certainly deserve a chapter to themselves.

Three main types of carbon–carbon bond may be distinguished at the outset, in all of which carbon exhibits its normal quadrivalence; the typical single bond occurs in ethane, the double bond in ethylene, and the triple bond in acetylene.

In ethane (Fig. 8.1 (a)) the bond angles are not far from the tetrahedral value ($109° 28'$) and the carbon atom may thus be regarded as in the tetrahedral (sp^3) valence state, essentially as in the methane molecule (p. 200). The C–C and C—H bonds may be perfectly well accounted for in terms of localized electron-pair wavefunctions, as discussed in Chapter 7; the two electrons of the C—C bond could occupy the bonding MO constructed from two overlapping sp^3 hybrids, while those of each C—H bond could occupy one formed from the overlap of its hydrogen 1s AO with a corresponding sp^3 hybrid (Fig. 7.8 (c)). All bonds are then single and of σ type. The considerations of §7.8 suggest (correctly) that the equilibrium geometry will be the 'staggered' or *trans* form in which one CH$_3$ group is rotated through 60° with respect to the other. The origin of the 'barrier' for twisting around the C—C axis has, however, been a matter of controversy for many years; we return later to this subject (§12.6). As usual, the electron-pair bonds may be described at least equally well by wavefunctions of Heitler–London form, and it does not matter much whether we use the language of VB theory or of MO theory.

In ethylene the bond angles are close to 120°, the molecule being planar, and similar considerations apply, but with trigonal (sp^2) hybridization instead of tetrahedral. The in-plane σ bonds involve the carbon sp^2 hybrids, but the odd 2p electron remaining on each carbon participates in a π bond, the two AO's forming a localized MO of π type—antisymmetric across the molecular plane (Fig. 8.1 (b)). The carbon–carbon bond is thus double, one σ bond and one π bond. Even without a more elaborate theory it is at once clear why the molecule is flat with the two CH$_2$ groups lying in the same plane, for if one group were twisted relative to the other the alignment of the two 2p AO's of the

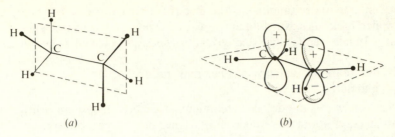

FIG. 8.1. Ethane and ethylene. (*a*) In ethane each carbon is in a tetrahedral valence state forming four σ bonds. The conformation is 'staggered', diagonally opposite CH bonds lying in the plane indicated. (*b*) In ethylene each is in a trigonal state, forming three σ bonds. The remaining electrons provide a π bond whose MO, formed by overlap of the 2p AO's shown, is antisymmetric across the molecular plane.

π-type MO would be destroyed and their mutual overlap would be substantially reduced; the σ-type MO's, and their corresponding energies, would not be much changed (assuming the sp² hybridization at each carbon to be maintained) but the bonding power of the π-type MO would be seriously impaired, i.e. its energy would rise. A plot of energy against angle of twist (θ) would thus show a *minimum* at $\theta = 0$, indicating a *planar molecule* in the equilibrium conformation. Indeed, the π classification of an orbital, strictly speaking, can only be applied when a molecule contains a plane of symmetry across which the wavefunction may be symmetric or antisymmetric; on twisting, the distinction between σ and π is blurred and finally lost.

In acetylene the atoms are collinear; the appropriate carbon hybridization is evidently digonal (sp) and the carbon-carbon triple bond resembles that in the nitrogen molecule (p. 101), consisting of one σ bond and two π bonds.

The two main categories of organic molecule are (1) *saturated* compounds, in which the carbon atoms each form *four* bonds of essentially σ type (i.e. the four valencies are 'saturated'), and (ii) *unsaturated* and *conjugated* compounds, in which one or more carbon atoms form only *three* σ bonds, the fourth valence electron being involved in somewhat weaker π bonding. The distinction we are making here is not 'chemical' so much as electronic; ethane is saturated and ethylene unsaturated, in terms of the nature and disposition of the electron pairs around the carbon atoms. There are, of course, corresponding differences of chemical behaviour between molecules of the two categories arising from their electronic differences.

Quantum theory has been most successful, so far, in accounting for the distinctive properties of unsaturated molecules, particularly conjugated hydrocarbons and other planar molecules in which the π bonding is 'delocalized' over a large region. Consequently, most of this chapter will be devoted to such π-electron systems. In recent years, however, with the increased availability of large computers, significant progress has also taken place in our understanding of the electronic structures and properties of saturated molecules, which in many ways provide a more exacting test of the theory. In later sections we shall give a brief account of such developments.

8.2. Heteroatoms, valence states, and molecular geometry

Before turning to the detailed treatment of π-electron systems a little more must be said about the underlying 'framework' of carbon and any other 'hetero' atoms over which the π electrons move. In ethylene, for example, the planar H_2C—CH_2 array is well described in terms of localized electron-pair bonds, essentially of σ type, as discussed in Chapter 7; the π electrons, which in ethylene provide the double bond, merely stabilize and 'stiffen' the σ-bonded framework and help to keep it flat. Although the framework bonds have rather constant properties, there are small variations, for example in the lengths of C—H bonds and the interbond angles, which are of considerable interest and are intimately related to the valence states of the atoms concerned. The appropriate valence states are largely determined by the geometry of the molecule and should always be established at the outset. In a saturated molecule, of course, there are no π electrons and the framework may not be planar, but in all cases it is important to establish at once the valence states of the atoms concerned. The following examples indicate common situations.

(i) *Pyridine.* One C—H of a benzene ring is replaced by a nitrogen atom (Fig. 8.2 (*a*)); all bond angles are close to 120°. The appropriate valence state is trigonal (sp^2) for both C and N, but the latter, with five valence electrons, must have one lone pair (sp^2).

(ii) *Pyrrole.* Again the molecule is planar (Fig. 8.2 (*b*)) and the angles around the nitrogen indicate roughly trigonal hybridization; the presence of three bonds in the plane (hence three singly occupied hybrids) shows that the lone pair is of π type, i.e. there are *two* electrons available for π bonding. As the ring is not hexagonal the carbon valence states, whilst basically trigonal, must be slightly deformed with two of the hybrids inclined at somewhat less than 120°.

(iii) *Amides.* The molecules are planar and are of the type

$$R-C \underset{\diagdown NH_2}{\overset{\diagup O}{}}$$

where the bond angles are all close to 120°. The appropriate valence states are as indicated in Fig. 8.2 (*c*); the nitrogen thus has two electrons available for π bonding, and the oxygen has one.

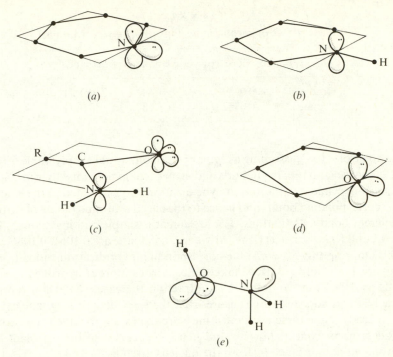

FIG. 8.2. Relationship of valence states to molecular geometry. (*a*) Nitrogen in pyridine forms two C—N bonds in the molecular plane, leaving one lone pair and one singly occupied π-type AO. (*b*) Nitrogen in pyrrole forms three coplanar σ bonds, leaving *two* π electrons. (*c*) Nitrogen and oxygen in an amide (planar) provide two and one π electrons respectively. (*d*) Oxygen in furan forms two C—O bonds in the molecular plane, leaving a σ lone pair and two π electrons. (*e*) Oxygen and nitrogen in hydroxylamine, in tetrahedral valence states suggested by non-planar molecular geometry and repulsion of lone pairs.

(iv) *Furan*. The five-membered ring (Fig. 8.2 (*d*)) again suggests roughly trigonal hybridization at every ring atom, and the absence of an external atom bonded to the oxygen implies an sp^2 lone pair, two singly occupied hybrids for the C—O bonds, and two electrons in the π-type 2p AO.

(v) *Hydroxylamine*. Here the molecule is not planar, bond angles suggesting (Fig. 8.2 (*e*)) tetrahedral valence states for both N and O. The lone pairs will then be as indicated, the equilibrium molecular geometry being largely dictated by their mutual repulsions.

In all the above examples, the geometry of the molecule suggests hybridizations which, although sometimes departing from regularity (some hybrids being inequivalent), are certainly compatible with the orthogonality constraints considered in §7.5. For example, if we point the carbon hybrids in pyrrole directly towards their neighbours, to achieve maximum overlap, the

TABLE 8.1

Covalent radius of carbon in various valence states (in nm)

	acetylene (sp)	ethylene (sp^2)	methane (sp^3)
Observed	0·0739	0·0764	0·0771
Calculated	0·0732	0·0758	0·0771

bond angle of $\sim 108°$ (assuming a regular pentagon) would simply imply 30 per cent 2s character in the ring hybrids and 40 per cent in the radial hybrids. Such geometry-dependent variations in hybridization may be related to the lengths of the corresponding bonds and hence to the effective atomic radius of carbon in different bonding situations. The argument is simple and ingenious.† The C—H bond lengths in acetylene, ethylene, and ethane are 0·1061, 0·1086, and 0·1093 nm, respectively, while the corresponding hybridization ratios (λ in $s + \lambda p$) are 1, $\sqrt{2}$, and $\sqrt{3}$. If we take the covalent radius of hydrogen in such bonds as 0·0322 nm we obtain by subtraction the effective carbon radii in Table 8.1. This adjusted value ensures that the C—C length in ethane comes out correctly. Now these radii must measure the extent to which the carbon hybrid projects from its nucleus and hence be dependent on λ. In fact, the centroid lies at a distance ($\bar{r}$) from the nucleus given by

$$\bar{r} = \frac{1 + (4/\sqrt{3})\lambda + (3/2)\lambda^2}{1 + \sqrt{3}\lambda + \lambda^2} \times \text{constant} \tag{8.1}$$

where the 'constant' depends on the orbital forms but not on λ. If we suppose the covalent radius (r_c) is proportional to $\bar{r}$, fixing the constant so as to obtain a perfect fit for ethane, we obtain the 'theoretical' values shown in the last line of Table 8.1. It is apparent that we can attribute shortening of the σ bond, with fair quantitative success, to an increase in the s character of the carbon hybrid concerned. It is also clear that the 'natural' lengths of C—C single bonds (before any allowance is made for the shortening effect of π bonding) will also be dependent on the carbon hybridization ratio; these natural lengths in typical compounds should be

ethane, 0·1542 nm   ethylene, 0·1528 nm   acetylene, 0·1478 nm

and we shall later be able to use these standard lengths in discussing the shortening which may arise from additional π bonding.

A final comment concerns molecules in which we apparently *cannot* satisfy the orthogonality constraints of (7.23) in the sense that we cannot find λ values for which the orthogonal hybrids at a given atom point towards its

† Coulson (1948). In the present discussion we have used improved values of the experimental bond lengths.

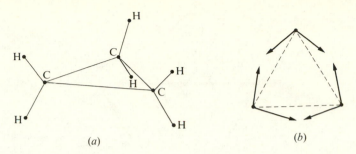

FIG. 8.3. Structure and bonding of cyclopropane: (*a*) carbons of the three CH_2 groups form an equilateral triangle, each group lying in a vertical plane; (*b*) directions of the hybrids used in forming the C—C bonds.

neighbouring atoms. Such molecules are said to be 'strained' and to contain 'bent' bonds. The most important example (Fig. 8.3 (*a*)) is cyclopropane, which consists of three carbon atoms at the vertices of an equilateral triangle with the three pairs of H atoms located symmetrically above and below the plane of the triangle, such that the H—C—H angle is approximately 115°. This is distinctly larger than the normal tetrahedral angle of 109° 28′.

It is clear that if we construct four equivalent tetrahedral hybrids at each carbon atom, the pair of hybrids used for each C—C bond will not point directly towards each other and the overlap will be reduced. If indeed we choose the hybrids such that two of them make an angle less than the tetrahedral one, they will overlap a little more effectively for the C—C bonds. Moreover, in order to preserve orthogonality the other two hybrids will make a bigger angle than the tetrahedral one. Now the fundamental equation (7.23) shows that by no real combination of s and p is it possible to obtain a valence angle less than 90°. Some compromise has to be reached. It appears that this occurs when the hybrids for the C—C bonds make a mutual angle of about 100°, and then the C—H bonds, which can remain straight, will make an angle of about 115°, just as found experimentally. The directions of the arrows in Fig. 8.3 (*b*) show the directions in which the in-plane hybrids point. As a result the bonds must be bent and their charge density must be different from that of a straight bond.

A most interesting verification of the existence of these bent bonds has been obtained by Hartman and Hirshfeld (1966), who studied the tri-cyano derivative of cyclopropane. The replacement of three hydrogen atoms by cyano groups will be expected to make very little difference to the electronic charge density in the C_3 plane. In this plane we have three bent C—C bonds. The overlap regions for the hybrids indicated in Fig. 8.3 (*b*) will not lie symmetrically along the lines joining the carbon atoms but outside the triangle formed by them. Thus whereas the difference density diagram for straight bonds should show (as it does in those cases that have been studied) a build-up

FIG. 8.4. Difference density map for tricyanocyclopropane in the C_3 plane. The important overlap regions have been emphasized by shading.

of charge along the line joining the atoms, in the case of bent bonds it should occur off this line. This is exactly what is found, as Fig. 8.4 nicely illustrates.

Thus Baeyer's theory of strain in a molecule has its modern interpretation in terms of an inability to achieve maximum overlap through steric restriction; this leads to a decrease in the overlap binding energy and hence to a smaller total binding energy (or heat of formation).

8.3. Conjugated and aromatic molecules: benzene

In practically all the molecules previously discussed it has been possible to localize the bonds; such delocalizations as were present were quite small. However, there is an extremely large and important class of molecule for which effective localization is completely impossible. These are the conjugated and aromatic molecules which form so extensive a part of organic chemistry.

As a first example, let us consider the most important of all such molecules, namely benzene C_6H_6. This is the prototype for all aromatic compounds. A large amount of evidence from X-ray crystallography and vibrational spectra shows that the carbon atoms lie at the vertices of a regular plane hexagon; the six hydrogen atoms lie in the same plane directed radially outwards from the carbons, in such a way that all the valence angles are $120°$. As we have noted in §8.2, this implies trigonal hybridization at each carbon atom. We must then arrange the directions of the hybrids as shown in Fig. 8.5 so that there is strong overlapping of pairs of orbitals to give localized C—C and C—H bonds, both of σ type. These bonds may be described either in MO or VB language; their

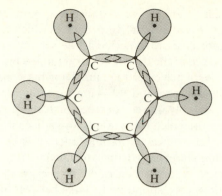

Fig. 8.5. Construction of σ bonds in the benzene molecule (the carbon hybrids have been highly stylized for clarity).

essential character is the same in either case. A simple counting of electrons shows that there are six still to be used; they occupy the unhybridized dumbbell-shaped 2p orbitals, one on each carbon atom. If the plane of the molecule is called the (x, y) plane, these orbitals are the $2p_z$ AO's, all directed perpendicular to the plane of the molecule as shown in Fig. 8.6 (a) and hence referred to as 'π-type' orbitals. It is precisely at this stage that our difficulty arises, for we look for one particularly favoured scheme of pairing these orbitals together, without which localization of the bonds is impossible, and clearly cannot find one.

The nature of our difficulty depends on whether we use the VB or the MO approach. In the former we couple the spins of electrons, which occupy pairs of strongly overlapping AO's, i.e. AO's on adjacent atoms, but this can be done in two equally satisfactory ways to give the VB 'structures' (§5.3) indicated in Fig. 8.6 (b, c). Both structures would apparently be equally eligible as components of the VB wavefunction describing the π electrons; they are the 'Kekulé structures', named after Kekulé (1865) who supposed the double bonds would oscillate between the alternative positions. The quantum-mechanical description, however, requires no oscillation between the two forms; indeed, in spite of the 'resonance' terminology (p. 124) the idea of an oscillation is foreign

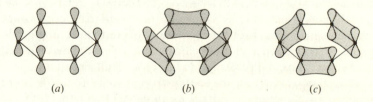

(a) $\qquad$ (b) $\qquad$ (c)

Fig. 8.6. (a) The π-type AO's in benzene and (b, c) two Kekulé pairing schemes.

to that of a stationary state. There is only *one* ground state, represented as some unique mixture of the alternative structures which can be set up, and the properties of the molecule are expected to lie somewhere in between those which could be expected of the individual structures. This is true, for example, of the C—C bond length, which is 0·139 nm, lying between the value 0·154 nm appropriate to a single bond and 0·134 nm for a double bond. The energy, however, lies considerably lower for the mixture than for either of the 'principal structures' (in which the π bonds join adjacent pairs of carbons), as would be expected for a linear variation function (§3.8); in VB theory the resultant stabilization, the energy drop associated with delocalization of the π bonds, is referred to as 'resonance energy'. The energy is lowered even further when other structures, corresponding to links between non-adjacent atoms, are admitted (Fig. 8.7), but the complexity of the calculations is then considerably increased and attention is more commonly confined to the principal structures.

FIG. 8.7. Simple representations of five VB structures for benzene. The first two are the Kekulé structures (Fig. 8.6); the others are 'Dewar' structures.

The MO approach also starts from the set of six π-type 2p orbitals of Fig. 8.6 (*a*), which we shall call $\phi_1, \phi_2, \ldots, \phi_6$, but, instead of setting up many-electron wavefunctions describing structures, it assumes that each electron can be described individually by a wavefunction (MO) represented as a combination of appropriate AO's. However, as the AO's do not fall naturally into pairs, each one overlapping *both* its neighbours, we are compelled to write our MO's in the form

$$\psi = c_1\phi_1 + c_2\phi_2 + \ldots + c_6\phi_6 \qquad (8.2)$$

where the coefficients $c_1 \ldots c_6$ are chosen to make the energy stationary. The significance of (8.2) is that the six electrons have to occupy MO's which extend over all six carbon atoms and are therefore completely delocalized. If we had set up localized MO's such as $c_1\phi_1 + c_2\phi_2$ to represent each of the electron pairs in a Kekulé structure (Fig. 8.6 (*b*, *c*)), the calculated energy would have been much higher. The MO interpretation of 'resonance energy' is thus unequivocally connected with the delocalization which allows each MO to spread over the whole molecule, and in future we shall almost always use the term delocalization energy for the energy drop in going from a (hypothetical) Kekulé-type wavefunction based on *localized* MO's to the MO function without any localization restrictions.

TABLE 8.2

Bonding π MO's in benzene†

	c_1	c_2	c_3	c_4	c_5	c_6
ψ_1	1	1	1	1	1	$1 \times (1/\sqrt{6})$
ψ_2	1	2	1	-1	-2	$-1 \times (1/2\sqrt{3})$
ψ_3	-1	0	1	1	0	$-1 \times (1/2)$

† Table entries in each row should be multiplied by the normalizing factor (extreme right).

To anticipate the results of the next section, there are six MO's of the form (8.2), three bonding and three antibonding. In the ground state the six π electrons fill the three bonding MO's and give the molecule great stability. Values of the coefficients $c_1, c_2, \ldots, c_6$ in these MO's are given in Table 8.2.

The MO of lowest energy, ψ_1, is depicted in Fig. 8.8. It has two lobes, one above the plane of the molecule and one beneath, each of 'doughnut' form; the wavefunction differs in sign in the two regions and vanishes in the molecular plane (which is thus a nodal surface). The other two MO's are apparently less symmetrical, but in fact they are degenerate and their forms are thus not unique—alternative linear combinations of the type $a\psi_2 + b\psi_3$ would also be wavefunctions of the same energy. The most symmetrical combinations of ψ_2 and ψ_3 arise when the coefficients are allowed to be complex numbers (all with the same *magnitude*, $1/\sqrt{6}$), but such MO's are less easy to visualize than those in Table 8.2. The three bonding MO's, in real form, are indicated schematically in Fig. 8.9.

To verify that all the C—C bonds are precisely equivalent, in spite of the apparent lack of symmetry of the MO's, we calculate the charge-density function for the π electrons. This takes the standard form (cf. (6.17))

$$P = P_{11}\phi_1^2 + P_{22}\phi_2^2 + \ldots + 2P_{12}\phi_1\phi_2 + 2P_{23}\phi_2\phi_3 + \ldots \qquad (8.3)$$

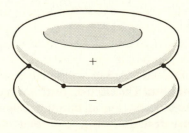

FIG. 8.8. Form of the lowest-energy bonding MO of π type in benzene. (This is *one* MO with two 'doughnut-shaped' lobes separated by a nodal plane which contains the carbon atoms.)

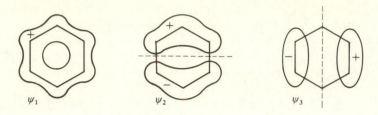

FIG. 8.9. Bonding MO's in benzene. ψ_1 is the MO with the three-dimensional form indicated in Fig. 8.8, while ψ_2 and ψ_3 are a degenerate pair of next-lowest energy. The nodes are indicated by broken lines.

where, since ψ in (8.2) makes a density contribution ψ^2, P_{rr} arises as a sum of squared coefficients (c_r^2) in all the occupied MO's (multiplied by 2 for a *doubly occupied MO*), while P_{rs} arises in a similar way from the products $c_r c_s$. In the context of π-electron theory it is usual to denote P_{rr} by q_r, calling it the π *charge* on atom r, and to denote P_{rs} by p_{rs}, calling it the π *bond order* of link $r-s$. The quantities P_{rs} are often set out in a square array—the 'charge and bond-order matrix' **P**; the charges are the diagonal elements, the bond orders are the off-diagonal elements. It is a simple matter to evaluate these quantities for benzene. Thus, adding up the values of c_1^2 in each of the MO's in Table 8.2 and multiplying by 2 (two electrons in each) we obtain charges

$$q_1 = 2(\tfrac{1}{6}+0+\tfrac{1}{3}) = 1$$

and similarly

$$q_2 = q_3 = 2[\tfrac{1}{6}+\tfrac{1}{4}+\tfrac{1}{12}] = 1.$$

In fact, $q_1, q_2, \ldots, q_6$ all have the same value, unity, and the amounts of π charge associated with all six carbon atoms are thus identical. In the same way, the weights of the overlap terms $\phi_1\phi_2$, $\phi_2\phi_3$, etc. are given by

$$p_{12} = 2\left\{\left(\frac{1}{\sqrt{6}} \times \frac{1}{\sqrt{6}}\right)+\left(0 \times \frac{1}{2}\right)+\left(\frac{1}{\sqrt{3}} \times \frac{1}{2\sqrt{3}}\right)\right\} = \frac{2}{3}$$

$$p_{23} = 2\left\{\left(\frac{1}{\sqrt{6}} \times \frac{1}{\sqrt{6}}\right)+\left(\frac{1}{2} \times \frac{1}{2}\right)-\left(\frac{1}{2\sqrt{3}} \times \frac{1}{2\sqrt{3}}\right)\right\} = \frac{2}{3}$$

etc. and each of the links 1–2, 2–3, 3–4, ... therefore has an electron-density term of exactly the same form. The electron distribution thus exhibits the full six-fold symmetry of the benzene ring, all bonds being precisely similar; there is no trace of a classical structure with the π bonds localized in particular positions.

Our use of the term 'bond order' in referring to the quantities p_{rs} takes on a special significance in dealing with π-electron systems. For a π bond between *two* overlapping 2p AO's, as in ethylene, the MO has the form $\psi = (\phi_1 + \phi_2)/\sqrt{2}$ and hence $p_{12} = 2 \times (1/\sqrt{2}) \times (1/\sqrt{2}) = 1$; there is *one* π bond. Benzene may thus be described as having *fractional* π bonds; each link has $\frac{2}{3}$ of a π bond (superimposed on the C—C σ bond) or a 'double-bond character' of $\frac{2}{3}$. This concept of (π) bond order (Coulson 1939) has proved to be of immense value in the electronic interpretation of organic chemistry and will be developed in later sections.

It should be apparent by now that the π charge q_r is simply the population of the π-type AO ϕ_r, while p_{rs} determines an overlap population for the pair of orbitals ϕ_r, ϕ_s:

$$q_r = P_{rr}, \qquad q_{rs} = 2S_{rs}P_{rs} = 2S_{rs}p_{rs} \qquad (r \neq s) \tag{8.4}$$

where S_{rs} is the overlap integral for ϕ_r and ϕ_s. The use of charges and bond orders may thus be regarded as one particular application of the very general technique (p. 154 and references cited) of electron population analysis. The prescription we are now using, to obtain these quantities in a simple MO approximation, is summarized mathematically by

$$P_{rr}(= q_r) = \sum_K n_K c_r^{(K)2}, \qquad P_{rs}(= p_{rs}) = \sum_K n_K c_r^{(K)} c_s^{(K)} \tag{8.5}$$

where $c_r^{(K)}$ is the coefficient of ϕ_r in the MO ψ_K, and n_K ($= 0, 1$ or 2) is the number of electrons ψ_K contains.

8.4. Hückel theory: conjugated chains and rings

We now turn to the actual determination of MO's, using approximations first introduced by Huckel (1931). First of all, we assume that the electrons of the atomic inner shells and of the localized σ bonds merely provide (along with the nuclei) an effective field in which the remaining π electrons move. Then we remember (cf. §4.2 and §2.5) that in the MO description the interactions among the electrons considered (i.e. the π electrons) may be taken into account, with fairly high accuracy, by an averaging technique; this is the essence of the self-consistent field approach. In the present context each π electron moves in a 'framework field', provided by the σ-bonded skeleton of the molecule supplemented by an average field due to the charge distribution arising from the remaining π electrons. In principle, this field, and hence the effective Hamiltonian operator for a π electron, could be calculated from first principles, and indeed, with the advent of large electronic computers, much progress in this direction has been made, but in simple MO theory it is customary to use a vastly simpler 'semi-empirical' approach. The secular equations for an electron in the MO (8.2), described by the (one-electron) Hamiltonian $\hat{h}$, are

$$(H_{11}-\varepsilon)c_1+(H_{12}-\varepsilon S_{12})c_2+\ldots$$
$$(H_{21}-\varepsilon S_{21})c_1+\quad(H_{22}-\varepsilon)c_2+\ldots \tag{8.6}$$

. . . .

where the AO's ϕ_1, ϕ_2, ... have been assumed normalized and

$$H_{rs} = \int \phi_r h \phi_s \, d\tau, \qquad S_{rs} = \int \phi_r \phi_s \, d\tau. \tag{8.7}$$

It is thus unnecessary to know the exact form of the operator $\hat{h}$ itself; what we need to know are the 'matrix elements' H_{rs}, whose numerical values may be either (i) calculated from first principles or (ii) identified by appeal to experiment. The second approach, which we shall follow, is termed 'semi-empirical'; the unknown quantities are chosen so that the results fit the experimental data for a few typical molecules and then used to predict results for other molecules. To keep the number of parameters (i.e. the required matrix elements) small, Hückel made the following approximations:

(i) $H_{rr} = \alpha$ for all conjugated carbon atoms (r)
(ii) $H_{rs} = \beta$ for any conjugated carbon link $r-s$ (adjacent atoms)
     $= 0$ otherwise
(iii) $S_{rs} = 0$, $r \neq s$ (i.e. all overlap integrals neglected).

These are by now familiar assumptions (cf. §7.3); Hückel was the first to introduce and successfully exploit them in a systematic treatment of conjugated systems. As in previous chapters, we call α a 'Coulomb integral' and β a 'bond integral', remembering that both are normally negative quantities.

Let us now write down the secular equations, with Hückel's approximations, for some typical systems encountered in organic chemistry—chains and rings (Fig. 8.10). The first few polyene chains are ethylene ($N = 2$), the allyl radical ($N = 3$), and butadiene ($N = 4$), while the most obvious ring is benzene ($N = 6$). For a chain or a ring with $N = 4$ the equations (8.6) take the form

$$\begin{array}{lllll}
(\alpha-\varepsilon)c_1 & + & \beta c_2 & [+\beta c_4] & = 0 \\
 & (\alpha-\varepsilon)c_2 & + & \beta c_3 & = 0 \\
 & \beta c_2 & + & (\alpha-\varepsilon)c_3 + & \beta c_4 = 0 \\
[+\beta c_1] & & + & \beta c_3 + & (\alpha-\varepsilon)c_4 = 0
\end{array}$$

where the terms within square brackets are absent for the chain (where atoms 1 and 4 are remote) but present for the ring (where 1 and 4 are adjacent). Such equations may be put in an even simpler standard form by dividing each line by β and putting

$$\frac{\alpha-\varepsilon}{\beta} = -x \qquad \text{(i.e. } \varepsilon = \alpha+\beta x) \tag{8.8}$$

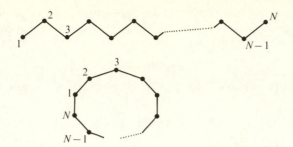

Fɪɢ. 8.10. Conjugated chains and rings. All molecules are assumed to be planar. Each atom has one π-type AO, $\phi_1, \phi_2, \ldots, \phi_N$.

for then they become

$$
\begin{aligned}
-xc_1 \;+\; c_2 && [+c_4] &= 0 \\
c_1 \;-\; xc_2 \;+\; c_3 && &= 0 \\
c_2 \;-\; xc_3 \;+\; c_4 &= 0 \\
[c_1] \;+\; c_3 \;-\; xc_4 &= 0
\end{aligned}
$$

Clearly, x may be described as the energy 'relative to α as datum' (i.e. $\varepsilon - \alpha$) 'in units of β' (i.e. divided by β), and by measuring the energy in this way we eliminate both the constants α and β. For a chain or a ring containing N conjugated carbon atoms the equations are entirely similar, but there are N of them instead of 4, i.e. we obtain

$$
\begin{aligned}
-xc_1+c_2 && [+c_N] &= 0 \\
\cdots\cdots\cdots\cdots\cdots\cdots\cdots\cdots\cdots\cdots\cdots\cdots\cdots\cdots \\
c_{m-1}-xc_m+c_{m+1} && &= 0 \qquad (8.9) \\
\cdots\cdots\cdots\cdots\cdots\cdots\cdots\cdots\cdots\cdots\cdots\cdots\cdots\cdots \\
[c_1] && c_{N-1}-xc_N &= 0.
\end{aligned}
$$

To solve (8.9) we need consider only the single general equation

$$
c_{m-1}-xc_m+c_{m+1} \qquad (8.10)
$$

which relates the coefficients of three adjacent AO's, as we now show. The details (small type) may be passed over on first reading, the results appearing in (8.18) and (8.19).

Let us consider first the case of a chain, deleting the terms in square brackets in the first and last equations of (8.9). These two equations then take the form (8.10), with $m = 1$ and $m = N$ respectively, provided we introduce two extra coefficients *and put both equal to zero*:

$$
c_0 = 0, \qquad c_{N+1} = 0. \qquad (8.11)
$$

These are 'boundary conditions' which tell us that there are no coefficients beyond those for $m = 1$ and $m = N$.

To satisfy the general equation (8.10) we look for a *periodic* solution, guided by the fact that for a particle in a box (§3.4) the wavefunctions are of sine and cosine form; a plausible dependence of c_m upon m (i.e. upon position in the chain) might thus be

$$c_m = \exp(im\theta) \tag{8.12}$$

where θ is a constant, which will determine the 'wavelength'. On substituting (8.12) in (8.10) we find

$$\exp\{i(m-1)\theta\} - x\exp(im\theta) + \exp\{i(m+1)\theta\} = 0$$

or, on removing a factor $\exp(im\theta)$,

$$e^{i\theta} + e^{-i\theta} - x = 0. \tag{8.13}$$

In other words, *all* the equations in (8.9) are satisfied when θ is related to the energy (x) by

$$x = 2\cos\theta. \tag{8.14}$$

Since changing the sign of θ does not change $\cos\theta$ it follows also that $c_m = \exp(-im\theta)$ satisfies the equations, and that

$$c_m = A\exp(im\theta) + B\exp(-im\theta) \tag{8.15}$$

will be a general solution, A and B being arbitrary constants. The allowed 'wavelengths' (θ values) are now fixed by the conditions (8.11); the first gives $c_0 = A + B = 0$, and hence

$$c_m = A\{\exp(im\theta) - \exp(-im\theta)\} = C\sin m\theta \tag{8.16}$$

while the second gives

$$c_{N+1} = C\sin(N+1)\theta = 0$$

and hence

$$(N+1)\theta = k\pi \qquad (k = 1, 2, \ldots). \tag{8.17}$$

The integer k is a *quantum number*; there is one solution for each k value, the MO and its corresponding energy being determined by putting (8.17) in (8.16) and (8.14) respectively.

Finally, then, the coefficient of the mth AO in the kth MO (ψ_k) is given by

$$c_m^{(k)} = C_k\sin\left(\frac{mk\pi}{N+1}\right) \tag{8.18a}$$

while the MO energy is determined by

$$x_k = 2\cos\left(\frac{k\pi}{N+1}\right). \tag{8.18b}$$

This is the required solution for *all* polyene chains, no matter how many atoms (N) there may be. The constant C_k is easily chosen so as to normalize the MO,

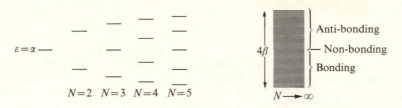

FIG. 8.11. Distribution of energy levels for a conjugated chain of N atoms. As N increases the levels become closer and closer, forming an 'energy band' of width 4β.

and then takes the value $C_k = \sqrt{\{2/(N+1)\}}$. In terms of α and β, (8.8) gives as the kth energy level

$$\varepsilon_k = \alpha + 2\beta \cos\left(\frac{k\pi}{N+1}\right) \qquad (k = 1, 2, \ldots, N). \qquad (8.19)$$

There are only N distinct solutions (for if we try $k = N+1, N+2, \ldots$ we simply find repetitions of the results for $k = 1, 2, \ldots$) and the distribution of energy levels is indicated in Fig. 8.11. The N levels are symmetrically disposed about $\varepsilon = \alpha$ (the energy of a π electron on a single carbon atom), some with lower energy and hence bonding MO's, and some with higher energy and hence antibonding MO's. We notice also that *when N is odd* there is always a solution with $x_k = 0$ $(k = \frac{1}{2}(N+1))$ and $\varepsilon_k = \alpha$; it is natural to refer to this orbital as the 'non-bonding MO'. As the chain becomes longer, we eventually obtain an almost continuous 'band' of levels of width 4β; the interaction between the atoms, and the consequent delocalization of the wavefunctions, has the effect of 'spreading' a typical atomic energy level into a band of N levels, and the greater the interaction β (depending on overlap between neighbours) the greater the band width.

The results obtained above are of great importance and point the way towards the energy band theory of the electronic structure of crystals, to which we return in Chapter 11: a long polyene, along with many other types of polymer chain, may in fact be regarded as a '1-dimensional crystal'.

For a *cyclic* polyene the terms in square brackets in (8.9) must be retained; they arise from the interaction between the end atoms, 1 and N, when the ends are brought together to make a ring (Fig. 8.10). To make the first and last equations fit the general form (8.10) we must then agree to choose

$$c_0 = c_N, \qquad c_{N+1} = c_1 \qquad (8.20)$$

In other words we simply replace (8.11) by new boundary conditions; these are often called 'periodic boundary conditions' because they ensure that if one goes on counting round the ring the $(N+1)$th atom (and its AO coefficient) must coincide with the first, the $(N+2)$th with the second, and so on. The new conditions lead to the solutions

$$c_m^{(k)} = A_k \exp\left(\frac{2\pi imk}{N}\right), \qquad x_k = 2\cos\left(\frac{2k\pi}{N}\right) \qquad (8.21)$$

instead of (8.18), where now positive and negative integer values of the quantum number k give distinct solutions,

$$k = 0, \pm 1, \pm 2, \ldots$$

though each pair of values (e.g. ± 2) gives the same energy x_k. The MO's are thus degenerate in pairs except for the first ($k = 0$) and, for an *odd*-membered ring, the last ($k = N$); for $|k| > N$ repetitions occur and there are no new solutions.

The complex MO's, with the AO coefficients given in (8.21), correspond to 'travelling-wave' solutions of the Schrodinger equation, positive and negative k values corresponding to the electron running round the ring in one direction or the other, with a momentum increasing with k (cf. p. 329). Very often they are combined in pairs, taking the sum and difference of MO's with k values of different sign, to give *real* MO's which still satisfy the secular equations but are easier to visualize; the two 'standing-wave' solutions for energy

$$\varepsilon_k = \alpha + 2\beta \cos\left(\frac{2k\pi}{N}\right) \qquad (8.22)$$

have coefficients which follow the sine and cosine pattern respectively:

$$a_m^{(k)} = C_k \sin\left(\frac{2\pi mk}{N}\right), \qquad b_m^{(k)} = C_k \cos\left(\frac{2\pi mk}{N}\right) \qquad (8.23)$$

where now only positive values of k need be considered and C_k is chosen as usual to normalize each ψ_k. It is easily confirmed, on putting $N = 6$, that the benzene MO's listed in Table 8.2, arise when $k = 0$ (b coefficients) and $k = 1$ (a and b coefficients giving the degenerate pair).

The distribution of the energy levels about the reference level $\varepsilon = \alpha$ is indicated in Fig. 8.12; for rings containing an even number of atoms the distribution is symmetrical, but for N odd it is not. Again, for $N \to \infty$, the levels form a continuous 'energy band' of width 4β, though each level now corresponds to a degenerate *pair* of states.

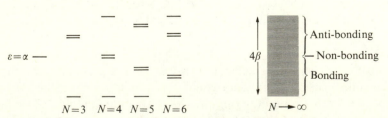

FIG. 8.12. Distribution of energy levels for a conjugated ring of N atoms. All the levels in the band are doubly degenerate except the lowest and, for N even, the highest.

Several interesting conclusions can be drawn from Fig. 8.12. In the first place the molecules for which the levels are symmetrically disposed about the line $\varepsilon = \alpha$ have a common property; the carbon atoms can be divided into two sets, for example by going round the ring and putting stars against alternate atoms, to obtain a 'starred set' and an 'unstarred set', and *no two stars fall in adjacent positions*. Such molecules are called *alternant hydrocarbons*. The chains (Fig. 8.10) are obviously alternants, and again give symmetrically disposed energy levels, but the *odd*-membered rings are *non-alternant* because the starring process would finish with two stars coming adjacent. The properties of alternant molecules are of great importance and are dealt with later.

Secondly, Fig. 8.12 shows the origin of a famous rule due to Hückel and known as 'Hückel's $4n+2$ rule'. The rule refers to the properties of cyclic conjugated molecules in which there are N π electrons; these depend on the nature of the integer N. If N is expressed in terms of a smaller integer n the rule may be stated as follows:

$N = 4n+2 \rightarrow$ the molecule is highly stable
$N = 4n+1 \rightarrow$ the molecule is a free radical
$N = 4n \qquad \rightarrow$ the molecule has a triplet ground state and is highly unstable.

The reasons are clear from Fig. 8.12. Because there is a single lowest orbital, while the others occur in degenerate pairs, the number of filled or partly filled levels must be of the form $1+2n$ (with, say, n degenerate pairs); they will be exactly filled, to give a closed-shell ground state, by twice this number of electrons, i.e. when $N = 4n+2$. If N is reduced by 1 the highest degenerate pair will contain only three electrons, the single occupied MO giving radical behaviour. If N is reduced by 2 the remaining two electrons will occupy the degenerate MO's singly with spins parallel (Hund's rules), giving a triplet state.

The $4n+2$ rule is illustrated by benzene, cyclopentadienyl, and cyclobutadiene. Benzene ($N = 4n+2$ with $n = 1$) is highly stabilized by π bonding. Cyclopentadienyl ($N = 4n+1$ with $n = 1$) has radical properties, but addition of an electron completes the closed shell and gives a stable anion, while withdrawal of an electron gives a highly unstable cation. Cyclobutadiene ($N = 4n$ with $n = 1$) should have a triplet ground state and high instability. For many years this molecule resisted all attempts at synthesis. It is now known (Emerson, Watts, and Pettit 1965) but has indeed proved to be exceptionally unstable.

In dealing with any ring, we must remember of course that stability depends on the *total* energy, not that of the π electrons alone, and that any departure from the 'natural' in-plane bond angles (120°, corresponding to trigonal hybridization) will raise the energy of the σ-bonded framework by an amount corresponding to the strain energy discussed in §8.2. In cyclobutadiene, for example, the overlapping hybrids could not point directly towards each other and would give 'bent bonds' unless their s content was diminished effectively to zero; in any event the reduced overlap would give diminished σ bonding and a higher σ-electron energy. The same effect would occur in cyclooctatetraene (C_8H_8), where planarity of the molecule would require bond angles of 135° instead of 120°; in this case it turns out to be energetically more favourable for the molecule to buckle out of plane, thereby lowering the σ-electron energy but sacrificing most of the π-electron delocalization energy—the bond integrals (β) being much reduced in value when the 2p AO's are no longer parallel. In such cases the planarity and π character of the molecule may be almost completely lost, the carbon–carbon links tending to alternate between single and double bonds. It must therefore be remembered that, although there is overwhelming evidence that many of the properties of conjugated molecules are associated with their π-electron systems, other factors must sometimes be considered, and that any deviation from 120° bond angles, or from planarity, is a sure indication that such factors may be important.

8.5. Alternant hydrocarbons. Charges and bond orders

In §8.3 we showed that the π charges on the carbon atoms in benzene were all unity and that the π electrons were thus distributed uniformly over the molecule. This turns out to be true also for all the molecules considered in §8.4, subject to Hückel approximations and the assumption that each carbon contributes only one π electron. In fact, with these assumptions uniformity of the π-electron distribution (every $q_r = 1$) is a characteristic property of *all alternant hydrocarbons*. Such molecules share many of the properties noted for alternant chains and rings. These properties may be summarized as follows.

(i) Every bonding MO has an antibonding 'partner', their energies lying symmetrically above and below that of a π electron on a single carbon atom ($\varepsilon = \alpha$).

(ii) If the number of conjugated atoms is odd there is at least one *non-bonding* MO with $\varepsilon = \alpha$.

(iii) When the MO's are filled in ascending energy order by electrons, one from each conjugated carbon, there is unit π charge on each atom. (Any non-bonding MO's are then singly occupied in accordance with Hund's rules).

The possible existence of non-bonding orbitals (NBMO's) is of great importance; as they hold the odd electrons, a knowledge of their forms allows us to predict the active centres in free radicals, and they may be used in various other ways as we shall discover presently.

First, however, we show how the total π-electron energy may be related to the charges and bond orders, for it is on this relationship that their value largely rests. The energy of the π electrons, with Hückel approximations, is the sum of the energies associated with the occupied orbitals (a term occurring twice if the orbital is doubly occupied). Now an orbital energy (ε) has the form

$$\varepsilon = \int (c_1\phi_1 + c_2\phi_2 + \ldots)\hat{h}(c_1\phi_1 + c_2\phi_2 + \ldots)\,d\tau$$

$$= c_1^2\alpha_1 + c_2^2\alpha_2 + \ldots + 2c_1c_2\beta_{12} + \ldots$$

where the α's and β's are the Coulomb and bond integrals (all α's the same for C atoms, all β's the same for C—C bonds), and we get the total electronic energy by summing the orbital contributions. On noting that the sums of squares and products give the charges and bond orders defined in (8.5) we thus find

$$E = q_1\alpha_1 + q_2\alpha_2 + \ldots + 2p_{12}\beta_{12} + \ldots. \tag{8.24}$$

There is thus an energy term $q_r\alpha_r$ from each atom r, and a term $2p_{rs}\beta_{rs}$ from each bond r—s; the latter term represents the decrease of electronic energy due to π bonding in the r—s link and is proportional to the bond order and also to the β factor (which in turn depends on the degree of overlap of the orbitals ϕ_r and ϕ_s).

The result (8.24) provides a more quantitative basis for some of our earlier conclusions about the effects of π bonding. In the first place it tells us that an energy term $2p_{rs}\beta_{rs}$, directly proportional to the bond order p_{rs}, arises from the effect of π bonding in link r—s; the larger the bond order the greater is the stabilization. Secondly, it allows us to discuss how the molecule responds to any *changes* that may occur (e.g. bending the molecule or stretching one of its bonds). The first-order change in π-electron energy may be calculated from (8.24) by making appropriate changes in the Coulomb and bond integrals, keeping the coefficients (i.e. the q's and p's) unchanged. This is a well known result of 'perturbation theory' in quantum mechanics. It is valid also in LCAO approximation provided the AO coefficients are variationally determined (i.e. by solving the secular equations). Thus, if the r—s bond length is changed, β_{rs} will change by $\delta\beta_{rs}$ and the change in π-electron energy will be

$$\delta E = 2p_{rs}\delta\beta_{rs}. \tag{8.25}$$

The π energy, computed assuming all bond lengths equal, may thus be lowered by allowing link r—s to shorten (increasing overlap increasing the *magnitude* of β_{rs}, which is negative, to give $\delta\beta_{rs}$ a negative value), thus compressing the corresponding σ bond of the framework, and the larger the bond order p_{rs} the greater the shortening will be. We therefore expect that each C—C bond in a conjugated system will be shortened, relative to the normal single bond, by an amount depending on the corresponding bond order p_{rs}. This conjecture is verified by observation; calculated bond orders and observed C—C bond

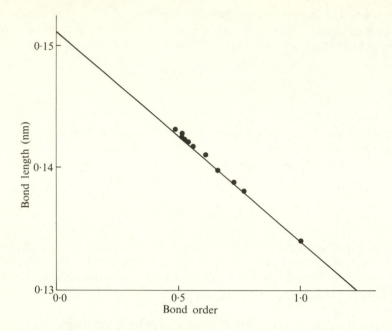

Fig. 8.13. Approximately linear relationship between C—C bond lengths and π bond orders. The points correspond to bonds in a considerable number of conjugated hydrocarbons, all with carbon atoms approximately sp^2 hybridized.

lengths for a wide range of conjugated systems give a fairly accurate linear relationship between bond length and bond order (Fig. 8.13).

There are two other immediate consequences of (8.24). If a conjugated molecule is 'folded' by bending it out of plane across two bonds r—s and t—u, the two bond integrals β_{rs} and β_{tu} will be drastically reduced in magnitude by the reduced overlap of the pairs of 2p AO's on which they depend, i.e. $\delta\beta_{rs}$ and $\delta\beta_{tu}$ will be positive. Thus δE will be positive, giving rise to a force tending to restore planarity; this is the basic reason that conjugated systems are planar. At the same time we note that the same degree of bending will be achieved more easily (δE smaller) when the affected bonds have small π bond orders, giving quantitative support to the idea that the planar conformation is 'stiffened' by the π bonding.

Finally, the delocalization ('resonance') energy associated with the π bonding is very easily computed from (8.24), once the charges and bond orders are known, for the 'reference energy' corresponding to one principal (Kekulé type) structure would contain a term $q_r\alpha_r + q_s\alpha_s + 2p_{rs}\beta_{rs}$ for each double-bonded link r—s, and with $q_r = q_s = p_{rs} = 1$ (and Hückel approximations $\alpha_r = \alpha_s = \alpha$, $\beta_{rs} = \beta$) this is simply $2\alpha + 2\beta$. We then obtain the following.

> The π-electron delocalization energy of a conjugated hydrocarbon is obtained by subtracting from the energy given by (8.24) an amount $2\alpha + 2\beta$ for each double bond in a single principal structure.
>
> (8.26)

Our earlier observations concerning possible changes of σ-electron energy due to bonding or compression of σ bonds tell us that 'resonance energy', about which much was written in the early days of quantum chemistry, is not a very satisfactory concept, nor one which is easily related to observable quantities. Even if an 'experimental' π-electron energy could be assigned to a single principal structure (using thermochemically inferred bond energies for C—C and C=C bonds), this structure would have bonds of alternating lengths, different from those in the actual molecule, and it would therefore always be necessary to introduce uncertain corrections for 'compression energy' in comparing experiment and theory. Nevertheless, there can be little doubt that the trends shown in the calculated delocalization energies of cyclic condensed systems give a useful guide to their stabilities. We return to the discussion of stability, and its chemical meaning, in Chapter 10. Hückel calculations have been performed, and the results tabulated (Streitwieser and Brauman 1965) for a vast range of conjugated systems, both alternant and non-alternant. The parallelism between calculated and 'experimental' delocalization energies is indicated in Table 8.3.

TABLE 8.3

Delocalization energies in selected hydrocarbons

Compound	Delocalization energy†	Observed‡ value (kJ mol^{-1})	Theoretical§ value (kJ mol^{-1})
Benzene	2.00β	155	155
Naphthalene	3.68β	314	285
Anthracene	5.32β	439	412
Phenanthrene	5.45β	460	422
Diphenyl	4.38β	330	339
Butadiene	0.47β	15	36
Hexatriene	0.99β	—	77

† From MO calculations with β values assumed equal for all C—C links. This is acceptable for the ring compounds, but less so for chains (where it exaggerates the delocalization energy).

‡ Taken mainly from Pauling (1960). See also Cottrell and Sutton (1947) for a critical discussion of resonance energy calculations.

§ Parameter β chosen to fit the experimental value for benzene.

8.6. Alternant radicals and ions

The form of the non-bonding orbital (NBMO), which occurs in an alternant hydrocarbon containing an odd number of centres and is associated with radical properties, is of special importance. It is also very easily determined, with nothing more than simple mental arithmetic! It is only necessary to divide the carbons into two sets, by 'starring' alternate atoms, and to apply the following rule.

> In the NBMO, the coefficient of every AO in the smaller of the two sets (usually taken as the unstarred set) is zero, and the *sum* of the coefficients of its neighbouring AO's is also zero.   (8.27)

The origin of this rule is in the secular equations; the one which connects the coefficient of ϕ_r with the coefficients of its neighbours (ϕ_s) is

$$-xc_r + \sum_{s(r-s)} c_s = 0. \tag{8.28}$$

where (r—s) beneath the summation sign means 'for s linked to r'. The non-bonding orbital has $x = 0$, by definition, and hence the sum over neighbours must vanish in the corresponding MO. A rigorous proof (Longuet–Higgins 1950) of the existence of at least one NBMO is a bit more difficult, and it should be noted that in some cases there may be more than one NBMO.

Let us take one example, the benzyl radical, starring alternant atoms as in Fig. 8.14 (*a*). If the AO coefficient on the left-hand carbon is taken to be 1, the remaining coefficients follow at once on application of the rule (8.27); they are -1 and $+2$ as shown in Fig. 8.14 (*b*). To normalize the NBMO we can multiply all coefficients by k and require the sum of the squares to be unity; hence $k^2(1+1+1+4) = 1$ and $k = 1/\sqrt{7}$. The full set of coefficients is shown in Fig. 8.14 (*c*), while their squares, which give the contributions to the charges q_1,

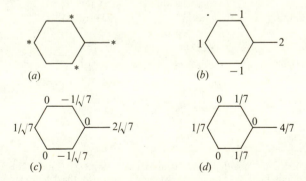

FIG. 8.14. Determination of the non-bonding MO in the benzyl radical: (*a*) definition of starred and unstarred sets; (*b*) AO coefficients (un-normalized); (*c*) normalized coefficients; (*d*) contributions to the π charges.

$q_2, \ldots, q_7$ associated with an electron in the NBMO, appear in Fig. 8.14 (*d*).

From Rule (i) (p. 244) we expect three bonding and three antibonding MO's, and in the ground state the seven π electrons will therefore doubly occupy all bonding MO's, leaving one π electron in the NBMO. The squared coefficients in Fig. 8.14 (*d*) therefore indicate where the odd electron is to be found; it is four times as likely to be found on the terminal carbon as on the ortho or para carbons and it will never be found at the meta positions. These results are in accord with the observed radical activity of the system, and are also supported by the observed hyperfine structure of the ESR signal, which arises from coupling with the spins of the protons attached to the predicted active centres.

It is interesting to turn briefly to the VB description of radical activity. This proceeds by drawing principal structures which, with an odd number of centres, always show one electron 'left over' (i.e. unpaired), corresponding to the odd electron in the MO approach. The benzyl structures are shown in Fig. 8.15 and suggest exactly the same active centres as the MO approach. What is more remarkable, however, and can be proved generally (Longuet–Higgins and Dewar 1952), is that *the number of principal structures with the unpaired electron at a particular centre is proportional to the magnitude of the AO coefficient, at that centre, in the NBMO*. Thus, counting the principal structures yields, apart from sign, the same AO coefficients (Fig. 8.14 (*b*)) as solving the MO secular equations. There is thus a very deep-seated parallel between the predictions of the MO theory and those derived from simple pictorial considerations based on possible resonance structures. This is not fortuitous but is a mathematical consequence of the topological properties of the diagrams; the relative numbers of structures with the dot in different positions satisfy equations similar to those for the NBMO coefficients. It is almost certainly for such reasons that, in spite of its defects from the point of view of quantum mechanics, the so-called 'theory of resonance' has had a fairly successful history as a qualitative tool in organic chemistry.

Knowledge of the form of the NBMO also allows us to discuss, with no

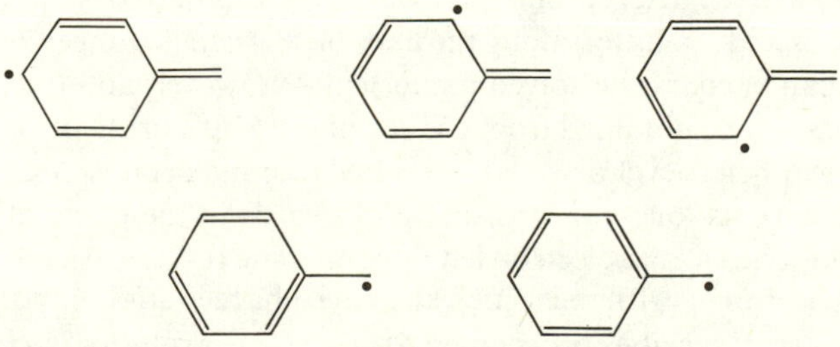

Fig. 8.15. Principal VB structures for the benzyl radical. Two structures show the odd electron on the terminal carbon; there is only one for each of the other active centres. The predictions of 'resonance theory' are thus similar to those of MO theory.

further calculation, the π-charge distribution in radical ions. In the case of benzyl the anion will have *two* electrons in the NBMO and the numbers in Fig. 8.14 (*d*) will thus represent the net negative charges on the various carbons, for before the electron is added the total π charge at each centre is unity (Rule (iii) on p. 244) and exactly balances the unit positive charge of the corresponding framework ion. The situation is reversed in the cation, loss of the single NBMO electron leaving the framework with net *positive* charges of 1/7 and 4/7 at the same positions.

8.7. Effects of heteroatoms and substituents

So far we have discussed only hydrocarbons; we must now turn to heterocyclic molecules, in which a carbon is replaced by a heteroatom (e.g. N, O), and substituted molecules in which a hydrogen atom is replaced either by another atom or by a substituent group (e.g. F, NH_2).

The essential feature which distinguishes a heterocyclic molecule from its 'parent' hydrocarbon is the changed electronegativity at the heteroatom. The matrix element H_{rr} ($=\alpha_r$) for the heteroatom will now differ from that for the remaining carbon atoms. The corresponding resonance integrals β_{rs}, where r is the heteroatom, will also differ slightly, but the difference, particularly when nitrogen replaces carbon, is small and it may be shown to exert only a second-order effect on the final charge distribution. Thus for pyridine, calling the N atom number 1, the secular equations will be approximately those for benzene except that the term α in the first line is altered. This alteration depends on the electronegativity change, and may be taken to be proportional to the difference of the electronegativities of the two atoms (§6.4). In the case of nitrogen it used to be customary to replace α for the N atom by $\alpha + 2\beta$, but later work suggests that $\alpha + \frac{1}{2}\beta$ is a rather more appropriate value. Many compilations of α values for heteroatoms have appeared in the literature. Some representative values are listed by Streitwieser (1961). In general, it is convenient to suppose $\alpha \rightarrow \alpha + k\beta$ for a heteroatom, and this means that in the 'x form' of the secular equations (see p. 238) the term $-x$ is replaced by $-x+k$ when it refers to the heteroatom. Unfortunately, this destroys the simplicity of the equations and a solution along the lines of §8.4 can no longer be found. The equations can of course be solved numerically, however, and the charges and bond orders can be obtained from the resultant MO's using (8.5).

In nitrogen heterocycles such as pyridine, the nitrogen is described as sp^2 hybridized with its lone pair projecting outwards in the plane of the ring; it therefore has one valence electron left to contribute to the π system. Pyridine is thus π-*isoelectronic* with benzene, and the charges and bond orders are calculated from three doubly occupied MO's. Some results for pyridine and for quinoline are shown in Fig. 8.16 for $k = 0.5$. They show how π-electron charge is drawn towards the more electronegative nitrogen atom at the expense of some of the carbon atoms. It also appears that, proceeding away from the

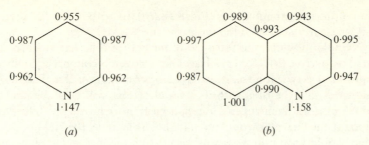

FIG. 8.16. π-Electron charges in pyridine and quinoline. All charges would be unity in the iso-electronic hydrocarbons (benzene and naphthalene).

heteroatom, it is only *alternate* atoms which are strongly affected. This conforms to an empirical 'law of alternating polarity', which can be shown to have a general foundation in MO theory (Coulson and Longuet–Higgins 1947).

Similar calculations can very easily be performed for other molecules using present-day computers, but it is interesting to note that the same qualitative predictions can often be obtained with virtually no calculation by using the properties of the NBMO introduced in the last section. To illustrate the approach let us take pyridine and imagine the nitrogen atom, with its one π electron, cut out of the framework. The 'residual molecule' is then essentially a pentadienyl radical (Fig. 8.17 (a)) for which the NBMO follows at once by the rules of the last section; the electron in the NBMO is divided equally over three centres (Fig. 8.17 (b)). Now we 'reconnect' the nitrogen and try to understand qualitatively what happens. The nitrogen provides an extra π electron, but being more electronegative than the carbons it keeps its own π electron and also tries to steal charge density from the remainder of the π system. Let us suppose it takes, say, 0·15 of an electron from the highest occupied orbital (the NBMO); this means 0·05 electron from each of the active centres in Fig. 8.17 (b). The resultant π charges should then be as in Fig. 8.17 (c), and these values are remarkably close to those calculated by solving the secular equations (Fig. 8.16 (a)). An exactly similar discussion of quinoline also leads to predictions in close agreement with those obtained by full

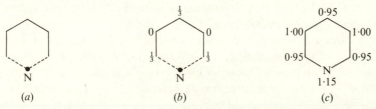

FIG. 8.17. Estimation of π charges in pyridine: (a) breaking two bonds leaves a pentadienyl radical; (b) distribution of electrons in the pentadienyl NBMO; (c) on re-connection some charge flows out of the NBMO towards the nitrogen, leaving total π charges as shown.

solution of the secular equations (Fig. 8.16 (*b*)). In both cases the 'alternating polarity' is clearly related to the nature of the NBMO.

An example in which the nitrogen atom is in a different valence state is provided by pyrrole (Fig. 8.2 (*b*)). Here the nitrogen is roughly sp^2 hybridized and thus has a π-type lone *pair* available for conjugation. The Hückel $4n + 2$ rule indicates that the ring, with its six π electrons, will be stable and this is indeed the case. If the nitrogen atom had been more nearly sp^3 hybridized as in ammonia, the lone pair would not have been of π type; the NH bond and the lone pair would have pointed above and below the molecular plane and the conjugated path would have been broken with a consequent loss of delocalization energy. The electron distribution in pyrrole may again be inferred qualitatively by considering the connection of the nitrogen, with its two π electrons, to the end carbon atoms of the residual molecule, a butadiene-like chain (Fig. 8.18 (*a*)). Delocalization can then lead only to a flow of electrons *away* from the N atom (which would have to have an infinitely negative α value to keep its π orbital filled), with the result that the nitrogen will carry a net positive charge instead of the net negative charge which occurs in pyridine. The results of a typical calculation (Miller *et al.* 1962) are shown in Fig. 8.18 (*b*). Similar considerations apply to the other heterocycles furan (Fig. 8.2 (*d*)) and thiophene (the oxygen and sulphur heteroatoms, respectively, having electronegativities of 3·4 and 2·6, compared with 3·0 for nitrogen), though in the latter molecule there is an added complication due to the possibility of d-orbital participation (Zauli 1960; see also §12.4).

There is an interesting confirmation of the difference between pyrrole and pyridine in a study of their dipole moments. The dipole moment of pyrrole (Nygaard *et al.* 1969) is 1·74D in a direction which makes the N atom positive. Not more than about 0·4D of this can be due to the σ electrons in the C—H, C—N, and N—H bonds,† so that approximately 1·3D arises from the π electrons. The value obtained from the π-electron distribution in Fig. 8.18 (*b*) is about 1·9D and in the correct direction. In pyridine, however, the observed dipole moment is 2·15D in a direction that makes the N atom negative (DeMore *et al.* 1952). This is precisely what is required from the charges in Fig. 8.17 (*c*), which lead to a π moment (sometimes called the resonance moment) of $\mu_\pi = 1\cdot2D$. When we add to this the C—H and C—N σ-bond moments and the moment of the lone-pair electrons on the nitrogen atom (Fig. 8.2 (*a*)), we obtain approximately the observed value.

The treatment of substituents, in which the heteroatom replaces a peripheral hydrogen and is thus external to the ring, follows exactly similar lines. We may either solve secular equations, giving the substituent a Coulomb integral $\alpha + k\beta$ with some appropriate value of k, or we may argue more qualitatively in terms of the substituent's capacity to donate or withdraw

† For some values of bond moments see Nash *et al.* (1968).

FIG. 8.18. π Charges in pyrrole: (*a*) breaking two bonds leaves butadiene; (*b*) on re-connection, nitrogen (with *two* π electrons) loses some charge to give total π charges shown.

electrons to or from the MO's of the 'parent' hydrocarbon. To illustrate the latter procedure, which has been developed at great length by Dewar (see Dewar and Dougherty (1975)), and which we discuss later in more quantitative form, let us consider aniline and nitrobenzene (Fig. 8.19), corresponding to benzene substitution with NH_2 and NO_2 respectively, assuming for the moment that both molecules are planar. The presence of seven conjugated centres would suggest that we regard the benzyl radical as the parent hydrocarbon and consider the effect of replacing the terminal CH_2 group by NH_2 or NO_2. If the NH_2 provides two π electrons, aniline will be isoelectronic with the benzyl *anion*, the extra electron in the NBMO going mainly onto the nitrogen (cf. Fig. 8.14(*d*)) but also being delocalized so that some net negative charge appears at the ortho and para positions. In nitrobenzene, in contrast, the nitrogen of the NO_2 group is electron deficient (its π electrons being strongly drawn to the more electronegative oxygens) and thus contributes *no* π electrons to the seven-centre conjugated framework; with only the six carbon π electrons the resultant system is isoelectronic with the benzyl *cation*, with net *positive* charges at the ortho and para positions. Again, there is an alternation of polarity as one goes away from the heteroatom, and again the expected charge distributions are confirmed by direct MO calculations.

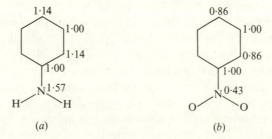

FIG. 8.19. π Charges in aniline and nitrobenzene: (*a*) aniline π system iso-electronic with benzyl anion, giving net negative charge (excess electrons) at ortho and para positions; (*b*) nitrobenzene iso-electronic with benzyl cation, giving net positive charge (electron deficiency) at ortho and para positions.

In the case of aniline it is a poor approximation to regard the molecule as planar, for the experimental evidence (Lister and Tyler 1966) shows that the valence angles at the nitrogen are closer to those in ammonia. Consequently, the lone pair is not accurately of π type and conjugation with the ring is inhibited. In spite of the correspondingly reduced value of β_{CN}, however, the above conclusions are qualitatively correct.

It is interesting to enquire *why* aniline is not planar. If it were so, it would be necessary to change the hybridization at the nitrogen atom from the approximately tetrahedral character found in ammonia to the trigonal type. This would involve energy, the amount of which may be estimated from a study of the vibration spectrum of ammonia. In the so-called 'inversion' spectrum the N atom passes right through the plane of the three hydrogens to a position of minimum energy symmetrically placed on the other side. It appears that about 0·25 eV is required to effect this inversion (Swalen and Ibers 1962); this is therefore the energy that is required to flatten the original pyramid into a plane. In the planar state the bonds are trigonal, so that we conclude that about 0·25 eV is needed to change the hybridization. Presumably much the same would be true for aniline, and this energy would have to be set against the gain in delocalization energy as a result of coplanarity. Now the observed delocalization energy in aniline is about 0·26 eV over and above the value for benzene, so in order that the molecule should be planar, the required delocalization energy would have to be at least 0·26 eV + 0·25 eV i.e. 0·51 eV greater than in benzene. It is not surprising that this is not achieved, for the two extra electrons would enter an orbital similar to the NBMO of the benzyl system. The balance of energy terms must be close, however, and in urea

$$
\begin{array}{c}
H_2N \\
\diagdown \\
CO \\
\diagup \\
H_2N
\end{array}
$$

the increase in delocalization energy favours a completely planar molecule (Vaughan and Donohue 1952; Waldron and Badger 1950).

Even though the delocalization energy in aniline (between ring and NH_2 group) is insufficient to ensure planarity of the molecule, it is evidently substantial; for acceptance of a proton to give $C_6H_5 \cdot NH_3^+$, which results in a fully tetrahedral nitrogen and consequent loss of the corresponding energy contribution, takes place much less readily than in the reaction $NH_3 + H^+ \rightarrow NH_4^+$ where no such loss occurs. This gives an immediate interpretation of the fact that aniline is a much weaker base than ammonia.

Experimental evidence for the effect of substituents comes from a variety of sources. We have noted already that shifts of π charge may lead to a substantial contribution to the dipole moment. This so-called 'resonance moment' may be estimated as the difference between the dipole moments of the corresponding

TABLE 8.4

Dipole moments (in Debye units) of some substituted hydrocarbons

X	F	Cl	Br	OH	NO$_2$	CN
$\mu(CH_3X)$	1·81	1·87	1·80	1·70	3·44	3·97
$\mu(C_6H_5X)$	1·58	1·72	1·77	1·45	4·23	4·42

aromatic and aliphatic molecules. Thus we may compare the dipole moments of CH_3X and C_6H_5X. Table 8.4 gives values for several distinct substituents X. It will be observed that for the first four molecules the resonance moment, determined as the difference of corresponding figures in the two rows, is always such as to make the substituent atom less negatively charged in the aromatic molecule than in the aliphatic. For the last two molecules, however, the situation is reversed. These results are in complete accord with our theoretical predictions that groups with π-type lone pairs surrender electronic charge to the ring while groups such as NO_2 withdraw charge.

For the first four molecules it is also verified experimentally that the charge which flows into the ring is concentrated at the ortho and para positions, for all four of these substituents are ortho–para directing,† i.e. they direct a positively charged approaching radical (e.g. NO_2^+) to substitution preferentially in the ortho or para positions. Without doubt this is chiefly due to the electrostatic attraction arising from the excess negative charge at these places which facilitates the approach of the ion. Again, the effect is reversed in the case of the last two molecules, which are rather unreactive towards electrophiles like NO_2^+ (*meta*-directed) but reactive towards nucleophiles such as OH⁻ or NH_2^- which are directed towards the *ortho* and *para* positions.

There is also evidence that the 'single' bond C—X acquires a partial double-bond character. First, there is a small contraction of the bond length. Secondly, as was first noted by Wilson (1950), accurate microwave spectroscopy shows a lack of cylindrical symmetry around the bond axis, which can most satisfactorily be interpreted as the acquisition of some π character. In the particular case of chloroethylene $CH_2 = CHCl$ the charge distribution around the C—Cl axis is distorted in such a way as to correspond to about 5 per cent double-bond character.

To conclude this section we distinguish two main types of substituent, which are commonly referred to as 'inductive' and 'mesomeric'. In MO terms an inductive substituent is one which, on account of its high electronegativity, changes the α value at the centre to which it is attached but does not extend the conjugated path in the parent molecule. An example would be fluorine in a

† The original work in this field is due largely to Ingold. For a review see Ingold (1934). This topic is considered in more detail in Chapter 10.

fluorobenzene, where the π-type lone pair on the fluorine is tightly bound in a compact AO which does not overlap much with the carbon $2p_\pi$ AO; the absence of a substantial β, connecting the fluorine to the ring, then precludes effective conjugation and there is negligible flow of π charge between ring and substituent. The C—F σ bond may, however, be very polar, and this induces an extra positive charge on the carbon ion of the framework (α becoming more negative corresponding to increased electron attraction), hence the use of the term 'inductive effect'. A mesomeric substituent, however, is one whose π-type orbital has a substantial overlap with the carbon $2p_\pi$ AO at the point of attachment, giving a fairly large β between ring and substituent and thus an extension of the conjugated path. A case in point would be nitrobenzene where the six-centre path in benzene is extended to include the nitrogen† of the NO_2 group, giving the seven-centre path characteristic of benzyl. Of course, the two effects are not strictly separable and both will generally occur to some extent simultaneously, but one or other is often dominant and the classification thus has a certain qualitative value.

8.8. Improvements of the simple theory

So far, our use of simple MO theory in its general form, i.e. as a means of calculating the completely delocalized MO's in large molecules, has been confined to π electron systems. For such systems it has been possible to make a number of not entirely unreasonable approximations in order to simplify the form of the secular equations; for example, the overlap between π-type AO's on adjacent atoms is much smaller (~ 0.25) than the 0.7–0.8 commonly encountered in the overlapping hybrids of a σ bond, and its neglect may therefore seem appropriate in a simple theory of Hückel type. In this section we take the first steps towards the improvement and extension of the theory, moving finally to the inclusion of σ electrons on the same footing as the π electrons. More detailed consideration of some of the methods of introducing the basic ideas of the self-consistent field (SCF) approach (§2.5), within the framework of the LCAO approximation, is postponed until Chapter 13. As this section and the next are concerned only with the nature of the approximations inherent in Hückel theory, they may be passed over on first reading without loss of continuity.

In trying to improve the theory we must first consider the approximation of putting all the overlap integrals equal to zero, for this apparently represents a serious inconsistenty—we have employed a 'principle of maximum overlap' to recognize the essential role played by overlap in chemical bonding and yet to simplify the mathematics we have systematically neglected all overlap integrals! To understand the general nature of this approximation we take first

---

† More accurately, of course, we should explicitly include the oxygens, in which case there would be a nine-centre conjugated path. In the simpler approximation the NO_2 group is imagined as a 'pseudo-atom' with an empty orbital consisting largely of a nitrogen $2p_\pi$ AO.

a π-electron system with N conjugated centres and replace the Hückel equations (§8.4) by the corresponding equations with overlap integrals included; we put $S_{rs} = S$ for all nearest neighbours and $S_{rs} = 0$ for more distant neighbours, and obtain (cf. (8.28))

$$(\alpha - \varepsilon)c_r + \sum_{s(r-s)} (\beta - \varepsilon S)c_s = 0 \qquad \text{(all } r) \qquad (8.29)$$

which means that there is one such equation for each conjugated atom (r) connecting its AO coefficient in any given MO with the coefficients of neighbouring AO's. If we divide every equation by $\beta - \varepsilon S$ we obtain

$$\frac{\alpha - \varepsilon}{\beta - \varepsilon S}c_r + \sum_{s(r-s)} c_s = 0 \qquad \text{(all } r). \qquad (8.30)$$

However, this is exactly the same set of equations we had with complete neglect of overlap (namely (8.28)) except that $-x$ $(=(\alpha - \varepsilon)/\beta)$ has been replaced by $(\alpha - \varepsilon)/(\beta - \varepsilon S)$. In other words, we may still use our original solution (overlap neglected), which for a particular MO gave $x = x_K$ and AO coefficients $c_1^{(K)}, c_2^{(K)}, \ldots, c_N^{(K)}$, provided (i) we calculate the energy using

$$\frac{\alpha - \varepsilon_K}{\beta - \varepsilon_K S} = -x_K$$

or, solving for ε_K,

$$\varepsilon_K = \frac{\alpha + \beta x_K}{1 + S x_K} = \frac{\varepsilon_K^0}{1 + S x_K} \qquad (8.31)$$

where ε_K^0 is the Kth orbital energy in zero-overlap approximation, and (ii) we use exactly the same AO coefficients in each MO, merely renormalizing the result to take account of overlap. Since the variational energy computed from a normalized orbital, with overlap neglected, is given by $\varepsilon_K = \int \psi_K h \psi_K \, d\tau$ and this must be divided by $1 + S x_K$ to admit overlap, the implication is that $\psi_K = \psi_K^0/\sqrt{(1 + S x_K)}$ (superscript zero again meaning 'with neglect of overlap') and hence that the AO coefficients are changed to

$$c_r^{(K)} = \frac{c_r^{(K)0}}{\sqrt{(1 + S x_K)}}. \qquad (8.32)$$

It is therefore possible to admit non-zero overlap with no further calculation in the present nearest-neighbour approximation merely by 'scaling' the energies and AO coefficients according to (8.31) and (8.32), respectively. This possibility was first pointed out by Wheland (1942) and the resultant MO's are sometimes called Wheland orbitals.

The effect of admitting overlap in this way is to destroy the symmetry across the line $\varepsilon = \alpha$ in the energy-level diagram, as shown in Fig. 8.20. This correctly shows, as already noted for a diatomic molecule (p. 98), that when a bonding

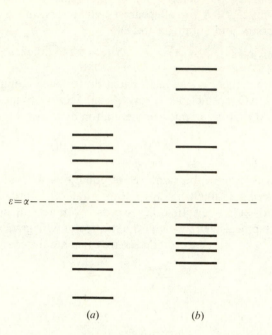

FIG. 8.20. Effect of overlap on π levels in naphthalene: (a) $S = 0$, (b) $S = 0.25$.

MO and its antibonding partner are both filled there is a net *anti*bonding effect.

A fundamental inconsistency in the use of orbital and overlap populations also disappears, namely the fact that although bonding has been associated with the build-up of charge in the overlap regions the overlap populations in (8.4) vanish identically when the overlap integrals are neglected. The contributions to the quantities P_{rr} and P_{rs} which appear in (8.3), arising from an electron in the MO ψ_K, follow from the coefficients in (8.32). Since the populations are defined by $q_r = P_{rr}$, $q_{rs} = 2P_{rs}S_{rs}$, the corresponding *partial* populations for an electron in ψ_K are

$$q_r^{(K)} = \frac{|c_r^{(K)0}|^2}{1 + Sx_K}, \qquad q_{rs}^{(K)} = \frac{2Sc_r^{(K)0}c_s^{(K)0}}{1 + Sx_K}. \qquad (8.33)$$

With *neglect* of overlap integrals, each orbital ϕ_r contains $q_r^{(K)0}$ $(= |c_r^{(K)0}|^2)$ electrons, while the number in each overlap region is zero. However, with inclusion of overlap each $q_r^{(K)0}$ is reduced by a factor $(1 + Sx_K)^{-1}$, and the overlap regions receive the charge taken away from the atoms; this is a physically much more satisfactory description of the bonding, conforming exactly to what we found in the two-centre case (p. 99) where overlap was included from the start.

It is also possible to derive an expression for the total electronic energy of exactly the same form as (8.24). The charges and bond orders are defined by

$$q_r = P_{rr} = \sum_K \frac{n_K |c_r^{(K)0}|^2}{1 + Sx_K}, \qquad p_{rs} = P_{rs} = \sum_K \frac{n_K c_r^{(K)0} c_s^{(K)0}}{1 + Sx_K} \qquad (8.34)$$

where n_K of the N electrons are in MO ψ_K. After some rearrangement of $\sum_K n_K \varepsilon_K$ we then find (with $\alpha_r = \alpha$ for each centre)

$$E = N\alpha + 2 \sum_{(rs)} p_{rs} \gamma_{rs} \qquad (8.35)$$

where (rs) means 'over all links r–s' and the new bond integral γ_{rs}, which replaces β_{rs} in (8.24), again has the same value for every bond, namely

$$\gamma_{rs} = \gamma = \beta - \alpha S. \qquad (8.36)$$

The introduction of overlap thus makes little difference; the energy, relative to the 'zero' $N\alpha$, is still a sum of bond orders times bond integrals and all we need do is to reinterpret the bond integral according to (8.36). If such quantities are treated as parameters, whose values are fixed by appeal to experimental data, it should make little difference which theory one uses.

The use of Wheland MO's has been discussed in a general way in a series of papers by Ruedenberg and collaborators,[†] and actual calculations fully support the above conclusions. Although numerical values of bond orders and orbital energies are affected somewhat by inclusion of the overlap integral, there is an extremely close parallel with Hückel theory; this is borne out by Fig. 8.21 which shows an accurate linear relationship between Hückel-theory bond orders and those calculated with overlap included using (8.34). The latter, since they measure the bond populations, may be expected to play a more fundamental role in, for example, discussions of the degree of π bonding and its effect on bond length, but since such correlations are essentially empirical the choice of which quantity to use is of little practical importance. Ham and Ruedenberg have shown that for alternant hydrocarbons the linear relationship in Fig. 8.21 is expected theoretically; they obtain (taking $S = 0.25$) the approximation

$$p_{rs} = p_{rs}^0 - 0.18$$

and this is good to within ± 2 per cent. The energy expression (8.35) may thus be written, substituting for p_{rs},

$$E = N\alpha' + 2 \sum_{(rs)} p_{rs}^0 \gamma_{rs} \qquad (8.37)$$

† See, in particular, Ham and Ruedenberg (1958). An alternative approach by Maslen and Coulson (1957) led to similar conclusions.

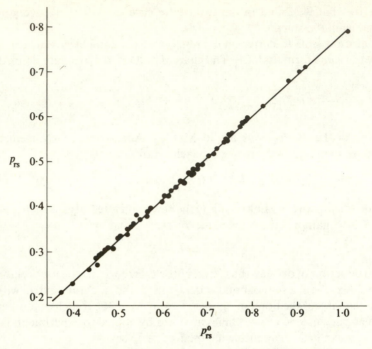

F<small>IG</small>. 8.21. Linear relationship between bond orders calculated with and without overlap. The points refer to a large number of alternant hydrocarbons.

where $N\alpha'$ represents a 'shifted' energy zero, independent of the distribution of bond orders. In other words, we may use the bond orders *computed with neglect of overlap* and take overlap into account simply by changing the values of the bond integrals.

For the reasons just discussed, Wheland's approximation is no longer much used in π-electron theory. It suggests, however, two further developments which are of importance in appreciating the nature of semi-empirical theories. In the first place, it suggests that Hückel theory may be applied no matter how large the overlap, provided we do not interpret the parameters literally but rather use the theory as a means of establishing numerical correlations with experimental data. The usual assumption of proportionality between bond integrals and overlap integrals (which are then neglected) no longer appears as a fundamental inconsistency, for with this assumption the modified bond integral γ defined in (8.36), remains proportional, like β, to the overlap integral S, though the energy formula (8.37) contains bond orders computed *as if S were zero*! Indeed it may be shown that Wheland's procedure is quite general, so long as β_{rs} is taken to be proportional to S_{rs} (for *all* pairs, not only nearest neighbours) and α_r is given a common 'mean value' for all the AO's admitted.

With this approximation there is no reason why we should not *extend* the Hückel theory to include all the valence electrons and thus to deal with a σ-bonded system on the same footing as the π electrons. The resultant 'extended Hückel theory' has been developed and widely applied by Hoffmann (1963) and many others, and, in spite of its theoretical limitations, often gives valuable insight into the forms of the MO's in complicated molecules.

Secondly, Wheland's procedure for formally 'eliminating' the overlap integrals suggests an alternative method of dealing with overlap, one which is rather less restrictive. We can illustrate this method most simply by considering the mixing of two AO's, ϕ_1 and ϕ_2, to form the MO's of a homonuclear diatomic. The secular equations with overlap included are

$$(\alpha-\varepsilon)c_1+(\beta-\varepsilon S)c_2 = 0$$
$$(\beta-\varepsilon S)c_1+(\alpha-\varepsilon)c_2 = 0.$$

Suppose, however, we started the calculation by forming two new 'atomic' orbitals, the first mainly ϕ_1 but with a little bit of ϕ_2 subtracted, and the second mainly ϕ_2 with some of ϕ_1 subtracted:

$$\bar{\phi}_1 = \phi_1-\lambda\phi_2, \qquad \bar{\phi}_2 = \phi_2-\lambda\phi_1. \tag{8.38}$$

Any LCAO MO formed from ϕ_1 and ϕ_2 can be equally well expressed in terms of $\bar{\phi}_1$ and $\bar{\phi}_2$; thus, in general

$$\psi = c_1\phi_1+c_2\phi_2 = \bar{c}_1\bar{\phi}_1+\bar{c}_2\bar{\phi}_2 \tag{8.39}$$

provided (equating coefficients of ϕ_1 and ϕ_2) we take

$$\bar{c}_1-\lambda\bar{c}_2 = c_1 \qquad -\lambda\bar{c}_1+\bar{c}_2 = c_2$$

which may be solved to give

$$\bar{c}_1 = \frac{c_1+\lambda c_2}{1-\lambda^2} \qquad \bar{c}_2 = \frac{c_2+\lambda c_1}{1-\lambda^2}. \tag{8.40}$$

In general, MO's formed from any basis set ϕ_1, ϕ_2, ϕ_3, ... can be written equally well in terms of a new basis set $\bar{\phi}_1$, $\bar{\phi}_2$, $\bar{\phi}_3$, ... (linear combinations of ϕ_1, ϕ_2, ϕ_3, ...), and the new expansion coefficients are expressible in terms of the old. Such a change of basis is called a *transformation*.

In the present example the point of the transformation is that, by choosing the numerical constant λ correctly, we can ensure that $\bar{\phi}_1$ and $\bar{\phi}_2$ are *accurately orthogonal*; the secular equations which determine the mixing coefficients of the *orthogonalized* AO's will then take the usual simple form. To determine λ, then, we require that

$$\int(\phi_1-\lambda\phi_2)(\phi_2-\lambda\phi_1)\,d\tau = S-\lambda-\lambda+\lambda^2 S = 0.$$

If S and λ are both small we can neglect the last term and then have $\lambda \simeq \frac{1}{2}S$; this approximation will serve our present purpose. The new OAO's (orthogonal-ized AO's)

$$\bar{\phi}_1 = \phi_1 - \tfrac{1}{2}S\phi_2, \qquad \bar{\phi}_2 = \phi_2 - \tfrac{1}{2}S\phi_1 \tag{8.41}$$

are also normalized if we neglect terms in S^2. In terms of $\bar{\phi}_1$ and $\bar{\phi}_2$ the MO's are of the form

$$\psi = \bar{c}_1 \bar{\phi}_1 + \bar{c}_2 \bar{\phi}_2 \tag{8.42}$$

and the new secular equations are

$$(\bar{\alpha} - \varepsilon)\bar{c}_1 + \bar{\beta}\bar{c}_2 = 0$$
$$\bar{\beta}\bar{c}_1 + (\bar{\alpha} - \varepsilon)\bar{c}_2 = 0. \tag{8.43}$$

The new Coulomb and bond integrals are

$$\bar{\alpha} = \int (\phi_1 - \lambda\phi_2)H(\phi_1 - \lambda\phi_2)\,d\tau = (1 + \lambda^2)\alpha - 2\lambda\beta$$

$$\bar{\beta} = \int (\phi_1 - \lambda\phi_2)H(\phi_2 - \lambda\phi_1)\,d\tau = (1 + \lambda^2)\beta - 2\lambda\alpha.$$

If we insert $\lambda = \tfrac{1}{2}S$ and discard terms in λ^2, this means

$$\bar{\alpha} \simeq \alpha - S\beta, \qquad \bar{\beta} \simeq \beta - S\alpha. \tag{8.44}$$

In other words, if we work throughout in terms of the new basis (8.41) we shall only need to solve the Hückel-type secular equations (8.43) (overlap integrals being accurately zero), obtaining the MO's (8.42), and at the end of the calculation reinterpret the α and β parameters according to (8.44). This procedure is not exactly equivalent to that of Wheland but is rather more general; it was first introduced by Löwdin (1950), and has been made the basis of many subsequent discussions of semi-empirical theories[†] and even as a convenient method of making *ab initio* calculations (see, for example, McWeeny and Ohno 1960). A great advantage of this transformation is that, besides making all overlap integrals vanish, it reduces the numerical values of many other troublesome integrals to the point at which they can safely be discarded; this gives a more convincing foundation for some of the current semi-empirical theories to which we turn in Chapter 13. At the same time, with the usual assumption $\beta \propto S$, the bond integral $\bar{\beta}$ over the orthogonalized AO's is also proportional to the overlap of the AO's *before* orthogonalization; this result is not inconsistent with the zero overlap of the orthogonalized AO's and provides useful guidance in assigning values to the bond integrals in semi-empirical work.

[†] See, for example, McWeeny (1954, 1955a) for VB theory; McWeeny (1964) for π-electron MO theory; and Cook *et al.* (1967) for all-electron MO theory.

Reference may be made to the literature for applications of Hückel-type MO theory to the σ bonds in organic molecules. The basis sets used may consist of the usual s and p AO's on the constituent atoms (see for instance Hoffmann 1963), in which case the resultant MO's, obtained by numerical solution of large secular equations, are completely delocalized. Alternatively, we may preserve contact with the qualitative picture (§8.1) by working with hybrids, chosen to give a 'natural' description of the localized σ bonds; this approach was pioneered by Sandorfy (1955) and was subsequently developed and used by Yoshizumi (1957), Pople and Santry (1964), and others. In both cases it is normally possible to find an equivalent set of localized combinations of the MO's, namely the bond orbitals discussed in §7.3, which recognize more clearly the near independence of electron-pair bonds upon which so much of qualitative valence theory is based. We shall therefore not pursue such extensions of Hückel theory at this point (see, however, Problem 8.14), though we shall need them later in discussing transition-metal complexes (§9.13).

8.9. Orbital energies and the total energy

Even when simple MO theory is improved and extended, as in the last section, it apparently provides only a semi-empirical scheme for calculating the *orbital* energies ε_K associated with an electron in orbital ψ_K moving in some hypothetical field (ideally the self-consistent field) due to the nuclei and all other electrons. These energies are of value because (in most cases) they correctly suggest the order of filling of the orbitals in the *aufbau* approach, and because they have a theoretical connection with ionization potentials ($I_K = -\varepsilon_K$) and spectra and therefore help us to understand ionization and other excitation processes. At the same time, as in (8.25), it is commonly assumed that

$$E_{\text{orb}} = \sum_K n_K \varepsilon_K$$

(namely the sum of the orbital energies, ε_K being counted n_K times if there are n_K electrons in ψ_K) may be used instead of the total energy E in discussing such questions as the stability of a molecule or the dependence of energy on geometry. In §7.3, for example, we used essentially E_{orb} in discussing the shape of the water molecule, and in §8.5 we have used E_{orb} (for the π electrons alone) in discussing the stabilizing effect of electron delocalization. The effect of electron–electron repulsions has been introduced only in connection with the purely qualitative interpretation of molecular shapes in terms of lone-pair and bond-pair interactions (e.g. in §7.4), and the inter*nuclear* repulsions have apparently been forgotten altogether. These are serious criticisms of the MO theory as we have developed it so far, and clearly require some comment.

First, it must be noted that the approximations implied are gross; the neglected electron repulsion energy in a molecule as small as N_2 amounts to about 34 000 kJ mol^{-1}, which is enormous compared with the energy required to change the shape of the molecule or to break a bond! Why, then, should we expect E_{orb} to have any predictive value at all? The answer can only be that for E_{orb} to be of any use its *variations*, at least, must in some way reflect those of the total energy E. If, for example, an increase of E_{orb} in a given geometry change, or a given substitution reaction, necessarily implied a corresponding increase of E (as it would if, for example, E were proportional to E_{orb}), then it would certainly be useful to study the simpler quantity, at least as an indicator of likely trends.

Now careful study of the results of accurate *ab initio* SCF calculations on molecules in their equilibrium geometries *does* suggest an approximate proportionality between E and E_{orb}, and this observation has provided a stimulus for a considerable amount of recent fundamental work. Why should there be any simple relationship between E and E_{orb}? The total energy, in the Hartree–Fock approximation, is given by

$$E = \sum_K n_K \varepsilon_K - V_{ee} + V_{nn} = E_{orb} + (V_{nn} - V_{ee}) \tag{8.45}$$

where V_{nn} is the internuclear Coulomb repulsion energy and V_{ee}, the electron–electron repulsion energy, is *subtracted* for reasons explained in §4.2 (see also Table 4.2 *et seq.*). The simplest rationalization for the use of E_{orb} instead of E would obviously be the hypothesis that internuclear and interelectron repulsion energies were roughly equal, so that (8.45) would reduce to

$$E \simeq E_{orb}$$

but quantitative studies show this assumption to be untenable.

The next step is to look for fundamental relationships among the various contributions to the total energy. There is, for example, a theorem to the effect that complete knowledge of the electron density function P (including its dependence on parameters such as nuclear charges and positions) is in principle sufficient to determine the ground-state energy. Since, however, P directly determines only the electron–nuclear term V_{en} this implies that it *indirectly* determines all the other terms. In fact, Politzer (1977) has given theoretical arguments for a relationship, for a molecule in its ground state and with equilibrium geometry,

$$E \simeq \tfrac{3}{7}(V_{en} + 2V_{nn})$$

in which V_{ee} and the electronic kinetic energy have been eliminated. As Ruedenberg (1977) has pointed out, this implies an approximate relationship (at the level of SCF theory) of the desired form:

$$E \simeq \tfrac{3}{2} \sum_K n_K \varepsilon_K = \tfrac{3}{2} E_{orb}. \tag{8.46}$$

There is considerable support for this relationship from actual *ab initio* calculations; it gives estimates of the total electronic energy (nuclear repulsions included as usual) which are usually within 2 or 3 per cent of the computed values.

There is no doubt that the existence of an approximate relationship of this kind considerably strengthens many of the pictorial and qualitative arguments which are characteristic of simple valence theory. At the same time it raises intriguing theoretical problems which lie at the frontiers of present-day molecular quantum mechanics.

Problems

8.1. Draw a curve to show the expected variation of electronic energy against angle of twist (θ) around the C–C bond in ethylene, taking $\theta = 0$ for the 'eclipsed' conformation. Discuss some of the factors which determine the equilibrium geometry.

8.2. Discuss *quantitatively* the variation of π-electron energy in ethylene and draw a curve similar to that in Problem 1. What would be the effect of exciting one π electron to the antibonding MO? Could this have photochemical implications? (Hint: use the vector property of p orbitals to find the dependence of β on angle of twist.)

8.3. Find any conjugated molecules you can (not already cited) to illustrate sp^2 nitrogen and sp^2 oxygen, respectively, providing (*a*) one π electron, and (*b*) two π electrons. (Refer to any textbook of organic chemistry.)

8.4. Nitramide ($H_2N{\cdot}NO_2$) is non-planar at the amine nitrogen. Propose suitable valence states for the nitrogen and oxygen atoms.

8.5. Predict the bond length in the CH radical by the method which led to Table 8.1. (The observed length is $0{\cdot}112$ nm.)

8.6. Perform a Hückel calculation for the allyl radical (polyene chain with $N = 3$) and compare your results with those obtained using equations (8.18) and (8.19). (Hints: write down the secular equations in the 'x form' and set up the secular determinant. Expand and equate to zero, obtaining a cubic equation which factorizes. Hence obtain three MO energies. Use each, in turn, to find the corresponding MO coefficients. (You may put $c_1 = 1$ and normalize at the end.)

8.7. Consider Problem 8.6 again but use symmetry about the central carbon atom to set up symmetric and antisymmetric combinations of the original AO's; then deal with the two symmetry types separately. (Hint: read §7.3 again, if necessary, and proceed in the same way.)

8.8. Verify that for $N = 6$ the general equations (8.23) lead to the benzene MO's in Table 8.2.

8.9. Calculate the two bonding π MO's in butadiene ($N = 4$) by the method used in Problem 8.7. Then calculate the total energy and the bond orders in the three C–C links and check your results by use of (8.24). Confirm the value of the delocalization energy given in Table 8.3. Show (by considering reflections or rotations as appropriate) that a similar argument applies for either the cis or trans form of the molecule.

8.10. Extend the method of Problem 8.7 to the case where there are *two* convenient planes of symmetry, as in the naphthalene molecule

Set up symmetry orbitals of four types SS, SA, AS, AA (symmetric across both planes, symmetric across the first but antisymmetric across the second, etc.) and solve separately for each type. How does the total π-electron energy compare with that of the cyclic polyene formed by removing the central C–C link?

8.11. Discuss whether the π-electron energy favours ring opening or ring closing for N = 4 up to $N = 10$. What do your results imply as regards the stability of conjugated rings and how do your conclusions compare with those provided by Hückel's $4n + 2$ rule? (Hint: use (8.19) and (8.22), for chains and rings, respectively.)

8.12. Strike out the nitrogen atom in quinoline (Fig. 8.16 (*b*)) and obtain the NBMO of the residual molecule by use of the rule (8.27). By considering reconnection of the nitrogen, shows that the charge distribution in Fig. 8.16 (*b*) is of the form to be expected.

8.13. The hyperfine structure of the ESR peak in the spectrum of the perinaphthenyl radical

indicates strong coupling between the protons attached at the starred positions and the odd π electron from which the signal originates. How do you account for this observation? (Hint: find the NBMO using the conditions given in the text. Whenever you cannot fix the value of a coefficient call it x (or y, or z, ...) and continue. Finally your conditions will determine all the unknowns.)

8.14. Set up the secular equations for an all-valence-electron MO treatment of the ethylene molecule, using three sp^2 hybrids and one $2p_\pi$ orbital on each carbon atom (all assumed orthogonalized). Show that with neglect of β's between hybrids on the same atom the equations are easily solved. How will the resultant MO's be occupied? By suitably combining the MO's which describe the C–H σ bonding show how four localized bond orbitals may be obtained, and hence justify the 'pairing' of strongly overlapping AO's as a useful approximate procedure for describing localized σ bonds. (Hints: use the same symmetry classification as in Problem 8.10, but with an additional σ or π factor. With an obvious notation you will need for the σ electrons $\alpha_C, \alpha_H, \beta_{CH}, \beta_{CC}$, β'_{CC} (intra-atomic), the last being small. Deal with each symmetry type separately in the usual way.)

9

Transition-metal compounds

9.1. d-Electrons in the transition elements

It is well known that the transition elements form a very wide variety of complexes. These vary from largely ionic compounds such as $[FeF_3]^{3-}$ to hydrated complexes such as $[Ti(H_2O)_6]^{3+}$. It seems almost inevitable that this wide variety of behaviour should be related in some way to the chief characteristic of the transition elements, that they all possess d electrons. Apart from Pauling's use of d orbitals to form hybrids, as discussed in §7.7, we have not so far needed to use them in discussing bond formation. This chapter, however, will be concerned with the behaviour of atoms having varying numbers of d electrons. It is important, therefore, that we should understand their distribution and properties in the gaseous atoms. After that we can see how this distribution is affected by the presence of surrounding groups, generally called ligands, placed around the atom in some definite pattern.

We saw in §2.4 that there are five distinct d orbitals, so that in any shell an atom can accommodate not more than 10 d electrons. As we pass from scandium to copper in the first long period we progressively fill up the 3d subshell. Thus Table 9.1 shows the electron distribution outside an argon-like core ($1s^2 2s^2 2p^6 3s^2 3p^6$) when the atom is in its ground state. At the beginning of this series the 4s electron (designated simply as s) is lower in energy than the 3d electron (designated simply as d), but as we move across the table the 3d electron drops relative to the 4s electron, though the energy differences are nowhere very large. Thus the mean size of the 4s and 3d charge clouds must be similar at the beginning of the row, but at the end of the row the d orbitals are more compact than the s. In neutral Fe, for example, self-consistent-field calculations show that the outer maximum of the 4s orbital is about 4·5 times as far out from the nucleus as the maximum of the 3d orbitals. These latter orbitals have approximately the same size as the 3p orbitals, so in many

TABLE 9.1

Atom	K	Ca	Sc	Ti	V	Cr	Mn	Fe	Co	Ni	Cu	Zn
Electron configuration	s	s^2	ds^2	d^2s^2	d^3s^2	d^5s	d^5s^2	d^6s^2	d^7s^2	d^8s^2	$d^{10}s$	$d^{10}s^2$

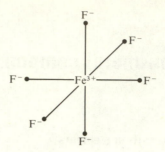

FIG. 9.1. Octahedral arrangement of ligands in $[FeF_6]^{3-}$.

respects they may be treated as if they belonged to an inner core, not greatly perturbed by chemical and other bonding. Thus (see Stern 1956) within the distance (about $2 \cdot 5a_0$) which represents one-half of the interatomic distance in metallic iron, no less than 95 per cent of the d-electron charge-cloud is to be found. In $[FeF_6]^{3-}$ (Fig. 9.1), therefore, it will be a reasonable approximation to think of a ferric ion Fe^{3+} with a total of five 3d electrons at the centre of an octahedron of six fluoride ions F^-. Similarly, in the hydrated ferrous ion there will be six 3d electrons around the central atom, and in $[Cr(NH_3)_6]Cl_3$ these will be the three d electrons of Cr^{3+} surrounded by six neutral ammonia molecules. In all such cases we are using a simple ionic model in which the electronegative atoms complete their octets, leaving a positive metal ion with a compact outer shell of d electrons.

9.2. Crystal-field, or electrostatic, splitting

In the isolated atom all five d orbitals have the same energy, but when the atom is perturbed by surrounding ligands this is no longer true. It becomes of the greatest importance to know how the degeneracy is split. If the perturbation can be represented by an electrostatic field we can speak of this as electrostatic splitting. Such situations arise particularly in ionic crystals (e.g. NaCl which may be regarded, anticipating Chapter 11, as a lattice of Na^+ and Cl^- ions), and it was in connection with this that Bethe (1929) first studied the splitting of orbitals of any symmetry type.† For this reason it is frequently referred to as crystal-field splitting. Different crystal types give rise to electrostatic fields which possess different symmetries. The way in which the degeneracy of the d electrons is split depends on this symmetry. A knowledge of group theory is desirable in order to understand the details of all this, but we shall find that considerable progress is possible with the aid of a few simple pictorial concepts. Some of the group theory is added in Appendix 3.

---

† The first *molecular* application appears to be by Van Vleck (1935).

9.3. A simple example: the square planar complex

Let us consider (Fig. 9.2) a central atom with d orbitals placed symmetrically between four ligands on the x and y axes. In all cases in which we shall be interested the ligands either possess a net negative charge (e.g. F^-) or align themselves so that a negative region of charge (e.g. an oxygen lone pair in H_2O) is nearest to the central atom. Thus the electrostatic field will be such as to repel electrons from the directions of the x and y axes.

We must now choose the five appropriate d orbitals. Symmetry suggests that the orientation shown in Fig. 9.2 is appropriate, for the orbitals then behave in the simplest possible way under the symmetry operations of reflection and rotation which characterize the symmetry of the square. The d_{z^2} AO, in particular, is uniquely fitted to lie along the z axis and is invariant against rotation around the axis and against reflection across any plane containing it.

Fig. 9.2 (a) shows that in the $d_{x^2-y^2}$ orbital all four lobes of charge are unfavourably placed relative to the ligands, shown as large dots. Therefore the energy of this orbital is raised very considerably by the electrostatic repulsion from the ligands. In Fig. 9.2 (b), however, we can see that the d_{xy} orbital energy is rather less raised. The remaining three orbitals are scarcely affected, since their greatest concentrations of charge lie in regions remote from the perturbation. Such considerations are not infallible, since the energy order depends on the detailed forms of the d orbitals and the perturbation, but from symmetry the d_{xz}, d_{yz} pair must always be degenerate. Thus the original five-fold degeneracy is split into one doublet level and three distinct singlet levels. The magnitudes of the splittings are proportional to the electrostatic field, but for fields of sufficiently simple form their ratios depend solely on the symmetry and therefore only need to be calculated once for each type of symmetry and each central atom. Figure 9.3 shows the usual order of the levels.

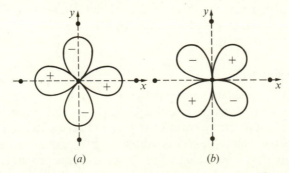

(a)   (b)

FIG. 9.2. Choice of d orbitals for a square planar complex. The dots denote the four ligands. With negatively charged ligands $d_{x^2-y^2}$, in (a), will lie at a higher energy than d_{xy}, in (b).

FIG. 9.3. d-orbital energies in a square planar complex.

One simple deduction can be made from Fig. 9.3. Since each orbital will accommodate two electrons with paired spins, there should be a tendency for systems with eight or nine d electrons to form square planar complexes, though this tendency should be much greater when there are eight electrons than when there are nine. This is because, with eight or nine electrons, we can leave the highly unfavourable orbital $d_{x^2-y^2}$ either completely empty or at least half empty. An example of this tendency is in the nickel cyanide ion $[Ni(CN)_4]^{2-}$ where, if we suppose it to be in the form $Ni^{2+}(CN^-)_4$, we see that there are eight d electrons. The ion is square planar. With fewer than eight electrons other symmetries may be preferred; with an octahedral arrangement of ligands there is a degenerate *pair* of highly unfavourable orbitals and both are left empty when only six electrons are present. This is the case if we put Fe instead of Ni, and in fact the ferrocyanide ion $[Fe(CN)_6]^{4-}$ is octahedral. It is a general rule, however, that Ni, Pd, and Pt tend to form planar complexes when there are eight d electrons.

9.4. Other symmetries

The most important geometrical symmetry found in molecular complexes is the octahedral type shown in Fig. 9.1. If we choose the directions from the central metal atom to the ligands as the x, y, and z axes, then our choice of d orbitals will be the same as in §9.3. A simple diagram, similar to Fig. 9.2, shows that now the splitting is such that d_{xy}, d_{zx} remain degenerate and considerably lower in energy than d_{z^2}, $d_{x^2-y^2}$, which also remain degenerate. The fact that d_{z^2} and $d_{x^2-y^2}$ are degenerate can be understood if we realize (cf. equation

(2.20)) that d_{z^2} is a linear combination of $d_{z^2-x^2}$ and $d_{z^2-y^2}$, which by symmetry must have the same energy as $d_{x^2-y^2}$.

The splitting into a triplet and a doublet can be established by group theory. In group-theoretical terminology (Appendix 3) we say that

d_{z^2} and $d_{x^2-y^2}$ are of symmetry e_g (sometimes called d_γ)

d_{xy}, d_{yz}, d_{zx} are of symmetry t_{2g} (sometimes called d_ε).

We shall frequently have occasion to refer to them by their symmetry names, in which the subscript g has the same significance as in §4.6 and in which we no longer need to distinguish between the *x*, *y*, and *z* axes.

It is desirable to have some scale on which the changes of energy can be represented. For this purpose it is useful to refer the splitting to an energy zero chosen as the centroid of the whole group of d levels. At this stage we are chiefly interested in the sequence of energy levels and their mutual separation. The whole group of levels may be shifted by the presence of the ligands but the *absolute* positions of the levels are not very important. In fact, the presence of negatively charged ligands raises all the levels by an amount which is usually much larger than the relative splittings, but in drawing energy-level diagrams it is customary to ignore this overall shift. Since there are three lower levels and two upper levels, it follows that their displacements from the centroid are in the ratio 2:3. It has become conventional to represent the overall splitting either by the symbol Δ, or (in older work) by $10Dq$, where $Dq > 0$. We therefore say that the e_g levels are raised $6Dq$ ($=\frac{3}{5}\Delta$) and the t_{2g} levels are lowered $4Dq$ ($=\frac{2}{5}\Delta$).

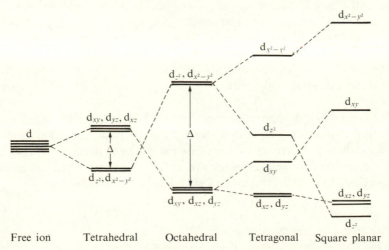

FIG. 9.4. Usual order of d-orbital energies. In octahedral symmetry the ligands are on the *x*, *y*, and *z* axes at the centres of six cube faces; in tetrahedral symmetry they are at four cube corners; in tetragonal symmetry the cube is stretched along the *z* axis (giving a square prism); in the square planar case the *z*-axis ligands are removed.

Similar calculations may be made for other types of symmetry. Figure 9.4 shows the results for the main symmetries (adapted from Griffith and Orgel 1957 and Pearson 1959). This is an exceedingly important diagram, on which much of the rest of our discussion will depend.

The question may be asked: What would have happened if we had not used the particular choices of axes, and hence of d orbitals, shown in Fig. 9.4? The answer to this question is that if we had chosen quite different directions for the five fundamental d orbitals, a full calculation of the effect of the perturbation would have yielded perturbed AO's which were linear combinations of the original set and were indeed new d orbitals, symmetrically aligned with respect to the ligands as we assumed at the beginning of this section. It is fortunate that when the ligands are used to define the x, y, and z axes, the corresponding d orbitals are appropriate and group theory is not required.

9.5. The spectrochemical series

Experimentally, most Δ values lie in the range

$$7000 < \Delta < 30\,000 \,\text{cm}^{-1}.$$

Since $8068 \,\text{cm}^{-1}$ is approximately 1 eV, this means that the usual values of Δ are in the range from 1 to 4 eV. It would be very nice if we could make absolute calculations of this crystal-field splitting. Unfortunately this has not yet been possible, since even now our knowledge of the sizes of d orbitals in complex atoms is still very incomplete and our ability to calculate the polarization of an orbital (in the field of the ligands) is yet smaller! It is reasonable enough to argue, for example, that, since the core in Na^+ is much more compact than in Au^+ the polarizability should be smaller, but less easy to distinguish between the d electrons in Fe and Ni, since in these atoms the d electrons lie well inside the outer shell of electrons. Nor, indeed, would it be sufficient to know the numerical value of Δ, as we shall see later, since other effects beside the purely electrostatic crystal-field splitting play a part in the complete electronic description.

It has been found empirically that many of the more frequent ligands can be arranged in a series such that Δ increases as we pass along this series. The precise values of Δ depend, of course, on the choice of the central metal atom, but the spectrochemical series, as it is called, is almost independent of this choice (Tsuchida 1938; Jørgensen 1962). The order of increasing Δ is

$$I^-, Br^-, S^{2-}, Cl^-, F^-, OH^-, C_2H_5OH, H_2O, NH_3, en, NO_2^-, CN^-, CO,$$

where en = ethylenediamine (see p. 276).

The most effective means of estimating Δ is from a study of the energy required to excite an electron from one of the d levels to another. The wavelength of light absorbed frequently lies in the visible region, a situation responsible for the colour of many of these complexes. Perhaps the simplest example is the aquo-complex of titanium, $[Ti(H_2O)_6]^{3+}$. Here the central metal atom has just one d electron, in an octahedral environment, so that the lowest state (see Fig. 9.4) should be the triply degenerate t_{2g} level. Excitation of the electron to the doubly degenerate e_g level should require energy Δ and be associated with an absorption frequency v, where $\Delta = hv$. Experiments show a band at $20\,000\,cm^{-1}$, which may be shown to arise from this transition $t_{2g} \rightarrow e_g$. Thus $\Delta = 20\,000\,cm^{-1}$ approximately. With more than one d electron there are other factors to be considered, as will appear later, but the essence of the method is the same.

It should be mentioned that, partly on account of their larger size and greater polarizability, values of Δ for the second and third transition series (involving the 4d and 5d orbitals respectively) are 40–80 per cent larger than for the first transition series (involving 3d orbitals).

9.6. Strong and weak fields: high and low spin

We now look at some of the additional factors that arise when more than one d electron is involved. For the present we neglect all interaction with the ligands other than that represented by a crystal field of appropriate symmetry. Later we shall relax this restriction.

Consider first the octahedral vanadium complex $[V(H_2O)_6]^{3+}$ in which the central atom is conventionally represented as V(III) or V^{3+}. Neutral vanadium has five valence electrons, so V^{3+} has two. According to Fig. 9.4 we must put these two electrons into the three lowest t_{2g} levels and according to Hund's rules they should go into different members of the degenerate set and should have parallel spins. In the corresponding chromium complex $[Cr(H_2O)_6]^{3+}$ with three d electrons we shall expect one to go in each of the three t_{2g} orbitals with parallel spin. Since, with $S = \frac{1}{2}$, each electron carries a magnetic moment of μ_B ($g\mu_B S \rightarrow \mu_B$), this means that these two complexes will appear to have resultant moments of $2\mu_B$ ($S = 1$) and $3\mu_B$ ($S = \frac{3}{2}$), respectively. However, when we come to the fourth d electron, as in $[Cr(H_2O)_6]^{2+}$ involving Cr(II), it is not clear whether we should put it in one of the t_{2g} orbitals, with opposite spin to the previous three electrons, or whether it is better to put it in one of the e_g orbitals. It is true that such orbitals lie higher than the t_{2g} orbitals by an amount Δ, but if Δ is not too big we may regain this energy by separating the electrons as much as possible and keeping their spins parallel, for then their mutual Coulomb repulsion is small, and their exchange interaction (p. 136) lowers the total energy. It all depends on the magnitude of Δ. If Δ is small we continue by putting up to two further electrons in the e_g orbitals, all with parallel spin, and then, when the number of d electrons exceeds five, we begin to complete the t_{2g} levels with opposed spins. If Δ is large we complete the t_{2g} subshell before starting the e_g subshell. In the first case the spin has a high value; in the second it has a low one. This is the origin of the terms high-spin and low-spin complexes. Table 9.2 shows the two situations

TABLE 9.2
Spins in octahedral complexes

No. of d electrons	High-spin (weak-field) arrangement		Resultant spin	Low-spin (strong-field) arrangement		Resultant spin
	t_{2g}	e_g		t_{2g}	e_g	
1	↑ — —		$\frac{1}{2}$	↑ — —		$\frac{1}{2}$
2	↑ ↑ —		1	↑ ↑ —		1
3	↑ ↑ ↑		$1\frac{1}{2}$	↑ ↑ ↑		$1\frac{1}{2}$
4	↑ ↑ ↑	↑ —	2	↑↓ ↑ ↑		1
5	↑ ↑ ↑	↑ ↑	$2\frac{1}{2}$	↑↓ ↑↓ ↑		$\frac{1}{2}$
6	↑↓ ↑ ↑	↑ ↑	2	↑↓ ↑↓ ↑↓		0
7	↑↓ ↑↓ ↑	↑ ↑	$1\frac{1}{2}$	↑↓ ↑↓ ↑↓	↑ —	$\frac{1}{2}$
8	↑↓ ↑↓ ↑↓	↑ ↑	1	↑↓ ↑↓ ↑↓	↑ ↑	1
9	↑↓ ↑↓ ↑↓	↑↓ ↑	$\frac{1}{2}$	↑↓ ↑↓ ↑↓	↑↓ ↑	$\frac{1}{2}$
10	↑↓ ↑↓ ↑↓	↑↓ ↑↓	0	↑↓ ↑↓ ↑↓	↑↓ ↑↓	0

and the resultant spin. In general, for a given metal atom high-spin complexes are associated with small Δ, and low-spin complexes with large Δ. For this reason the two types of complex have often been called weak-field and strong-field complexes. In the case of $[Cr(H_2O)_6]^{2+}$, which we mentioned earlier, the fourth electron goes into the e_g orbital ($\Delta = 14\,000\,\text{cm}^{-1}$) leading to a high-spin complex. In the case of $[Co(H_2O)_6]^{3+}$ with six 3d orbitals in Co(III), however, we have a low-spin complex represented by t_{2g}^6. Table 9.2 shows that it is only for the cases of d^4–d^7 that the high-spin and low-spin situations differ. This magnetic criterion for distinguishing these two situations was recognized very early as being of the greatest importance (Pauling 1960; see Figgis and Lewis 1964 for a review of the magnetochemistry of transition metal complexes).

A particular example of the distinction between high-spin and low-spin complexes is illustrated in Fig. 9.5 (taken from Nyholm 1958) which shows the crystal-field splittings in two different octahedral complexes of the ferric ion. In the fluoride, with small Δ, we have a high-spin complex, but in the cyanide, with large Δ, we have a low-spin complex.

There are two other effects of a purely electrostatic crystal field which must be briefly mentioned. Both have the result of complicating the simple picture just described. The first, called spin–orbit coupling, arises because there is an interaction between the magnetic effect of the electron in its orbit (orbital magnetic moment) and the spin. In the case of a single electron, such as our previous example of Ti(III) as in $[Ti(H_2O)_6]^{3+}$, spin–orbit interaction lifts the

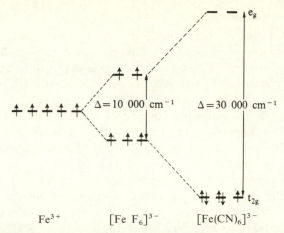

Fe³⁺   [Fe F₆]³⁻   [Fe(CN)₆]³⁻

FIG. 9.5. Crystal-field splitting in octahedral complexes of the ferric ion. Small Δ gives a high-spin complex; large Δ gives a low-spin complex.

degeneracy of the t_{2g} levels. Fortunately, however, this splitting is not usually large except for heavy atoms (see Slater 1960). The second effect is known as the Jahn–Teller effect (Jahn and Teller 1937; for a simpler approach see Clinton and Rice (1957)): a symmetrical non-linear molecule cannot remain in equilibrium in an orbitally degenerate state, but will inevitably distort in such a way that the degeneracy is lifted. The case of Ti(III) will serve as an illustration (Carrington and Longuet–Higgins 1960). Here the single electron is in a triply degenerate orbital level t_{2g}. The result is an unstable system, which distorts in such a way that the octahedron is compressed or extended along the x, y, or z direction. Figure 9.6 shows how one of the three t_{2g} levels separates out below the others. Instability of this kind arises whenever the orbitals of a degenerate group are unequally filled, for then the charge cloud lacks the full symmetry of the molecule and the resultant asymmetric pattern of forces on the nuclei produces the distortion. This effect is usually quite small.

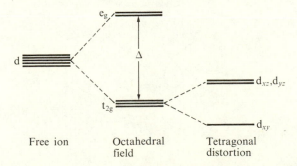

Free ion   Octahedral field   Tetragonal distortion

FIG. 9.6. Jahn–Teller splitting of the t_{2g} level (octahedral field) by tetragonal distortion (compression or expansion along the z axis).

9.7. Ligand-field theory

Our discussion so far has been entirely in terms of the crystal field provided by the ligands and its perturbing effect on the energies of the d-shell electrons. We have supposed the ligands to be held by simple electrostatic attractions, and no other participation of ligand electrons has been considered. This is a serious omission, however, since there are many cases where electrons from the ligands take part in some form of covalent bonding to the central metal atom. There are two ways in which this bonding takes place, via σ bonds and via π bonds. Since the σ-bond contribution is usually the more important we shall discuss it first.

Our problem may be put in the following way. Given that there are five d orbitals on the metal atom and one orbital on each of the ligands, what molecular orbitals can be formed from their linear combination? The ligand orbitals will be taken to have axial symmetry relative to the directions joining each ligand to the metal atom. In the case of the fluoride ion, as in $[FeF_6]^{3-}$ represented in Fig. 9.1, the σ orbital is the $2p_\sigma$ fluorine orbital whose axis lies along the line joining the Fe and F nuclei. In the case of the cobaltammines, such as $[Co(NH_3)_6]^{3+}$, the σ orbitals are the lone-pair hybrids of each ammonia molecule, as indicated in Fig. 7.18. In the case of a chelating ligand such as ethylenediamine $NH_2 \cdot CH_2 \cdot CH_2 \cdot NH_2$, which occupies two of the co-ordinating sites (see Fig. 9.7), the ligand orbitals are the lone-pair orbitals of the two end nitrogen atoms. The MO's that we are now going to form will obviously be completely delocalized, so that, from the mathematical point of view, we have a situation similar to that dealt with in Chapter 8 where we formed delocalized MO's for aromatic and conjugated molecules. The difference now is that each separate ligand orbital has local σ character and not π character.

Let us return to the case of a square planar complex represented in Fig. 9.2. We want to know how to combine four ligand orbitals in such a way that these combinations may have the right symmetry to combine with suitable s, p, and

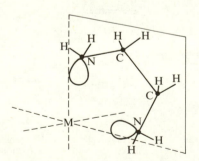

FIG. 9.7. The chelating property of ethylenediamine.

d orbitals on the central atom. Group theory provides the general method of solving this problem, but again we can use a simple pictorial approach, as in §7.3 and in several of the problems on Chapters 7 and 8. References are given in Appendix 3. What we must do is combine the ligand orbitals into *symmetry orbitals* which behave in the same way as the AO's on the central atom when subjected to the various reflections or rotations which describe the symmetry of the molecule.

Let us define the ligand orbitals in such a way that each of them is positive in the region where they approach the central atom. Then if, as in Fig. 9.8, we refer to these ligand orbitals as A, B, C, D, we can see that the allowed combinations are as follows.

(i) $A+B+C+D$, which has the full symmetry of the complex and will combine with a central s orbital.

(ii) $A-B$, which has the same symmetry as a p_x orbital at the centre and will therefore combine with it.

(iii) $C-D$, which has the same symmetry as a p_y orbital at the centre and will therefore combine with it.

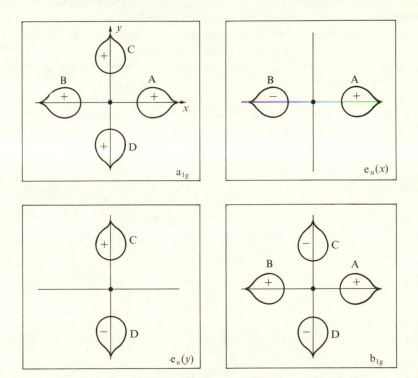

Fig. 9.8. Appropriate combinations of ligand σ orbitals for a square planar complex, labelled according to symmetry type.

(iv) $A + B - C - D$, which has the same symmetry as a central $d_{x^2-y^2}$ orbital (see Fig. 9.2 (a)) and will combine with it.

On account of the almost zero overlap between A, B, C, and D the energies of these four combinations of $A, \ldots, D$ will all be about the same and effectively equal to that of A alone. The correct group-theory designations of these four combinations, which will also apply to the MO's formed by them, are shown in Fig. 9.8. We must also take into account the remaining d-type AO's on the central atom (see Fig. 2.9) namely $d_{z^2}(a_{1g})$, $d_{xy}(b_{2g})$, and d_{yz} (an e_g pair). The resultant MO's will be as follows:

$$\{A+B+C+D\} + \lambda_1 s + \lambda_2 d_{z^2} \qquad \text{Symmetry } a_{1g}$$

$$\left.\begin{array}{l} \{A - B\} + \mu p_x \\ \{C - D\} + \mu p_y \end{array}\right\} \qquad \text{Symmetry } e_u$$

$$\{A + B - C - D\} + \nu d_{x^2-y^2} \qquad \text{Symmetry } b_{1g}$$

$$\left.\begin{array}{l} d_{xz} \\ d_{yz} \end{array}\right\} \qquad \text{Symmetry } e_g$$

$$d_{xy} \qquad \text{Symmetry } b_{2g}$$

The numerical values of the coefficients λ, μ, ν must be found by the use of the variation method. Solution of the secular equations for each symmetry type will give three a_{1g} levels, two degenerate pairs of e_u levels (x type and y type), two b_{1g} levels, a single degenerate e_g pair, and a single b_{2g} level. As usual (§3.9) the levels corresponding to the orbitals which are mixed behave as if they repelled one another.

The octahedral situation can be dealt with in a similar way. On account of its great importance we have listed in Table 9.3 the orbitals on the central

TABLE 9.3
Molecular orbitals for octahedral complexes

Central atom AO's	Ligand σ- group orbitals	Ligand π- group orbitals	Symmetry name
s	$(A+B+C+D+E+F)$	None	a_{1g}
p_x	$(A-B)$	$(C_{\pi x}' + D_{\pi x} + E_{\pi x} + F_{\pi x})$	
p_y	$(C-D)$	$(A_{\pi y} + B_{\pi y} + E_{\pi y} + F_{\pi y})$	t_{1u}
p_z	$(E-F)$	$(A_{\pi z} + B_{\pi z} + C_{\pi z} + D_{\pi z})$	
$d_{x^2-y^2}$	$(A+B-C-D)$	None	e_g
d_{z^2}	$-(A+B+C+D)+2(E+F)$		
d_{xy}		$A_{\pi y} - B_{\pi y} + C_{\pi x} - D_{\pi x}$	
d_{yz}	None	$C_{\pi z} - D_{\pi z} + E_{\pi y} - F_{\pi y}$	t_{2g}
d_{zx}		$A_{\pi z} - B_{\pi z} + E_{\pi x} - F_{\pi x}$	

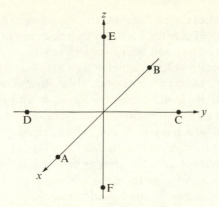

FIG. 9.9. Labelling of octahedral ligand orbitals.

atom, and the particular combinations of the ligand orbitals $A, B, \ldots, F$ (Fig. 9.9) with which an LCAO molecular orbital can be built. We may now form these linear combinations. A typical situation is shown in Fig. 9.10, where the metal orbitals are on the left, the ligand group orbitals are on the right, and in the middle there are the full molecular orbitals. Since the ligand groups are usually more electronegative than the metal itself, we have drawn the right-hand energy level below any of those on the left.

We can now see how the new diagram in Fig. 9.10 modifies the picture previously presented in Fig. 9.5. Formerly we considered only the d electrons

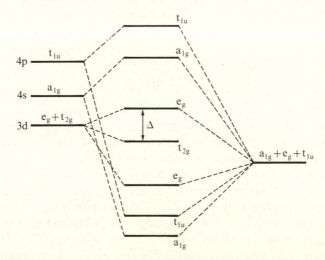

FIG. 9.10. Energy levels of MO's in an octahedral complex (centre) related to those of the metal AO's (left) and the ligand group orbitals (right). The e- and t-type orbitals are doubly and triply degenerate respectively. (Only the order of the levels is significant.)

of the central atom, but now we include the ligand electrons and also s and p electrons of the metal. There are normally two electrons in each ligand orbital $A \ldots F$ (in the ionic picture), and according to Fig. 9.10 they will completely fill the three lowest levels, which comprise six $(1+3+2)$ MO's, and can thus accommodate all 12 electrons. However, these 'ligand' electrons have now partly gone back into the metal-atom orbitals 3d, 4s, and 4p, thus reducing the purely ionic character of the bonding in the crystal-field model. Of course, the previously purely metal orbitals such as d_{z^2} and $d_{x^2-y^2}$ (i.e. the e_g pair) will now involve a fraction of the ligand orbitals. Thus the separation Δ, which is marked in the figure, is not quite the same as before. It is much more difficult to calculate this Δ, since both metal-atom and ligand orbitals are involved. However, if, as often happens, the ligand orbitals lie a long way below the metal-atom d orbitals, these e_g and t_{2g} orbitals will correspond quite closely to those calculated on the electrostatic crystal-field basis.

Diagrams such as that of Fig. 9.10 may be drawn for each type of symmetry represented in Fig. 9.4. They enable the allocation of electrons to be made to the various orbitals. Further, the distinction between high-spin and low-spin complexes is unaffected by the inclusion of the ligand orbitals.

The theory just described is usually known as ligand-field theory, to distinguish it from the less accurate but simpler electrostatic crystal-field theory with which we started.

9.8. Relation with Pauling's octahedral hybrids

We are now in a position to relate our ligand-field theory description of an octahedral complex to the hybridization theory of Linus Pauling, which may be applied as indicated in §7.7. Let us consider the ferricyanide $[Fe(CN)_6]^{3-}$ as an example in which the central atom is a transition metal. In Pauling's approach it is assumed initially that the negative charge resides on the metal atom, which will then have the electron configuration

$$Fe^{3-}[KL3d^9 4s^2].$$

Each CN group possesses an odd electron and we therefore set up six singly occupied octahedral hybrids (d^2sp^3) to form six σ-type electron-pair bonds. That leaves five electrons to occupy the lower-energy d orbitals, which remain essentially atomic. If the bonds were described by a wavefunction of VB type, without admitting ionic structures, each bond would associate one electron with the metal and one with the ligand, and the system would have to be described as $Fe^{3-}(CN)_6$. If they were described by a purely ionic VB structure, each electron pair would be associated with the ligand and we should have $Fe^{3+}(CN^-)_6$. In view of the highly electronegative character of the CN group as compared with Fe we expect the latter description to be more nearly correct. If it were indeed absolutely correct, we should have the case of an electrostatic crystal field, and it would not be very reasonable to begin our description in terms of octahedral Fe hybrids and then find them to be of no use. The true situation is of course somewhere in between; each bond is partly covalent and partly ionic—in other words *polar*. The VB approach of Pauling cannot readily describe such polarity without including a large number of ionic structures. The ligand-field approach, in which the relative weights of the metal and ligand AO's in each MO are determined in principle by the variation method (i.e. by solving secular equations), automatically allows for the right amount of polarity, which will usually be fairly high. It seems, therefore, the more sensible description. Indeed, the ionic description $Fe^{3+}(CN^-)_6$ puts a most improbably high charge on the central atom, and the covalent description $Fe^{3-}(CN)_6$ seems even more outrageous.

The effect of polarity in the bonds is to 'even out' the charge distribution, and Pauling embodied this observation in a 'postulate of neutrality' to the effect that admission of ionic structures would minimize the deviations from electrical neutrality of the constituent atoms in the molecule. As an example we may consider the hexammino-cobalt ion $[Co(NH_3)_6]^{3+}$ which occurs in the cobaltammines and again has an octahedral structure. To form six electron-pair bonds we should have to take an electron from each NH_3 group, leaving a singly occupied tetrahedral hybrid pointing towards the metal atom. The cobalt would then be Co^{3-}, with $6-3$ excess electrons. However, the electronegativity difference between Co and N ($x_{Co} = 1\cdot8$, $x_N = 3\cdot0$) leads to an ionic character for each bond of about 50 per cent (according to (6.35)), so about $6 \times \frac{1}{2} = 3$ electrons would leave the cobalt and go back to the nitrogens. The cobalt would thus become approximately neutral, and the greatest deviation from neutrality would not exceed about $\frac{1}{2}$ electron on each of the six ligands.

In retrospect, Pauling's imaginative use of hybridization failed only because it could not be conveniently implemented for highly polar bonds within the framework of VB theory. There is no similar objection to the use of the hybrids in a localized MO (or 'bond-orbital') description of each electron pair along the lines indicated at the end of §8.8. Certainly, the main deficiencies of Pauling's approach could be corrected in this way. However, in cases of high symmetry the calculation of completely delocalized MO's as mixtures of appropriate symmetry orbitals is quite straightforward (as will be seen in §9.13) and this is the form of ligand-field theory that has won the widest acceptance.

Finally, we note that whichever description is used, there are usually some non-bonding d electrons left occupying the t_{2g} set (d_{xy}, d_{yz}, d_{zx}). These are responsible for the paramagnetism of such compounds. The Pauling approach is not sufficiently general to explain the slight delocalization of these 'magnetic' electrons onto the six ligands (observed in spin resonance experiments), but according to Table 9.3 such delocalization is possible if we invoke a small degree of π-type bonding. The Pauling approach is also too inflexible to give a simple description of excitation processes, such as the transfer of an electron between the central t_{2g} and e_g levels in Fig. 9.10, and is therefore not well adapted to the interpretation of spectra. For such reasons the ligand-field theory description is generally preferred (Griffith and Orgel 1957).

9.9. Nature of the π-bonding

We have so far dealt only with σ bonding between the metal atom and the ligands, but quite evidently there will be π bonding also. Fig. 9.11 shows one type of π bonding in the planar square complex. It is clear from this diagram that the combination of ligand orbitals

$$A_{\pi y} - B_{\pi y} + C_{\pi x} - D_{\pi x}$$

is of precisely the right symmetry to overlap strongly with d_{xy} of the central atom. Reference to p. 278 will show that d_{xy} was one of the atomic orbitals, with symmetry b_{2g}, not previously used in a delocalized MO, but now it becomes at least partly delocalized. Similar descriptions apply to the other orbitals. We show, in the last column of Table 9.3, those combinations of π-ligand orbitals which combine with p and d central-atom orbitals in octahedral complexes. Since, however, this π bonding complicates the analysis without in this case adding anything essentially new, we shall not describe any further the nature of the resulting modifications to Fig. 9.10. Some of the most important applications so far are to the cyanide and carbonyl complexes. For further details the reader is referred to the books by Orgel (1960), Griffith (1961), and Figgis (1966).

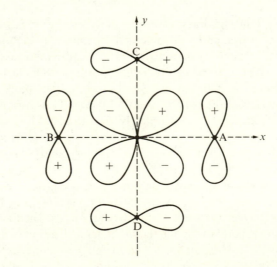

FIG. 9.11. An example of π bonding in a square complex.

9.10. Experimental evidence for electron delocalization

One of the most significant features of the ligand-field picture is the prediction of electron delocalization. This occurs from the ligands to the metal atom in the low-lying a_{1g}, t_{1u}, and e_g levels, all being doubly filled, and it occurs from the metal atoms to the ligands in the partly filled subshells t_{2g} and e_g (Fig. 9.10). As Table 9.3 shows, a t_{2g} MO consists mainly of a metal d orbital and an e_g borrows only σ orbitals. Again (cf. §6.6) the techniques of physics provide information about the form of the electron distribution.

In this case, confirmation of electron delocalization comes from three main sources.

(i) *g values.* We have seen in §3.10 that application of a magnetic field (flux density B) to a system with one odd electron gives rise to a small interaction energy $\Delta E_{spin} = g\mu_B m_s B$, where $m_s = \pm\frac{1}{2}$ is the electron spin quantum number, and that transitions between the two levels of such a 'spin doublet' are observed in ESR experiments. More generally,

$$\Delta E_{spin} = g\mu_B M_s B$$

where (with several unpaired electrons) M_s is a *resultant* spin component in the field direction, and if there is any 'unquenched' orbital angular momentum the value of g differs somewhat from the value of 2 for a free electron. Now if the d electrons are drawn away from the central atom their contributions to the orbital angular momentum will diminish and g will tend towards its free-electron value. By measuring the g value we can work back to the distribution of the unpaired d electrons. In the case of $[IrCl_6]^{2-}$, for example, it is found (Cipollini *et al.* 1962) that the central atom fails to complete its t_{2g}^6 shell by about 0·68 electrons and that there is about 0·05 of an unpaired spin on each of the six ligands.

(ii) *Hyperfine coupling with the ligands.* Since each chlorine atom in the above example fails to complete its closed shell (by $\sim 0·05$ electrons), there will be a resultant electron spin on each ligand; this will interact with the spin of the chlorine *nucleus* to give a hyperfine structure to the iridium ESR signal. When this situation is properly unravelled (Griffiths *et al.* 1953) the results quite closely substantiate the values found from g-value measurements.

(iii) *Hyperfine coupling with the metal.* If unpaired electrons migrate from the metal to the ligands, they will give a smaller magnetic coupling with the metal nucleus, and the corresponding hyperfine splitting should become less marked. Again this effect is confirmed by experiment, providing a technique complementary to that in (ii).

The above techniques are essentially physical, but they amply confirm the picture provided by our chemical arguments, particularly in the ground state of these complexes. For further discussion of these powerful methods see Carrington and McLachlan (1967).

Finally, further evidence of delocalization is obtained from a study of excited states, where, as Jørgensen (1959) has shown for a series of hexahalide complexes, there are strong absorption bands in the visible and near ultraviolet. These may be interpreted as being due to charge-transfer transitions, where an electron originally in an MO mainly on the six halide ions jumps into an orbital of d or s type chiefly concentrated on the metal atom. Figure 9.10 shows that there will normally be vacant orbitals in one or both of the e_g and a_{1g} symmetries. Detailed analysis supports this assignment.

Before turning to other types of transition-metal complex, it is worth remarking that the three concepts of d-orbital splitting, high- and low-spin complexes, and delocalization of ligand orbitals have brought a sense of order and pattern which was previously lacking in this important area of inorganic chemistry; yet it is doubtful whether any one of these three concepts would have been developed without the simplifying ideas of wave mechanics. It must be regarded as one of the major triumphs of quantum chemistry in the period 1945–60 that it provided this enormous degree of understanding.

9.11. π-Electron ligands

Our previous complexes have been chiefly dependent upon the use of σ electrons for co-ordination, but in the last 20 years many new complexes have been discovered in which the π electrons of molecules such as butadiene and benzene are directly involved in co-ordination. Thus there is the ethylene complex of platinum (*a*), the acetylene complex of dicobalt hexacarbonyl (*b*), the butadiene complex (*c*), and the para-benzoquinone complex of $Fe(CO)_3$ (*d*). One of the most interesting examples of these π-electron ligands is dicyclopentadienyl iron (*e*), now called ferrocene (Kealy and Pauson 1951; Miller *et al.* 1952). This substance forms orange crystals, which are sufficiently stable to permit vaporization without decomposition. The two cyclopentadienyl rings are not necessary for co-ordination as is shown by the existence of (*f*). There is obviously a close parallel between the ability of π electrons in these molecules to act as ligand orbitals in a complex, and the formation of so-called π-complexes (Dewar 1946, 1969), such as that between ethylene and Br^+ (*g*), which are postulated as intermediates in many chemical reactions.

There is no great difficulty in understanding some of these molecules (Dewar 1951; Chatt 1953). Thus in (*a*) and (*g*) we may suppose that the orbital which describes the π bond (Fig. 9.12 (*a*)) is distorted in such a way (Fig. 9.12 (*b*)) as to overlap effectively with an empty metal orbital. The whole ethylene molecule then behaves just like any other ligand, the distorted π orbital

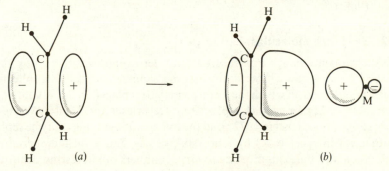

participating in a bond of roughly σ symmetry when it points towards the metal as in Fig. 9.12. There will be certain small compensating changes, since the sharp division into σ and π electrons in ethylene was dependent upon the plane of symmetry, now removed. Thus, in the butadiene (Mills and Robinson 1960) complex (*c*) all three C—C distances appear to be almost equal, with a value of 0·145 nm corresponding to relatively little double-bond character. Figure 9.12 represents donation of an electron from the olefin to the metal

FIG. 9.12. Change of ethylene bonding (π) MO on formation of a π complex. M denotes an empty σ-type orbital on the metal atom. The bond formed is of essentially σ type.

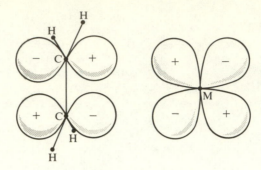

FIG. 9.13. Overlap between ethylene antibonding (π) MO and metal d orbital. The bond formed is of π type, the metal acting as a donor.

atom. There is also evidence that in many such complexes the butadiene geometry is closer to that of the first excited state, with one electron promoted to the first antibonding orbital. This suggests an interplay of donation and back-donation (McWeeny *et al.* 1969), which can be understood by reference to the case of ethylene; Fig. 9.12 represents donation of an electron from the olefin bonding MO to the metal. As Fig. 9.13 shows, however, there may also be migration back to the empty antibonding π orbital of the double bond, which overlaps strongly with the appropriate d orbital on the metal atom. Such donation and back-donation appears to be very common, and clearly provides a mechanism for evening out the charge distribution in a manner conforming to Pauling's postulate of electrical neutrality (p. 281). Evidently the binding of π-electron ligands is formally very similar to that which we described earlier in this chapter for octahedral and other similar molecular complexes. We shall not therefore discuss it further. It should be noted, however, that the whole field of 'organo-metallic chemistry', in which organic ligands are bonded to metal atoms, has by now established itself as one of the important and rapidly developing areas of chemistry (see, for example, Coates *et al.* (1968)).

9.12. Sandwich molecules. Ferrocene

The dicyclopentadienyl iron molecule (*e*), ferrocene, is not by any means unique.† There is now a large family of these 'sandwich' molecules, collectively called *metallocenes*, in which the Fe atom may be replaced by Ti, V, Cr, Mn, Co, Ni, Ru, Os, or Mg, or even U, Np, or Pu (Streitwieser and Muller–Westerhoff 1968; Hayes and Edelstein 1972), and the five-membered rings may be replaced (though not in every case) by four-, six-, seven-, and eight-membered rings C_nH_n. Such molecules all give rise to large numbers of interesting derivatives.

† For a survey of the properties of this and similar molecules see, for example, Huheey (1972); see also Orgel (1960).

FIG. 9.14. Superposition of relevant sections of three-dimensional Fourier maps to give picture of ferrocene electron density.

It was originally supposed that ferrocene could be represented as an ionic complex $(C_5H_5)^-Fe^{2+}(C_5H_5)^-$ in which each five-membered ring completed its 'aromatic sextet' to achieve the stability recognized in the Hückel $4n+2$ rule (p. 243), but this cannot be correct since the molecule behaves chemically as if the groups were essentially neutral. In nickelocene $Ni(C_5H_5)_2$, however, where the central atom has two more d electrons, there is NMR evidence (McConnell and Holm 1957) for some charge migration, leading to an excess of $0·14$ electrons on each carbon and a nickel atom with positive charge $1·4e$. To allow for both covalent and ionic character in the bonding we shall therefore use ligand-field theory, forming MO's of appropriate symmetry by combining the π-electron MO's of the attached groups with the s, p, d orbitals of the metal atom.

The precise structure of ferrocene was for many years a subject of controversy. The X-ray analysis of Dunitz, Orgel, and Rich (1956) pointed clearly to a sandwich in which the C_5H_5 rings were parallel, with the metal atom lying symmetrically between them, with the staggered conformation shown in Fig. 9.14. Later work (Bone and Haaland 1966; Haaland and Nilsson 1968) suggested that the rings were eclipsed. It is now generally accepted (for further discussion see Churchill and Wormald 1969 and Palenik 1969) that the free molecule (i.e. in the vapour phase) has the eclipsed conformation but that the barrier height for rotation of one ring relative to the other is very small $(\sim 4\,\mathrm{kJ\,mol^{-1}})$. As a consequence, the conformation adopted in the crystal is determined essentially by packing considerations, which favour the staggered form. In what follows, we assume the two rings are staggered and choose the co-ordinate system as in Fig. 9.15 (*a*).

Let us start† with the π-type MO's of the two rings, disregarding the antibonding orbitals which lie several eV above the others. These MO's appear in Fig. 9.15 (*b,c,d*), their forms being determined from the results of §8.3. For each ring there is the totally symmetric bonding MO (e.g. ϕ_1), followed by a

† For other representative MO discussions see Moffitt (1954a); Robertson and McConnell (1960); Sohn *et al.* (1971).

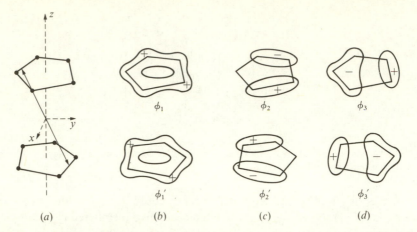

FIG. 9.15. Staggered conformation of C_5H_5 rings in ferrocene: (a) inversion symmetry (metal ion at the origin); (b) (c) (d) inversion-related MO's on top and bottom molecules.

degenerate pair (ϕ_2, ϕ_3) in which each MO has one node. With respect to rotations around the z axis (principal axis) of symmetry, these orbitals resemble roughly a σ-type MO of a diatomic molecule and a pair of π-type MO's (π_x, π_y). Since the whole system has a centre of symmetry, we shall be looking for MO's of g and u symmetry as usual, and we therefore begin by setting up g and u combinations of the ligand orbitals; these are clearly $\phi_1 \pm \phi_1'$, $\phi_2 \pm \phi_2'$, and $\phi_3 \pm \phi_3'$, where (with the orbitals defined as in Fig. 9.15) the upper signs give the g combinations and the lower signs the u combinations. The AO's of the metal are already symmetry orbitals; the s and d orbitals are of g type, the p orbitals are of u type. The orbitals to be combined in forming the MO's are collected below and given their group-theoretical names.

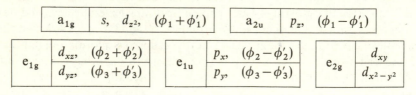

a_{1g}	s, d_{z^2}, $(\phi_1 + \phi_1')$		a_{2u}	p_z, $(\phi_1 - \phi_1')$	

e_{1g}	d_{xz}, $(\phi_2 + \phi_2')$	e_{1u}	p_x, $(\phi_2 - \phi_2')$	e_{2g}	d_{xy}
	d_{yz}, $(\phi_3 + \phi_3')$		p_y, $(\phi_3 - \phi_3')$		$d_{x^2-y^2}$

It will be remembered from §9.7 that the label a applies to non-degenerate orbitals which are unchanged under the symmetry operation of rotation around the principal axis (e.g. d_{z^2}), while e designates a degenerate pair (e.g. p_x, p_y) which, under a similar rotation, turn into new linear combinations of each other; the further classification, by means of subscript 1 or 2, is explained in Appendix 3. As always, the classification is important because in forming the MO's we must mix only functions of the same type, i.e. those which appear on any one line in the above boxes. Fig. 9.16 shows how three bonding MO's (one

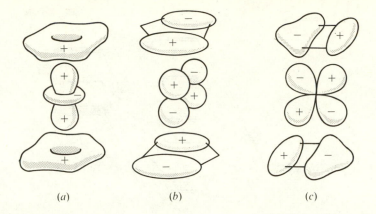

(a) (b) (c)

FIG. 9.16. Ferrocene MO's (schematic) formed by combination of ring orbitals (Fig. 9.15) and metal d orbitals. (Horizontal nodes in the ring orbitals have been omitted for clarity).

non-degenerate and two of a degenerate pair) arise from the a_{1g} and e_{1g} functions. We can also infer the likely relative positions of the resultant energy levels by considering the extent of the overlap between the metal orbitals and the ligand-group orbitals. For example, the d_{z^2} AO points into the 'hole' in each cyclopentadienyl MO of the combination $\phi_1 - \phi_1'$, giving small overlap and relatively weak metal–ligand bonding (even though the energy may be low); in contrast the d_{xz} strongly overlaps the lobes of the $\phi_2 + \phi_2'$ combination to give one MO of a strongly bonding pair.

The most likely energy order of the resultant MO's is in most cases that shown in the correlation diagram of Fig. 9.17. The details are uncertain and may vary somewhat from molecule to molecule, but the correlation diagram does account for the main facts. Thus, in ferrocene there are nine bonding (or low-energy non-bonding) MO's, and these are exactly filled by the available electrons $(5+5+8)$ of the two π-electron rings and the iron d^6s^2 configuration to give a closed shell of 18 electrons. Indeed, in many complexes of this kind an '18-electron rule' appears to be obeyed, i.e. that when d orbitals are included in the valence shell of a central atom the electron configuration of maximum stability corresponds formally to a krypton-like $d^{10}s^2p^6$ structure. It was, in fact, on the basis of such considerations that Longuet–Higgins and Orgel (1956) first proposed that the highly unstable molecule cyclobutadiene (p. 243) might exist in the complex $(C_4H_4)MX_2$, where X is a univalent ligand and M is Ni, Pd, or Pt. This prediction was subsequently verified by Criegee and Schröder (1959), who prepared the dimer of $(C_4H_4)NiCl_2$ and showed it had the expected structure.

The correlation diagram (Fig. 9.17) also allows us to understand the magnetic properties of the metallocenes. Ferrocene, with nine doubly

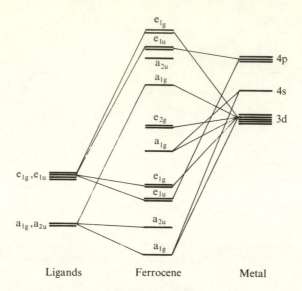

Fig. 9.17. Ferrocene orbital energies related to metal and ligand orbital energies.

occupied MO's, should be diamagnetic, but the cobalt analogue with its extra electron in the first antibonding orbital should be paramagnetic and less stable. Nickelocene, with two additional electrons, appears to have one electron in each of a degenerate pair (probably e_{1u} or e_{2u}) with spins parallel (Hund's rules), again giving paramagnetism. However, the manganese, chromium, and vanadium analogues should show a progressive reduction of the number of electrons in the highest bonding pair (e_{2g}), again leading to paramagnetism. Of course our theory is very primitive (we have even ignored the antibonding MO's on the rings) and we cannot expect Fig. 9.17 to apply unchanged to all the metallocenes. There is indeed evidence for variations in the ordering of the levels. Broadly speaking, however, this simple treatment gives a satisfactory picture of the bonding and properties of such molecules.

9.13. Simple semi-empirical theories

So far we have been content with a purely qualitative discussion of bonding in transition-metal compounds, based on the combination of metal AO's and ligand-group orbitals of similar symmetry. Very often the number of functions of given symmetry type is small, and simple considerations of metal–ligand overlap indicate the nature of the resultant MO's and the form of the correlation diagram. In fact, we are able to exploit the high symmetry of such molecules to avoid direct solution of secular equations which, in the absence of symmetry, would be of quite high order. It is natural, however, to carry the approach a step further and to attempt a more quantitative calculation of the

MO's using approximations parallel to those of Hückel theory in organic chemistry. There are firm grounds (§8.8) for believing that, though we are no longer dealing with simple π-electron systems, similar methods should be applicable. In this section we outline the procedure. Of course, it is possible to attempt more elaborate SCF calculations, even at the *ab initio* level (cf. §6.1), but reference to these will be postponed until Chapter 13.

In the applications of Hückel theory in Chapter 8 the matrix elements H_{rs} in the secular equations were given values α_r (for $s = r$) and β_{rs} ($s \neq r$). These Coulomb and bond integrals were treated as disposable parameters and determined empirically. It was supposed that α_r, representing the expectation energy of an electron in the AO ϕ_r, was somewhere near to $-I_r$ (I_r being the ionization energy needed to remove an electron from ϕ_r of the corresponding free atom in its valence state) while β_{rs} was roughly proportional to the overlap of the pair of AO's ϕ_r and ϕ_s. In the secular equations themselves the overlap integrals were usually neglected.

On turning to highly symmetrical transition-metal complexes, where MO's of given symmetry type are constructed by combination of metal AO's and ligand *group* (or symmetry) orbitals of the same type, the matrix elements have a somewhat different interpretation. For example, if ϕ_r is a metal AO and ϕ_s is a ligand-group orbital then S_{rs} will be a 'group overlap'. The simplest form of semi-empirical theory is that due to Wolfsberg and Helmholz (1952) which requires only the calculation of group overlaps. Their assumptions are as follows.

(i) The values of $-H_{rr}$ are assumed close to corresponding valence state ionization energies for the metal or ligand orbitals. As the ligands are not bonded to each other, the IP for a ligand *group* orbital does not differ much from that for an individual ligand orbital.

(ii) The values of H_{rs} are proportional to the group overlaps S_{rs} but also depend on the electronegativities of the atoms connected, a suitable approximation being

$$H_{rs} = \tfrac{1}{2}kS_{rs}(H_{rr} + H_{ss}) \tag{9.1}$$

where k is a numerical constant with values ranging from about 1·5 to 2·0 (for σ and π overlap respectively).

(iii) The group overlap integrals, which vary widely in magnitude, are retained explicitly in the secular equations.

It is not difficult to calculate the group overlap S_{rs} for any geometrical arrangement of ligands in terms of diatomic overlap integrals for specific pairs of Slater orbitals, the latter in turn being calculated from standard formulae. To obtain the overlap integral for two p_x orbitals, for example, one on the metal and one on a ligand (Fig. 9.18), we remember the vector properties of the

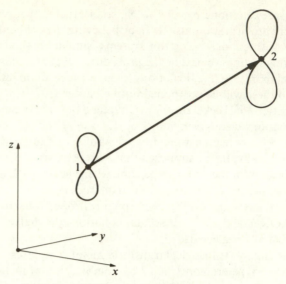

FIG. 9.18. Globally defined p_z orbitals on centres 1 and 2.

p orbitals (p. 185) and express each in terms of three alternative p orbitals pointing along and transverse to the internuclear axis (p_σ, p_π, and $p_{\bar\pi}$, say); the overlap between $p_{x1} = lp_{\sigma 1} + mp_{\pi 1} + np_{\bar\pi 1}$ and $p_{x2} = lp_{\sigma 2} + mp_{\pi 2} + np_{\bar\pi 2}$ where l, m, and n are the appropriate direction cosines. Since p orbitals of different symmetry (σ and π) have zero overlap we obtain, with an obvious notation,

$$S(p_x, p_x) = l^2 S(pp, \sigma) + m^2 S(pp, \pi) + n^2 S(pp, \bar\pi)$$

Now the π and $\bar\pi$ integrals are identical (the π AO's differing only by rotation through 90° around the axis), and also $l^2 + m^2 + n^2 = 1$. The result is thus

$$S(p_x, p_x) = l^2 S(pp, \sigma) + (1 - l^2) S(pp, \pi). \qquad (9.2)$$

The overlap integrals for pairs of AO's with axes defined in the complex may thus be expressed in terms of standard diatomic overlaps of σ, π, δ, ... type, and it is then a simple matter to assemble such integrals to obtain the overlap between a metal orbital and a group orbital.

The most frequently needed overlap integrals for s, p, and d orbitals of given 'global' orientation (Fig. 9.18) are given in Table 9.4 in terms of the direction cosines of the internuclear axis ($1 \rightarrow 2$) and the basic diatomic overlaps, an obvious abbreviation being used. Integrals not shown can be inferred by reinterpreting the axial directions. The group overlap integrals for any particular symmetry combinations of ligand orbitals are easily derived from these results and have also been tabulated (see, for example, Wolfsberg and Helmholz (1952) and Gray (1965) for further details).

TABLE 9.4

Overlap integrals between globally defined s, p, d orbitals†

$$S(s, s) = S(ss, \sigma)$$
$$S(s, p_x) = l S(sp, \sigma)$$
$$S(s, d_{z^2}) = \tfrac{1}{2}(2n^2 - l^2 - m^2)S(sd, \sigma)$$
$$S(s, d_{x^2-y^2}) = \tfrac{1}{2}\sqrt{3}\,(l^2 - m^2)S(sd, \sigma)$$
$$S(s, d_{xy}) = \sqrt{3}\,lm\,S(sd, \sigma)$$
$$S(p_x, p_x) = l^2 S(pp, \sigma) + (1 - l^2)S(pp, \pi)$$
$$S(p_x, p_y) = lm\,S(pp, \sigma) - lm\,S(pp, \pi)$$
$$S(p_x, d_{z^2}) = \tfrac{1}{2}l(2n^2 - l^2 - m^2)S(pd, \sigma) - \sqrt{3}\,ln^2\,S(pd, \pi)$$
$$S(p_x, d_{x^2-y^2}) = \tfrac{1}{2}\sqrt{3}\,l(l^2 - m^2)S(pd, \sigma) + l(1 - l^2 + m^2)S(pd, \pi)$$
$$S(p_y, d_{z^2}) = \tfrac{1}{2}m(2n^2 - l^2 - m^2)S(pd, \sigma) - \sqrt{3}\,mn^2\,S(pd, \pi)$$
$$S(p_y, d_{x^2-y^2}) = \tfrac{1}{2}\sqrt{3}\,m(l^2 - m^2)S(pd, \sigma) - m(1 + l^2 - m^2)S(pd, \pi)$$
$$S(p_z, d_{z^2}) = \tfrac{1}{2}n(2n^2 - l^2 - m^2)S(pd, \sigma) + \sqrt{3}\,n(l^2 + m^2)S(pd, \pi)$$
$$S(p_z, d_{x^2-y^2}) = \tfrac{1}{2}\sqrt{3}\,n(l^2 - m^2)S(pd, \sigma) - n(l^2 - m^2)S(pd, \pi)$$
$$S(p_x, d_{xy}) = \sqrt{3}\,l^2 m\,S(pd, \sigma) + m(1 - 2l^2)S(pd, \pi)$$
$$S(p_x, d_{yz}) = \sqrt{3}\,lmn\,S(pd, \sigma) - 2lmn\,S(pd, \pi)$$
$$S(p_x, d_{zx}) = \sqrt{3}\,l^2 n\,S(pd, \sigma) + n(1 - 2l^2)S(pd, \pi)$$

† The s, p, d orbitals are as given in Tables 2.3, 2.4, with 'global' Cartesian axes defined in Fig. 9.8; d orbitals are admitted on one centre only. The cosines of the angles between $1 \to 2$ and the co-ordinate axes are denoted by l, m, n, and Table entries express any $S(\phi_1, \phi_2)$ in terms of σ- and π-type overlaps between similar functions defined *locally* with $1 \to 2$ as z axis. Thus $S(pd, \sigma)$ is the overlap between p_z and d_{z^2} orbitals pointing along a common z axis, while $S(pd, \pi)$ is that between p_x and d_{xz} lying transverse to the axis.

Entries not shown in the table may be obtained by cyclic permutation of x, y, z and l, m, n. For example, the expression for $S(p_y, p_z)$ follows from $S(p_x, p_y)$ on replacing x, y, z by y, z, x and hence l, m, n by m, n, l.

Although the results of calculations along the above lines are generally encouraging, it soon becomes clear that further refinements are necessary. In particular, the order of the levels in the resultant correlation diagram (e.g. Fig. 9.17) is sensitive to the choice of parameter values (including the orbital exponents, which control expansion or contraction and hence the overlap values) and even small variations may lead to an 'instability' in which there is a 'landslide' of electronic charge towards either the metal or the ligands. The reason for such behaviour is not hard to find. It is clearly unrealistic to give the parameters H_{rr} *fixed* numerical values when we know from our discussions in §6.4 that the electronegativities to which they are closely related are dependent on the charge distribution; if electrons flow towards the metal, increasing the population of the valence shell, its power to attract further electrons will diminish and the magnitude of H_{rr} will be reduced. If this effect is allowed for, the 'landslide' will not occur because the reduction of $|H_{rr}|$ will ensure that, on solving the secular equations, the corresponding atom does *not* take an unreasonably large share of the electron density.

The simplest extension of the Wolfsberg–Helmholz approach is that due to Ballhausen and Gray (1962; 1965) who assumed a simple polynomial dependence of H_{rr} for a metal orbital (ϕ_r) on the net charge carried by the metal. If the total population is Q (which, as in §6.4, will be a gross population), this amounts to assuming

$$H_{rr}(\text{metal}) = a + bQ + cQ^2 \tag{9.3}$$

where the constants can be chosen to give a good fit to successive ionization potentials for the free atom (or its computed valence states). To incorporate this dependence in the calculation it is necessary to proceed as in Hartree's self-consistent field method (§2.5). We make a guess at the most likely distribution of electrons between metal and ligands, set up the secular equations for each symmetry type, using (9.3) with the corresponding population values, and solve to obtain the MO's. If the populations calculated from the MO's disagree with those assumed initially, we use them to obtain revised estimates of the matrix elements H_{rr}, and then go on repeating this cycle until 'self-consistent' values are finally obtained. The results of calculations based on this procedure (see, for example, Viste and Gray (1964), Dahl and Ballhausen (1968), and the many references cited therein) are certainly much more satisfactory, and the method has been applied, with minor variations, to a large number of interesting problems.

It will be realized that the introduction of a population dependence into the parameters is a step towards the introduction of self-consistency in the Hartree–Fock sense (§2.5); the extended Wolfsberg–Helmholz method may thus be viewed as a very crude form of SCF theory. Had it not been for the rapid development of computers, this kind of semi-empirical analysis would probably have had a long future—comparable with that of Hückel theory in organic chemistry. However, the simple theories have to some extent been overtaken by events, for it is now possible to perform much more ambitious calculations which, even though they may not reach the near-Hartree–Fock accuracy achieved for smaller molecules, permit an altogether more realistic allowance for electron interaction effects. We shall turn to the development of semi-empirical SCF theory in Chapter 13.

Problems

9.1. Turn back to Problem 3.4 where you found a set of three degenerate orbitals resembling d_{xy}, d_{yz}, d_{zx}. Give an interpretation of the behaviour of the d_{xy}, d_{yz}, and d_{zx} energies in Fig. 9.4 as the axial ligands on the octahedral complex are withdrawn.

9.2. Make 3-dimensional sketches of the σ-type ligand group orbitals in Table 9.2, similar to those in Fig. 9.8, noting their resemblance (from the point of view of symmetry) to the corresponding s, p, and d-type AO's of the central atom.

9.3. Confirm that the naming of the orbitals in Problem 9.2 (see Table 9.2) is correct, using the group theory conventions summarized in Appendix 3.

9.4. Make 3-dimensional sketches of σ-type group orbitals which could mix with central-atom s and p AO's in a tetrahedral complex. Name the orbitals as in Problem 9.3 and construct a table analogous to Table 9.2 (π type ligand AO's are needed in constructing group orbitals to match those of the central atom d AO's: the results are less easy to visualize). Try to understand the pictures given in other textbooks, e.g. Fig. 6.42 in Karplus and Porter 1970. (Hint: represent the ligand AO's (A, B, C, D) by four spheres, at the appropriate corners of a cube, centred on the origin; choose the signs to match those on the central atom AO's).

9.5. It seems likely that the bonding MO's in ferrocene, whose energies are indicated in Fig. 9.17, have the following composition (in ascending energy order):

a_{1g}: mainly $(\phi_1 + \phi_1')$ and 4s, with a little $3d_{z^2}$
a_{2u}: mainly $(\phi_1 - \phi_1')$ with a little $4p_z$
e_{1u}: mainly $(\phi_2 - \phi_2')$ and $4p_x$ (similarly for the y-type partner)
e_{1g}: mainly $(\phi_2 + \phi_2')$ and $3d_{xz}$ (similarly for the y-type partner)
a_{1g}: mainly $(\phi_1 + \phi_1')$ and $3d_{z^2}$, with a little 4s
e_{2g}: mainly $3d_{xy}$ (with partner $3d_{x^2-y^2}$).

Try to rationalize these suggestions, without numerical calculations. Sketch the anti-bonding orbitals of the ligands, set up corresponding symmetry combinations, and discuss the probable effects of admitting them into the calculation.

9.6. Calculate a few of the overlap integrals between p and d orbitals on a metal and the ligand group orbitals of corresponding symmetry, assuming first octahedral and then tetrahedral geometry. You may assume σ-type ligand orbitals and may neglect their mutual overlap. (Hint: use the combinations given in Table 9.2 and the Table constructed in Problem 9.4. List the direction cosines of the ligands and obtain their individual overlap integrals from Table 9.4. Sum, with appropriate coefficients, to get the group overlaps.)

9.7. The vanadyl ion $VO(H_2O)_5^{2+}$ has a distorted octahedral structure with a strongly bonded oxygen and weakly bonded H_2O above and below the vanadium, respectively, and four H_2Os at the equatorial sites. Discuss the probable forms of the MO's (using the reduced, square planar, symmetry) and name their symmetry types; propose a correlation diagram relating the MO energies to the energies of the metal and ligand orbitals; and indicate a likely electron configuration for the ion. (Hints: use the 3d, 4s and 4p AO's of vanadium, σ-type hybrids on the water oxygens, and both σ and π AO's on the oxide oxygen. Assume metal-orbital energies (ascending order 3d, 4s, 4p), higher than ligand-orbital energies ($\sigma(O)$, $\sigma(N_2O)$, $\pi(O)$). Remember that pairs of orbitals (including group orbitals) which overlap most interact most; and that overlap may be increased by setting up hybrids. You should sketch all the orbitals concerned (cf. Figs. 9.2 and 9.11).)

(A full treatment of this system involves most of the main ideas presented in Chapter 9. When you have done all you can turn to Ballhausen and Gray (1962); or to their book (1965).)

10

Chemical reactivity

10.1. Basis of reactivity theories. Substitution reactions

To account for the electronic structure of a molecule is one thing; to predict its properties, and in particular its chemical reactivity, is a vastly more difficult matter. Much progress has been made, however, and a brief review of the main ideas—in so far as they relate to the theory of valence—is therefore desirable. These ideas have been developed mainly through applications in organic chemistry, and parallel applications elsewhere are still in their infancy; the emphasis of this chapter will therefore fall on organic reactions.

The earliest ideas were formulated mainly by Vorlander, Lapworth, Robinson, and Ingold, during the period 1900–1935, and were based essentially on classical electrostatics; in a heterolytic reaction, involving the approach of a charged reagent, the point of attack should be determined mainly by the net charges at the various positions in the molecule. We have in fact already used this idea (p. 256) to account for the directing effects of substituents.

To proceed in a more quantitative way we must go back to thermodynamics. The equilibrium constant (K) in a reversible reaction is determined by the (molar) free-energy increase

$$\Delta G = G_{\text{products}} - G_{\text{reactants}}$$

according to

$$-RT \log K = \Delta G. \tag{10.1}$$

For an irreversible reaction the rate constant (k) is determined by a similar equation

$$-RT \log k = \Delta G^{\dagger} \tag{10.2}$$

where $\Delta G^{\dagger}$ is the free-energy increase in going to a *transition state*, from which the products are formed irreversibly. The accurate calculation of a free-energy change ΔG is clearly impossible, for, since $G = U - TS + PV$, ΔG includes an entropy change (and hence requires detailed knowledge of vibrational partition functions etc.) as well as the energy of *all* the electrons, together with solvent interactions etc. In many cases, however, especially in organic chemistry, we study not *absolute* reaction rates but the rate of one reaction *relative* to that of another, rather similar, reaction and, if we are prepared to assume that similar changes of entropy, solvation energy, etc. occur, then the difference between the free-energy changes, ΔG for one reaction and ΔG_0 for a 'standard' reaction, may be assumed to arise principally from the difference between ΔU and ΔU_0. Since ΔU is simply ΔE for 1 mol we may then write, for a typical irreversible reaction,

$$-RT \log (k/k_0) \simeq \Delta E - \Delta E_0 \qquad (10.3)$$

where ΔE is the total *electronic* energy change (i.e. the relevant part of the internal energy change). For many classes of organic reaction (notably the substitution reactions of aromatic molecules) the difference $\Delta E - \Delta E_0$ arises principally from the difference of π-*electron energy changes*. It is for reactions of this kind that the simplest theories have been most successful. We shall therefore start by confining attention to the substitution reactions of conjugated systems.

The most widely adopted model of the transition state is that due to Wheland (1942), the substitution being imagined to proceed through the three stages indicated in Fig. 10.1. The conformation (*b*), in which both X and H are

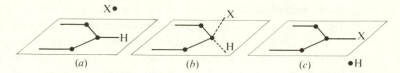

FIG. 10.1. Wheland's model of a substitution reaction: (*a*) approach of substituent X; (*b*) roughly tetrahedral transition state; (*c*) expulsion of hydrogen.

attached (roughly tetrahedrally) to the carbon atom, represents the hypothetical 'transition state' from which the system proceeds spontaneously to (*c*). X may be a radical or an ion (e.g. if it is a positive ion a *proton* is expelled instead of H), but in any case the carbon is removed from conjugation and 0, 1, or 2 π electrons go into the *localized* C—X bond. When the same substitution occurs at different positions the changes in σ-electron energy may be assumed to be similar (involving similar *local* changes in the molecular framework), but the changes in π-electron energy may be very different, depending on the shape and size of the whole conjugated system. In such cases, therefore, the interpretation of each ΔE in (10.3) as a change of π-electron energy is reasonable, the other terms cancelling.

Let us now represent the π-energy variation, as a given substituent approaches two alternative reaction sites, by the two curves in Fig. 10.2. The aim of all theoretical discussions of aromatic substitution is to estimate the forms of such curves for alternative positions of substitution. There are two main approaches, depending on whether we look at the initial parts of the

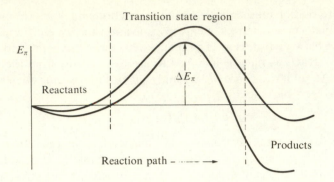

FIG. 10.2. Variation of π-electron energy for two substitution positions.

curves (the 'isolated molecule approximation') or at the transition state region (the 'localization approximation'), and in both approaches we shall find that the energy changes are related to the charges and bond orders.

10.2. Isolated molecule theory

We start from the formula (8.24) which refers to the π-electron system of the isolated molecule before it is appreciably perturbed by the approaching substituent X, and which expresses the total π-electron energy in terms of charges and bond orders. We have noted already that the first-order effect of a perturbation can be calculated using the q's and p's for the isolated molecule, simply allowing the α's and β's to change as the Hamiltonian is modified by the perturbation. If a charge (heterolytic reagent) comes close to, say, centre r its main effect will be to change the value of α_r, the additional potential energy term in the Hamiltonian being greatest for an electron in ϕ_r. The first-order change in E may then be calculated as

$$\delta E = q_r \delta \alpha_r. \tag{10.4}$$

If the reagent is electrophilic, electrons will be pulled towards centre r, which means that $\delta \alpha_r$ is negative; the lower curve in Fig. 10.2 will thus be the one for the centre where q_r has the greater numerical value. We conclude that

$$\boxed{\text{Positions of high } q_r \text{ are most prone to attack by electrophilic reagents, and those of low } q_r \text{ to attack by nucleophilic reagents.}} \tag{10.5}$$

This is merely a quantum-mechanical statement of Ingold's ideas: electrophilic or positively charged reagents are attracted towards positions of high negative charge (i.e. electron density), and *vice versa*.

Rule (10.5) is, of course, of no value when all the q's are the same, as for an alternant hydrocarbon where $q_r = 1$ at every centre. Formula (10.4) can be extended, however, to give the change to *second* order and is then written

$$\delta E = q_r \delta \alpha_r + \tfrac{1}{2} \pi_r \delta \alpha_r^2 \tag{10.6}$$

where π_r is called the 'self-polarizability' or 'mutability' of atom r. If two positions have the same q values it is necessary to look at their π values to see which is the more susceptible to attack; π_r is usually negative, and since $\delta \alpha_r^2$ is positive, this means that

> Of two positions r and s with $q_r = q_s$, the one more prone to attack by heterolytic reagents (electrophilic *or* nucleophilic) is the one with the numerically larger π value. (10.7)

The π values are quite awkward to calculate, and the expressions for them derived by formal perturbation theory have no simple physical interpretation. It is, however, possible to obtain an excellent approximate formula which gives at the same time immediate chemical insight (McWeeny 1956). When the β_{rs}, connecting r with its neighbours are all equal ($\beta_{rs} = \beta$) this formula becomes

$$\pi_r = \frac{2q_r - q_r^2}{2N_r \beta}. \tag{10.8}$$

Here N_r is the total number of (π) bonds connecting r with its neighbours, where (using, as in Chapter 8, the notation $s\!-\!r$ to mean 's linked with r')

$$N_r = \sum_{s(s-r)} p_{rs} \tag{10.9}$$

and is sometimes called the 'Dewar reactivity number' since it has been widely used, in a completely different context, by Dewar. Thus, for benzene $N_r = 2 \times \tfrac{2}{3}, q_r = 1, \pi_r = 3/(8\beta)$; this compares with $0.398/\beta$ from perturbation theory, a difference of only ~ 5 per cent. Equation (10.8) shows that the most easily 'polarizable' centres are those which are *least* involved in bonding (N_r small), a result which gives substance to the intuitive ideas formulated long ago by Thiele. The reason for the name 'polarizability' is that π_r describes also the change of q_r due to the change of α_r (i.e. the polarization of the charge distribution)

$$\delta q_r = \pi_r \delta \alpha_r. \tag{10.10}$$

Thus in electrophilic attack ($\delta \alpha_r$ negative, π_r negative), δq_r is positive and electronic charge flows towards the position of attack, making attack still easier. Similarly, in nucleophilic attack $\delta \alpha_r$ is positive and δq_r is negative; electronic charge flows *away* from the point of attack, but again (negative reagent) attack is made easier. In both cases polarization *reduces* the energy and there is thus no discrimination between electrophilic and nucleophilic substituents.

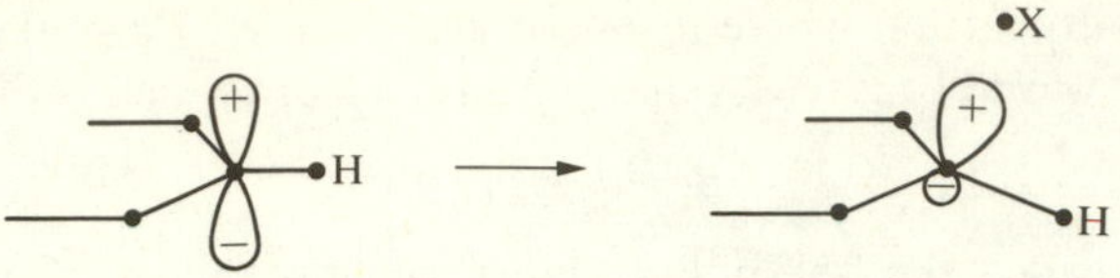

FIG. 10.3. Change of hybridization during the approach of a substituent.

If the attacking reagent is *uncharged* (homolytic reaction), there are no strong electric fields and the α's are not much affected. We then consider what happens in the very early stages of orbital overlap (Fig. 10.3). As X approaches, the carbon begins to extend its bonding to all *four* neighbours (X included) by adjusting the s–p mixing to give a hybrid overlapping with X; the pure 2p AO of the conjugated system is contaminated with some 2s character, and begins to overlap its neighbouring 2p AO's less favourably. As a result $\beta_{rs} \rightarrow \beta_{rs} + \delta\beta_{rs}$, for each neighbour s, with $\delta\beta_{rs}$ *positive* (i.e. β_{rs} becomes less negative). If we assume that each C—C bond is equally affected (by $\delta\beta$, say) the energy formula (8.24) gives a first-order change

$$\delta E = 2 \sum_{s(s-r)} p_{rs}\delta\beta = 2N_r\delta\beta \qquad (10.11)$$

where again N_r is the total number of bonds, defined in (10.9), between r and its neighbours (s). Thus, in a homolytic reaction the position of lowest 'bond number' (N_r) gives the lowest curve in Fig. 10.2. This is often expressed differently by introducing N_{max} (with a theoretical value $\sqrt{3}$) and defining (Coulson 1947)

$$F_r = N_{max} - N_r \qquad (10.12)$$

as the 'free valence' at position r. This quantity, which represents the maximum conceivable increase in N_r, gives a numerical measure of Thiele's 'partial valence' or Werner's 'residual affinity'. Our conclusion is that

> Positions of high free valence are prone to attack by homolytic reagents. (10.13)

The indices we have defined all refer to the *early stages* of a substitution reaction before extensive geometry changes have occurred; they can never be very reliable but are easy to use and correlate reasonably well with experiment.

Sometimes the charges, bond orders, and free valences are displayed in a molecular diagram (Fig. 10.4) which shows the net charge at each atom r (i.e. $1 - q_r$), the total bond order $(1 + p_{rs}$, counting the σ bond) against each link r–s, and the free valence (F_r) at the end of an arrow originating on atom r. Such diagrams give a good overall picture of molecular properties: the bond orders

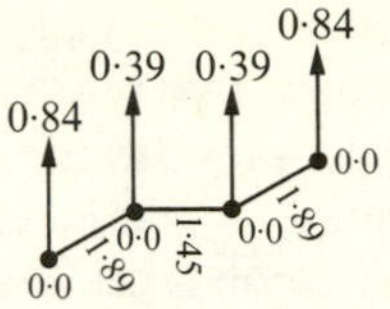

FIG. 10.4. Examples of molecular diagrams: (a) butadiene; (b) naphthalene.

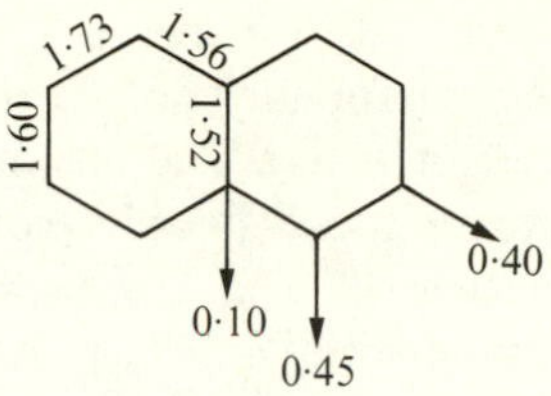

indicate the long and short bonds, and also indicate whether there is any appreciable 'bond fixation' with the double bonds located as in a classical structure; the net charges (when they are non-zero) indicate the most likely points of attack by ions such as NO_2^+ or NH_2^-, and also allow a rough estimation of dipole moments (i.e. of the 'resonance moment' p. 252); the free valences indicate the positions which, initially at least, are prone to attack by neutral free radicals. It is worth noting that F_r values greater than 1·0 are usually associated with free radicals, values around 0·8 with terminal carbons, values around 0·4 with aromatic carbons (as in benzene), and very small values with the internal atoms of a polycyclic condensed molecule.

It has been noted that the indices used so far, in particular the charges, provide a quantitative framework for Ingold's ideas in so far as they determine (at least roughly) the electric field outside the isolated molecule: the value of the charges for this purpose has been reviewed by Julg (1975). It is therefore not surprising that carefully computed *maps* of the electrostatic potential outside a molecule have also been used to discuss reactivity, the directions of attack by charged reagents being suggested by 'valleys' in the potential energy surface. This development (Scrocco and Tomasi 1973) may be regarded as the culmination of Ingold's approach: it has been applied with considerable success even to large molecules, for example to the protonation reactions of nucleic acid bases (see Pullman and Pullman 1973 for many references).

10.3. Localization theory

In this approach (Wheland 1942) we consider a 'transition complex', intended to represent a conformation through which the system is likely to pass near the peak of the energy curve. In such a complex the substituent is actually attached to the molecule, as for example in Fig. 10.1 (b). For different positions of attachment the peak heights may differ, the lower peak indicating the more reactive position.

We consider the substitution reaction and assume that in the transition state shown in Fig. 10.1 (*b*) the centre under attack (*r*, say) is taken out of conjugation as the tetrahedral bonding develops; the 'residual molecule' is then an open chain, with π energy E_r say. Sometimes the conjugated system keeps all its electrons (a nucleophilic reagent does not want them); sometimes it loses one electron (in a radical reaction one electron is needed to form a new σ bond); sometimes it loses two electrons (when the reagent is electrophilic and needs both electrons). The number of electrons *localized* in this way is indicated by a superscript n ($=0, 1, 2$) and the *localization energy* is defined as

$$L_r^{(n)} = E_r^{(n)} - E. \tag{10.14}$$

This may be used as a measure of the π component of the activation energy ΔE in a rate equation such as (10.3); it is not of course the whole of the localization energy, for it disregards the energy of the electrons (if any) that have been localized in forming the new σ bond C—X, but it does take account of any change in the number of π electrons. As usual, we assume that in *comparing* rates for two alternative positions of attack the same σ-bond changes occur. The *difference* of two activation energies should then be determined by the difference of (π) localization energies:

$$\Delta E_r - \Delta E_s = L_r^{(n)} - L_s^{(n)} \tag{10.15}$$

for two positions r and s. It should be noted that for alternant hydrocarbons $L_r^{(n)}$ must have the same value for $n = 0, 1, 2$, because the highest MO of the residual molecule is the NBMO (the conventional 'zero' of energy), but for non-alternants and heterocycles this is not so and care must be taken to choose the right index.

Calculation of localization energies is straightforward; it is only necessary to perform two calculations (usually of Hückel type), one for the molecule and one for the residual molecule, and to take the difference. Many such calculations have been made and again there is a fairly good correlation with results based on experimental rate constants. In particular, the concept of localization energy has a direct link with that of 'aromaticity'. Originally, the term 'aromatic' meant 'having a chemistry like that of benzene' but with the development of quantum theory it came to be associated with the possession of a high resonance (i.e. delocalization) energy. Now it is clear from Fig. 10.2 that the two ideas are not necessarily equivalent; chemistry has to do with the *activation* energy (an energy *difference* between molecule and transition complex) while resonance energy refers only to the molecule in its ground state, and high resonance energy means only that the level on the left of the hump is low. The localization energy thus appears to provide a more satisfactory theoretical index of aromaticity and calculations seem to support this conclusion. For example, naphthalene and azulene both have a much higher resonance energy than benzene, but they are *less* aromatic; in contrast, naphthalene has $L = -2{\cdot}30\beta$ (lowest at the α position) and azulene has $L^{(1)} =$

$-2 \cdot 24\beta$, $L^{(0)} = L^{(2)} = -1 \cdot 98\beta$, correctly indicating less aromatic character than benzene ($L = -2 \cdot 54\beta$) and a much higher reactivity towards electrophiles and nucleophiles. Roughly speaking we expect aromatic character when all localization energies exceed about -2β. The 'conventional' definition of aromaticity (i.e. possession of a high resonance energy) is nevertheless useful in characterizing reactivity in processes which involve ring opening and closing, as we shall see in a later section.

Various ways of estimating localization energies, without recourse to full calculations on the molecule and residual molecule, have been proposed. Thus, provided we know the bond orders in the molecule itself, we may use (10.11) to obtain a rough estimate of the energy change on breaking the conjugation with atom r; we need only put $\delta\beta = -\beta$ to obtain

$$L_r = -2\beta N_r. \tag{10.16}$$

This gives $L = -2 \cdot 58\beta$ for the naphthalene α position (Fig. 10.4 (*b*)) and $L = -2 \cdot 78\beta$ for the β-position; both values are some 10 per cent too large but their relative order is correctly predicted.

An even simpler method has been used extensively by Dewar (Dewar and Dougherty 1975). This is based on a quantitative form of the argument used in §8.7 to discuss the effect of heteroatoms; instead of pulling a carbon atom out of conjugation by breaking two π bonds, we consider the molecule to arise from the *union* of the residual molecule and the single carbon with its π electron. For an even alternant hydrocarbon, the residual molecule is then odd and is a radical, the orbital energies being as in Fig. 10.5. Instead of building up the MO's of the complete molecule as LCAO's, let us now imagine them to be built up as linear combinations of the molecular orbitals of the two separate parts; this LCMO method is completely equivalent to the usual LCAO method, provided we use *all* the MO's of the two fragments. However, we know from general principles (§4.7) that the orbitals which mix most heavily are those whose energies are equal or nearly equal, and Dewar has therefore simplified the method by supposing that only one such pair of orbitals need be considered. In the present case (Fig. 10.5) we expect that connection of the two

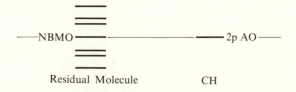

FIG. 10.5. Localization energy by Dewar's method. An even alternant hydrocarbon may be formed by the union of a radical and a CH fragment, with orbital energies shown left and right respectively.

fragments will lead to bonding and antibonding combinations of the NBMO of the residual molecule with the carbon AO, both having exactly the same energy ($H_{11} = H_{22} = \alpha$). As usual, the energy change is $\pm H_{12}$ where H_{12} is an off-diagonal matrix element for the two orbitals concerned and is thus

$$H_{12} = \int \phi_r h(c_1\phi_1 + c_2\phi_2 + \ldots + c_s\phi_s + \ldots)\,d\tau$$

the coefficients being those appropriate to the NBMO of the residual molecule (r not appearing). With the usual nearest-neighbour approximation, contributions only occur for AO's ϕ_s adjacent to ϕ_r, giving an interaction energy (two electrons going into the new bonding MO)

$$E_{int} = 2H_{12} = 2\beta \sum_{s(s-r)} c_s^{NBMO}. \tag{10.17}$$

This is negative (interaction lowering the energy) and reversal of the sign gives the energy needed to separate the fragments

$$L_r = -2\beta \sum_{s(s-r)} c_s^{NBMO}. \tag{10.18}$$

The value of this approximate result is that the coefficients in the NBMO can be worked out in a few seconds, using only simple mental arithmetic. Comparison of (10.18) with (10.16) suggests that we are, in effect, making use of the NBMO to estimate the orders of the bonds connecting r with the residual molecule; very roughly in fact

$$p_{rs} \simeq c_s^{NBMO} \qquad (s \text{ linked with } r).$$

This approximation is easily tested. If we remove a 1-carbon from naphthalene we obtain the residual molecule shown in Fig. 10.6, with the NBMO coefficients shown (determined as in §8.6). The resultant bond-order estimates of 0·30 and 0·60 do not compare very well with the values (0·56 and 0·73) in Fig. 10.4 (which is not surprising since mixing with all the lower MO's in Fig. 10.5 has been ignored), but at least they appear in the correct order. The localization energy given by (10.18) is $-1·81\beta$, some 20 per cent lower than the calculated value $-2·30\beta$, but, as Dewar has stressed, even such rough values reflect surprisingly well the general trends in going from one position, or one molecule, to another.

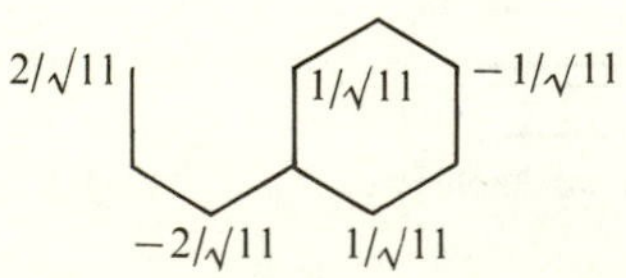

$2/\sqrt{11}$ $1/\sqrt{11}$ $-1/\sqrt{11}$

$-2/\sqrt{11}$ $1/\sqrt{11}$

FIG. 10.6. Residual molecule when the 1-position in naphthalene is removed from conjugation. The numbers give the coefficients in the NBMO.

10.4. Some applications to substitution reactions

The two approaches just indicated are only intended to be representative of the many attempts made during the last 40 years to obtain a simple, but reasonably realistic, theory of reactivity, and we have discussed only substitution reactions, neglecting for example, oxidation–reduction reactions where other important concepts arise. In particular the 'charge-transfer' mechanism in which the highest occupied (HO) MO of the donor and the lowest unoccupied (LU) MO of the acceptor are of central importance. A comparison of many of the different indices has been made by Greenwood and McWeeny (1966) and more recent developments have been reviewed by Fujimoto and Fukui (1972). It is even possible, using present-day computers, to discuss simple substitution (and other) reactions by non-empirical calculation of the variation of total electronic energy along a postulated reaction path.

Instead of carrying the theory further, however, we devote this section to showing how the simple indices so far introduced can be applied in the laboratory, using nothing more than pencil and paper. The approach, due largely to Longuet–Higgins (1950) and Dewar, consists basically of using differences of localization energy to measure differences of activation energy for given positions of substitution, and asking how such differences are likely to be affected by the presence of heteroatoms or other substituents. We give three examples to indicate a vast field of applications:

(i) A classic problem concerns the orientation of substituents in the benzene ring. How is the ease of substitution by Y affected by the presence of an existing substituent X? This amounts to asking how the localization energy at the Y position (Fig. 10.7 (a)) depends on the $\delta\alpha$ change induced at the centre to which X is attached. Let us denote the energies of the molecule and the residual molecule, in the absence of substituents, by E and E'; the localization energy for Y substitution, in the absence of X, is then $L_0 = E' - E$. Now if X is attached at position r, these energies will become $E + q_r \delta\alpha$ and $E' + q'_r \delta\alpha$, respectively, accord-

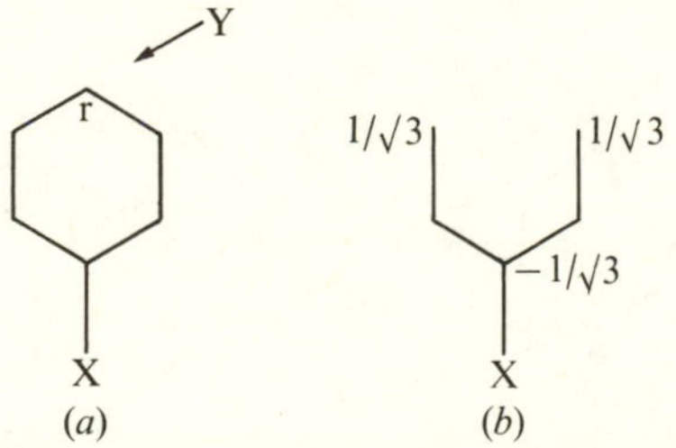

FIG. 10.7. Orientation of substituent Y by presence of X: (a) substituent approaching position r; (b) coefficients in the NBMO of the residual molecule.

ing to (10.4). The localization energy in the presence of X at position r will thus be

$$L = (E' + q'_r \delta\alpha) - (E + q_r \delta\alpha) = L_0 + (q'_r - q_r)\delta\alpha. \qquad (10.19)$$

It is therefore only necessary to know the charge at centre r in the molecule and in the residual molecule for Y substitution. Now in benzene $q_r = 1$ at every position, while in the residual molecule (the 5-carbon chain) the charges q'_r follow from the form of the NBMO; at positions ortho, meta, and para to the position of Y they are as follows (Fig. 10.7 (b)).

	ortho	meta	para	
4 electrons	$\frac{2}{3}$	1	$\frac{2}{3}$	(electrophilic)
5 electrons	1	1	1	(radical)
6 electrons	$1\frac{1}{3}$	1	$1\frac{1}{3}$	(nucleophilic)

If the change $\delta\alpha$ is negative (e.g. X being a halogen) the ortho and para positions should thus be deactivated towards electrophilic substitution, while for $\delta\alpha$ positive (e.g. X being NO_2, CN, CH_3) they should be activated. The effect of an existing substituent on radical reactions should be much less marked. These conclusions provide a slightly more sophisticated interpretation of the directing effects already predicted (p. 256) on the basis of a simple electrostatic argument applied to the isolated molecule. The implication is that the two curves in Fig. 10.2 retain their relative positions in proceeding from the isolated molecule to the transition state.

(ii) A similar argument applies to heterocyclic systems. Suppose we consider the hydrolysis of 1-chloronaphthalene (Fig. 10.8 (a)) and of the aza-derivative with a nitrogen at position r. The hydroxyl radical is nucleophilic and leaves all 10 π electrons in the 9-centre residual molecule; the NBMO is thus doubly occupied to give the charges shown in Fig. 10.8 (b). By the argument of the last example, the localization energy is increased by $(q'_r - q_r)\delta\alpha$ when C at position r is

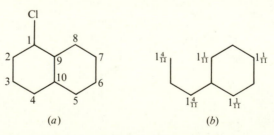

(a)   (b)

FIG. 10.8. Hydrolysis of 1-chloronaphthalene: (a) conventional numbering system; (b) charges in the 10-electron residual molecule.

replaced by N; this amounts to $(4/11)\delta\alpha$ at the 2 and 4 positions, and $(1/11)\delta\alpha$ at the 5, 7, and 9 positions. Hydrolysis of the chlorine atom should thus be strongly favoured by aza substitution at the 2 or 4 positions.

(iii) Finally, let us try to understand the properties of some nitrogen-containing aromatic bases. Here the lone pair of the heterocyclic nitrogen atom acts as a proton acceptor, the reaction is reversible, and the analogue of (10.3) is

$$-RT \log (K/K_0) \simeq \Delta E - \Delta E_0. \qquad (10.20)$$

In other words the pK_a value should be linearly related to ΔE, the change in π-electron energy on protonation. If now q_r is the charge at position r in the parent molecule (i.e. with C instead of N), the energy change on going to the base will be $q_r \delta\alpha$ and on going to the protonated base it will be $q_r \delta\alpha^+$, where $\delta\alpha$ represents the change of α associated with $C \to N$ and $\delta\alpha^+$ that associated with $C \to NH^+$. Thus, if we know q_r we can estimate ΔE on protonation as

$$\Delta E \simeq q_r(\delta\alpha^+ - \delta\alpha) \qquad (10.21)$$

which is negative since $\delta\alpha^+$ will be more negative than $\delta\alpha$. For a hydrocarbon parent, $q_r = 1$ everywhere and the basicity should not depend much on the nitrogen position. In, say, an aza-aromatic amine such as 4-aminoquinoline (Fig. 10.9 (a)), however, q_r at the nitrogen position may be considerably enhanced by the presence of the NH_2 group. The simplest way to estimate the enhancement is once again by use of the NBMO; the nitrogen supplies two π electrons and the excess charges in the (11-centre) isoelectronic carbanion are shown in Fig. 10.9 (b). Disregarding the electronegativity difference, we expect an extra $\frac{1}{5}$ electron at the 4-position; $q_r = 1\frac{1}{5}$ compared with 1 in the absence of the NH_2 group. From (10.21) it follows at once that 4-aminoquinoline (observed pK_a 9·08) should be more basic than quinoline (pK_a 4·85).

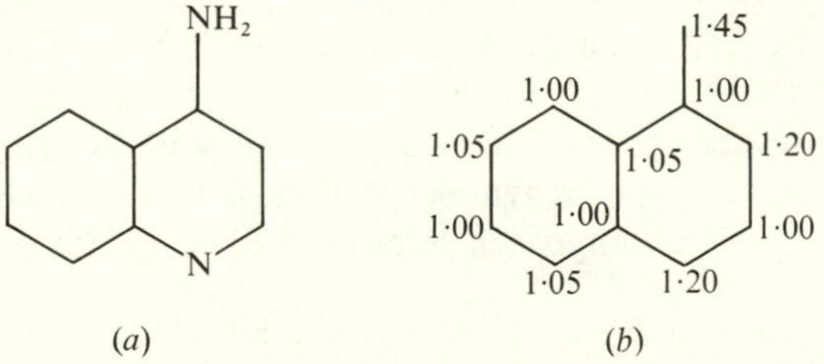

FIG. 10.9. Effect of aza substitution on basicity: (a) 4-amino quinoline; (b) π charges in isoelectronic carbanion.

A remarkable amount of progress has been achieved using ideas no more sophisticated than those used in this section.† In view of the chemical insight they provide and the ease with which they are applied, such methods, primitive as they are, still have much to offer.

10.5. Ring closing and opening.
Woodward–Hoffman rules

In §8.8 we looked at the basis of the Hückel method and began to understand the reasons for its success. In particular, the Coulomb and bond integrals (α's and β's) play the part of *parameters* and the exact interpretation of the underlying AO's is crucial neither to the structure of the theory nor to the form of the secular equations; the energy order and overall shapes of the MO's are determined not so much by the parameter values as by the geometry of the molecule. Indeed, with a nearest-neighbour approximation even the geometry of the molecule is of secondary importance; what really matters is which AO's are adjacent, for if ϕ_r and ϕ_s come together there will be a non-zero bond integral β_{rs}, and it is therefore the *topology*‡ of the molecule which determines the pattern of β's in the secular equations and hence the forms and energies of the LCAO MO's. Our analysis also suggested that essentially similar approximations could be extended with equal validity to the study of the σ electrons, using appropriately orthogonalized AO's, with or without hybridization, to construct the σ MO's. The resultant 'extended Hückel theory' (EHT), with its many sweeping approximations (even beyond its disregard of electron interactions), clearly cannot be relied upon in any way as a means of making accurate numerical predictions, and yet calculations even at this level have led to a surprising advance in our understanding of the changes of electronic structure which occur in chemical reactions. Such changes are often much more severe than in the case of substitution reactions, involving a complete change of geometry in which the σ–π classification is lost, and it then becomes imperative to consider all the valence electrons, at least in that part of the system where major changes are occurring.

To show how much can be obtained from purely qualitative arguments, we may consider one of the most important developments in recent years; we shall take a few examples of cycloaddition and electrocyclic reactions to illustrate the origin of the now famous Woodward–Hoffmann rules. Although the formulation of the rules initially leaned heavily on experience gained from EHT calculations, they are now most commonly presented in purely pictorial form with an emphasis on the symmetry of the orbitals undergoing the most significant change. This approach is developed in great detail, with many

---

† See for example Dewar (1969) and Dewar and Doughery (1975) for many stimulating and provocative examples.

‡ That is, the way in which the numbered centres (or more precisely orbitals) are *connected* to each other—thus, *cis*- and *trans*-butadiene have the same topology but different geometry.

applications, by Woodward and Hoffmann (1970). The use of symmetry arguments was first stressed by Longuet–Higgins and Abrahamson (1965). It has also been pointed out by Dewar (Dewar and Dougherty 1975) that similar conclusions, though in a more limited context, were reached many years ago by M. G. Evans. We shall start with the Woodward–Hoffmann approach, turning to that of Evans and Dewar in the next section.

Example 1

As a first example of an addition reaction let us take the dimerization of two ethylene molecules to form cyclobutane, choosing the maximum-symmetry approach (Fig. 10.10) in order to simplify the discussion. In Fig. 10.10 (*a*) the two molecules have approached to the point at which their $2p_\pi$ AO's begin to overlap, but without appreciable change of hybridization; this is essentially a simplified model of a transition state in which intramolecular π bonds are being weakened and intermolecular σ bonds are developing. In Fig. 10.10 (*b*) the reaction has gone to completion, increasing σ overlap being favoured by a hybridization change from trigonal to distorted tetrahedral, yielding bent bonds in the cyclobutane ring with the CH_2 groups directed radially outwards (four-fold symmetry axis). In the Woodward–Hoffmann approach attention is focused on the changing orbital energies of the electrons most affected in the formation of the transition complex (Fig. 10.10 (*a*)) and the discussion is thus similar to that which leads to Walsh's rules (p. 191). Initially, there are two pairs of π electrons, and finally two pairs of σ electrons, and most of the change can be accomplished without greatly disturbing the electron pairs responsible for the C—C and C—H single bonds.

If we suppose that the four carbons in Fig. 10.10 (*a*) form a rectangle, it is easy to set up MO's of correct symmetry (Problem 8.10); they should either be unchanged or change sign under reflection through either of the two planes of

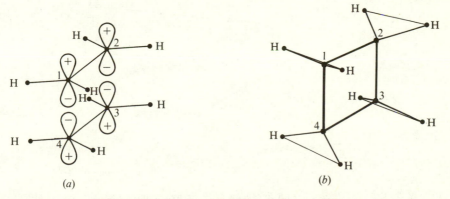

(*a*)   (*b*)

Fig. 10.10. Dimerization of two ethylene molecules to form cyclobutane: (*a*) maximum-symmetry approach, slight intermolecular overlap of $2p_\pi$ AO's; (*b*) increasing overlap and change of hybridization, leading to σ-bonded C_4 ring.

symmetry—the vertical plane which bisects the π bonds and the horizontal plane which bisects the incipient σ bonds. There is of course a third plane of symmetry, the C_4 plane, but all the orbitals considered are automatically symmetric across that plane. The appropriate symmetry orbitals, classified SS, SA, etc. according to symmetry or antisymmetry across vertical and horizontal reflection planes respectively, are

$$
\begin{aligned}
\text{SS:} \quad & \psi_1 = N_1(\phi_1 + \phi_2 + \phi_3 + \phi_4) \\
\text{SA:} \quad & \psi_2 = N_2(\phi_1 + \phi_2 - \phi_3 - \phi_4) \\
\text{AS:} \quad & \psi_3 = N_3(\phi_1 - \phi_2 - \phi_3 + \phi_4) \\
\text{AA:} \quad & \psi_4 = N_4(\phi_1 - \phi_2 + \phi_3 - \phi_4).
\end{aligned}
\tag{10.22}
$$

There is only one orbital of each type and therefore there are no secular equations to solve; each orbital is an MO. At the beginning of the approach, when there is negligible overlap between the two ethylenes, ψ_1 and ψ_2 will be nearly degenerate, low-energy orbitals, describing the two π bonds. This is easily seen if we use π_{12}, for example, to denote the localized π-type bonding orbital between centres 1 and 2. The (unnormalized) π-type bond orbitals on the two ethylenes are then, using a star to indicate antibonding character,

$$
\begin{aligned}
\pi_{12} &= \phi_1 + \phi_2 & \pi_{43} &= \phi_4 + \phi_3 \\
\pi_{12}^* &= \phi_1 - \phi_2 & \pi_{43}^* &= \phi_4 - \phi_5
\end{aligned}
\tag{10.23}
$$

and thus, from (10.22),

$$
\psi_1 = N_1(\pi_{12} + \pi_{43}) \qquad \psi_2 = N_2(\pi_{12} - \pi_{43}).
\tag{10.24}
$$

When the off-diagonal matrix element connecting π_{12} and π_{43} is negligible (small overlap) the energy associated with each MO will be that of an ethylene π electron in the bonding orbital (π_{12} or π_{43}). The electron configuration of lowest energy is thus $[\psi_1^2 \psi_2^2]$, as shown on the left in Fig. 10.11 which also

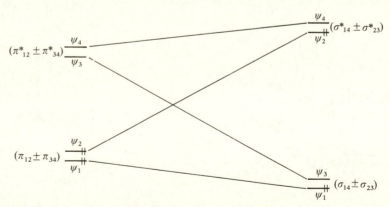

FIG. 10.11. Correlation diagram for approach of two ethylene molecules as in Fig. 10.10 (a). Initially, the MO energies are close to those of bonding and antibonding ethylene π MO's (left); finally, they are nearer to those of intermolecular bonding and antibonding σ MO's (right). Configuration $\psi_1^2 \psi_2^2$ then gives a highly excited state of the transition complex in Fig. 10.10 (a).

shows the energies associated with ψ_3 and ψ_4—roughly that of a π electron in an *anti*bonding orbital (π_{12}^* or π_{43}^*).

What happens as the two ethylenes approach? The sides 1–4 and 2–3 of the rectangle eventually turn into (bent) σ bonds (Fig. 10.10 (*b*)), but first let us try to make use of the localized combinations

$$\sigma_{14} = \phi_1 + \phi_4 \qquad \sigma_{23} = \phi_2 + \phi_4$$
$$\sigma_{14}^* = \phi_1 - \phi_4 \qquad \sigma_{23}^* = \phi_2 - \phi_3 \qquad (10.25)$$

which would describe two separate σ bonds formed from the axially directed 2p AO's prior to any change of hybridization. In this (imaginary) penultimate stage of the reaction, the MO's would still have the forms (10.22), but it would be more natural to express them in terms of the σ-type bond orbitals (10.25); instead of (10.24) we could equally well write, using (10.25) in (10.22),

$$\psi_1 = N_1(\sigma_{14} + \sigma_{23})$$
$$\psi_2 = N_2(\sigma_{14}^* + \sigma_{23}^*). \qquad (10.26)$$

It is now clear that as the reaction proceeds the electrons in ψ_2 are carried into an *excited* MO of the σ-bonded complex, namely a delocalized mixture of two *anti*bonding σ^* orbitals. The other MO's, ψ_3 and ψ_4, may be treated in the same way and expressed alternatively as

$$\psi_3 = N_3(\pi_{12}^* + \pi_{43}^*) = N_3(\sigma_{14} - \sigma_{23})$$
$$\psi_4 = N_4(\pi_{12}^* - \pi_{43}^*) = N_4(\sigma_{14}^* - \sigma_{23}^*) \qquad (10.27)$$

On the left of Fig. 10.11 the interaction between the strong π bonds is weak and the 'in-phase' combination ψ_3 is only slightly lower in energy than ψ_4, both being close to that of an antibonding π orbital; on the right, the interaction between the strong σ bonds is weak, but while the ψ_3 energy is roughly that of a bonding orbital the ψ_4 energy is roughly that of its antibonding partner and lies very much higher. A one-determinant wavefunction (see p. 135) based on the configuration $[\psi_1^2\psi_2^2]$ would therefore describe the ground state of two ethylene molecules quite well, but as the reaction proceeded it would go towards a highly excited state of the dimer. To describe the ground state of the dimer we should clearly need to assume a different occupation of the orbitals, taking the configuration $[\psi_1^2\psi_3^2]$.

The correlation diagram of Fig. 10.11 clearly suggests that the proposed addition of two ground-state ethylene molecules would be energetically 'forbidden', for the energy required to reach the electronically excited product would greatly exceed the thermal energy normally available. However, if the pair of ethylene molecules could be excited first to a state with electron configuration $[\psi_1^2\psi_2\psi_3]$ then the reaction might proceed spontaneously, the rising energy of the electron in ψ_2 being more than compensated by the falling energy of that in ψ_3. This is exactly what is observed; the cycloaddition of two

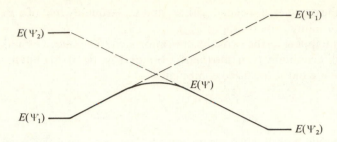

FIG. 10.12. Variation of energy with allowance for configuration interaction. When Ψ_1 and Ψ_2 are allowed to mix, there is an 'avoided crossing' as Ψ changes from mainly Ψ_1 to mainly Ψ_2.

ethylenes to form cyclobutane is 'thermally forbidden' but 'photochemically allowed'.

The argument has been deliberately oversimplified, but the conclusion is unaffected by refinements. For example, the one-determinant wavefunction based on the initial electron configuration, let us denote it by $\Psi_1[\psi_1^2\psi_2^2]$, would correspond to the same energy at the 'crossing point' in Fig. 10.11 as would the function $\Psi_2[\psi_1^2\psi_3^2]$. To obtain a better description we should allow the two functions to mix, taking

$$\Psi = c_1\Psi_1 + c_2\Psi_2 \tag{10.28}$$

and determining the coefficients in the usual way from the secular equations. That these functions are *allowed* to mix follows because both are of SS character† for reflections in the two planes of symmetry. We shall then obtain an 'avoided crossing' when we plot the *total* electronic energy (Fig. 10.12) as in the discussion of §3.9. The system can then pass smoothly from $[\psi_1^2\psi_2^2]$ on the left to the ground-state configuration $[\psi_1^2\psi_3^2]$ on the right, c_1 rapidly diminishing at the crossing point as c_2 rapidly grows. The modified conclusion is thus that a transition from the ground state of the reactants to the ground state of the products is *possible* but only by climbing a very high potential hill; the reaction is evidently still thermally 'forbidden', as the original discussion suggested. A similar argument applies in principle to the excited states, but in practice the wavefunction $\Psi_3[\psi_1^2\psi_2\psi_3]$ is so far removed in energy from the next of similar symmetry (namely AA) that the avoided-crossing situation does not arise (Fig. 10.13) and mixing is unimportant.

Example 2

A contrasting example is provided by the $[4+2]$-cycloaddition of butadiene to ethylene (Fig. 10.14) to give cyclohexene, a Diels–Alder reaction. Here there is only one plane of symmetry, but similar considerations apply. The energy

† Note that ψ_3 appears *twice* in Ψ_2, and under reflection in the vertical plane *both* factors are multiplied by -1, giving the many-electron wavefunction S character for this reflection.

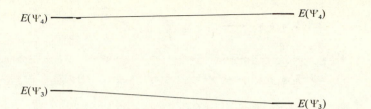

$E(\Psi_4)$ ————————————————— $E(\Psi_4)$

$E(\Psi_3)$ ————————————————— $E(\Psi_3)$

FIG. 10.13. Negligible effect of configuration interaction for excited-state functions which differ greatly in energy throughout the geometry change.

levels for chains with $N = 2$ and $N = 4$ are given by (8.19) and the symmetries of the corresponding MO's (AO coefficients following the sinusoidal pattern given by (8.18 (a)) are alternately S and A, the lowest energy MO being S type. Again, the bonds in cyclohexene will be described most conveniently in terms of a rehybridization accompanying the geometry change, but if we consider the penultimate stage of the reaction we may disregard such changes and use the original set of $2p_\pi$ AO's.

The two lowest energy MO's of the reactants are shown on the left in Fig. 10.15 and are of S type. They may be combined to give

$$\psi = a(\phi_1 + \phi_2) + b(\phi_3 + \phi_4) + c(\phi_5 + \phi_6) \qquad \text{(S)} \qquad (10.29)$$

where the coefficients a, b, c will be positive in the lowest-energy combination. As the 1–3 and 2–4 interactions develop the first two terms will grow at the expense of the third, and as the molecules approach we therefore expect the lowest MO of the product to become

$$\psi_1^P \simeq (a\phi_1 + b\phi_3) + (a\phi_2 + b\phi_4). \qquad (10.30)$$

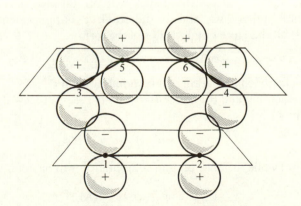

FIG. 10.14. Approach of ethylene and butadiene in [4+2] cycloaddition. There is a plane of symmetry, bisecting the bonds 1–2 and 5–6. (The numbering conveniently recognizes the symmetry—odd numbers on one side, even on the other.)

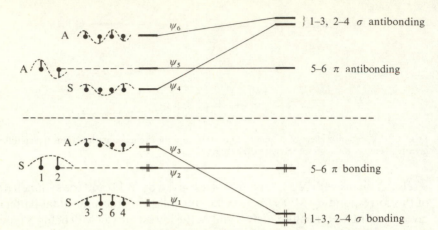

Fig. 10.15. Correlation diagram for approach of ethylene and butadiene as in Fig. 10.14. Energy levels and MO's (vertical strokes indicating AO coefficients on the numbered atoms) of the reactants are on the left; corresponding levels as the product (cyclohexene) is approached are on the right. The formation of σ bonds and the development of a 5–6 π bond should be noted.

This is a symmetric combination of localized MO's representing two σ bonds, 1–3 and 2–4, in the complex. The second MO of type (10.29) must stay orthogonal to the first and will therefore stay away from the σ-bond region, becoming mainly

$$\psi_2^P \simeq c(\phi_5 + \phi_6) \tag{10.31}$$

representing a localized π bond in the butadiene central link. The third MO will initially (i.e. on the left in Fig. 10.15) be mainly the first A-type butadiene MO. It can, however, mix with other A-type orbitals, the nearest being the ethylene antibonding MO, and as the ethylene–butadiene interaction increases we shall therefore obtain MO's of the form

$$\psi = a'(\phi_1 - \phi_2) + b'(\phi_3 - \phi_4) + c'(\phi_5 - \phi_6) \quad \text{(A)}. \tag{10.32}$$

The lowest-energy combination will be bonding with respect to the new 1–3 and 2–4 interactions, and the first A-type MO of the product will thus become predominantly

$$\psi_3^P \simeq (a'\phi_1 + b'\phi_3) - (a'\phi_2 + b'\phi_4) \tag{10.33}$$

(a', b' positive) while the antibonding combination will be mainly

$$\psi_4^P \simeq c'(\phi_5 - \phi_6) \quad \text{(A)}. \tag{10.34}$$

We now know how the first few MO's develop during the course of the reaction, and see that the ground-state electron configuration of the non-interacting molecules (on the left in Fig. 10.15) can pass smoothly into that for cyclohexene

$$[(\psi_1^R)^2(\psi_2^R)^2(\psi_3^R)^2] \rightarrow [(\psi_1^P)^2(\psi_2^P)^2(\psi_3^P)^2] \qquad (10.35)$$

and there is no promotion of electrons into the *antibonding* MO's of the product.

The complete correlation diagram is shown in Fig. 10.15, which should be compared with Fig. 10.11. The conclusion is exactly reversed; the [4+2] cycloaddition reaction of butadiene and ethylene should be favoured *thermally* rather than *photochemically*. Again, this result is in complete accord with experiment. It is not difficult to generalize the argument, which depends essentially on the alternating S and A characters of the MO's to the addition of a polyene of N_1 carbons to one of N_2 carbons. The result is that if $N_1 + N_2 = 4n$ (*n* being an integer) the reaction will be thermally forbidden and photochemically allowed, while if $N_1 + N_2 = 4n + 2$ the preference is reversed.

Example 3

Woodward and Hoffmann introduced the term 'electrocyclic reaction' to describe the formation of a single bond between the end carbon atoms of a conjugated chain, or the reverse process in which the single bond is broken and the ring opens. As an example, we may consider the ring opening in which cyclobutene is converted, by heating, into butadiene.

$$\begin{array}{ccc} \text{HC}\!=\!\text{CH} & & \text{HC}\!-\!\text{CH} \\ | \quad\; | & \longrightarrow & \diagup \qquad \diagdown \\ \text{H}_2\text{C}\!-\!\text{CH}_2 & & \text{H}_2\text{C} \qquad\quad \text{CH}_2 \end{array}$$

The connection with our previous examples is clearest if we consider the reverse reaction; the conversion of cyclobutadiene to cyclobutene is then seen to be an *intra*molecular cycloaddition reaction. There are two possible ways of bringing about this change in a symmetrical fashion. The reaction need not, of course, proceed in a perfectly symmetrical way, but by considering the most symmetrical cases we can simplify the discussion and distinguish very easily the two main *types* of process. In the first (Fig. 10.16 (*a*)) the CH_2 groups are twisted in the opposite (*disrotatory*) sense, while in the second (Fig. 10.16 (*b*)) they are twisted in the same (*conrotatory*) sense. The disrotatory process preserves a *plane* of symmetry and the conrotatory process preserves a twofold *axis* of symmetry; in each case the orbitals may be classified as S or A.

Let us denote the 2p AO's by $\phi_1, \ldots, \phi_4$ and keep ϕ_1 and ϕ_4 perpendicular to the CH_2 groups. The lowest butadiene π MO is of the form (see Fig. 10.15)

$$\psi_1 = a(\phi_1 + \phi_4) + b(\phi_2 + \phi_3) \qquad \text{(S)} \qquad (10.36)$$

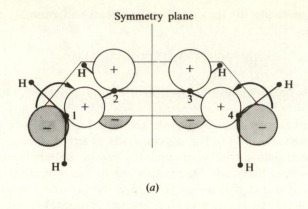

(a)

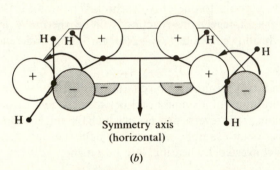

Symmetry axis
(horizontal)

(b)

FIG. 10.16. Two modes of twisting the butadiene CH_2 groups: (a) disrotatory (one clockwise, one anticlockwise); (b) conrotatory (both anticlockwise). In (a) the positive lobes of the $2p_\pi$ orbitals approach; while in (b) there is a sign difference.

which is preserved during disrotatory twisting. As the twist approaches 90°, however, the positive lobes of ϕ_1 and ϕ_4 overlap (overlap integral positive, β_{14} negative), and the $\phi_1 + \phi_4$ combination begins to look like a localized σ bond while $\phi_2 + \phi_3$ remains essentially an ethylene-like π-bond MO. Therefore we may rewrite ψ_1 as

$$\psi_1 = a\sigma_{14} + b\pi_{23}. \tag{10.37}$$

The next butadiene MO of S symmetry is ψ_3 (denoted by ψ_4 in Fig. 10.15 where the sequence included the ethylene MO's) which may be written similarly as

$$\psi_3 = c\sigma_{14} + d\pi_{23}.$$

The lowest-energy MO of the twisted system (i.e. our model of the transition complex) will be a mixture of ψ_1 and ψ_3 in which σ_{14} is favoured by the growing value of the bond integral β_{14}. The lowest MO of the product is expected to be mainly σ_{14}, the upper combination (still bonding) being mainly π_{23}:

$$\psi_1 \to \psi_1^{\mathrm{P}} \simeq \sigma_{14}, \qquad \psi_3 \to \psi_2^{\mathrm{P}} \simeq \pi_{23}. \tag{10.38}$$

In a similar way the A-type butadiene MO's ψ_2 and ψ_4 are of the form

$$\psi = a'(\phi_1 - \phi_4) + b'(\phi_2 - \phi_3) \qquad \text{(A)} \tag{10.39}$$

and may be rewritten in terms of localized *anti*bonding orbitals, σ_{14}^* and π_{23}^*. The next lowest energy MO of the product, ψ_3^{P}, is therefore expected to be predominantly π_{23}^* (the least *anti*bonding), while ψ_4 should approach σ_{14}^*:

$$\psi_2 \to \psi_3^{\mathrm{P}} \simeq \pi_{23}^*, \qquad \psi_4 \to \psi_4^{\mathrm{P}} \simeq \sigma_{14}^*. \tag{10.40}$$

It is now clear that the correlation diagram must have the form shown on the left in Fig. 10.17 (butadiene levels in the centre).

In the conrotatory process (Fig. 10.16 (*b*)) the combinations (10.36) and (10.39) again have appropriate symmetry but are, respectively, A and S for rotation around the symmetry axis, so the labelling is reversed. As the twist approaches 90°, $\phi_1 + \phi_4 \to \sigma_{14}^*$ and therefore ψ_1 and ψ_3 (A-type) are of the form

$$\psi = a\sigma_{14}^* + b\pi_{23} \qquad \text{(A)}.$$

The lowest-energy combination must be mainly π_{23} (π bonding), which is

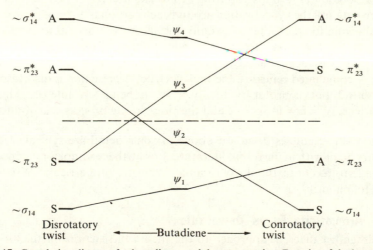

FIG. 10.17. Correlation diagram for butadiene–cyclobutene reaction. Energies of the butadiene MO's are in the centre. On the left, the product is formed by disrotatory twisting (Fig. 10.16 (*a*)); on the right, it is formed by conrotatory twisting (Fig. 10.16 (*b*)). Symmetry symbols on the left are for reflection; those on the right are for rotation around the symmetry *axis* (interchanging the CH_2 groups).

expected to lie higher in energy than σ_{14}

$$\psi_1 \to \psi_2^P \simeq \pi_{23} \tag{10.41a}$$

while the second combination must give the highest level (σ antibonding)

$$\psi_3 \to \psi_4^P \simeq \sigma_{14}^*. \tag{10.41b}$$

In a similar way we find for the symmetric orbitals

$$\psi_2 \to \psi_1^P \simeq \sigma_{14}, \qquad \psi_4 \to \psi_3^P \simeq \pi_{23}^*. \tag{10.42}$$

The change in orbital energies during the conrotatory twisting is thus likely to be as shown in the right-hand half of the correlation diagram (Fig. 10.17).

It is now clear that if we start from the butadiene ground-state configuration $[\psi_1^2 \psi_2^2]$, disrotatory twisting would lead towards a product with the excited electron configuration $[(\psi_1^P)^2(\psi_3^P)^2]$, while conrotatory motion would yield $[(\psi_2^P)^2(\psi_1^P)^2]$ in which the two lowest energy MO's remain filled. As in Example 1, further analysis of the many-electron wavefunction would show that the excited state would not actually be reached during the disrotatory twist, configuration interaction yielding an avoided crossing, but there would be a high potential energy hill which is not encountered in the conrotatory process. Experiment shows that the reaction is indeed conrotatory. As in cycloaddition, the rule is reversed when the reaction proceeds photochemically.

Again, the discussion has been reduced to its essentials. As the geometry changes (in ring closing, say) so will the hybridization at the carbons, the p orbitals of the CH_2 groups acquiring s character and the C—C bonds bending somewhat until the hybridization around each carbon is distorted tetrahedral, but allowing for such changes would not affect the essential features of our argument or in any way deepen our understanding of the process.

As in the case of cycloaddition reactions, the argument is easily generalized. With N conjugated centres, a thermal electrocyclic reaction is disrotatory for $N = 4n+2$ but conrotatory for $N = 4n$, n being an integer. Thus for hexatriene, $N = 6 = (4 \times 1)+2$ and the ring-closing process is *dis*rotatory.

The three examples discussed above in some detail are typical of a vast number dealt with in the recent literature. For further examples reference may be made to recent textbooks of organic chemistry (in particular Woodward and Hoffmann 1970).

10.6. Aromaticity. Evans–Dewar rules

An altogether different approach to the theory of reactions of the kind just discussed has been developed at great length by Dewar, starting from a principle which was first enunciated by Evans (1939). This approach is so dependent on the concept of aromaticity that before we can use it we must go back and extend our earlier discussion (p. 302). We noted that aromaticity was usually taken to mean 'having a large resonance energy (i.e. bond de-localization energy)', but that this property, in itself, did not have immediate *chemical* implications; to discuss the chemistry of a molecule, and hence its aromaticity in the 'old-fashioned' sense, we should consider the bearing of its resonance energy on the reactions it can undergo. From this point of view the high stability of benzene against electrophilic substitution (i.e. its high 'chemical' aromaticity) is a consequence of the high localization energy (i.e. activation energy) of the Wheland transition state, and this in turn results from the loss of resonance energy when the ring is broken and replaced, in effect, by an open chain whose resonance energy is much smaller. The resonance energy is thus important only in so far as it bears on an energy *difference*, the activation energy, and the resonance energy of the transition state is just as important as that of the molecule itself. It is not surprising, then, that cycloaddition and electrocyclic reactions can also be discussed by asking how the resonance energy *changes* as the transition state is approached, for the more aromatic the transition complex (in the conventional energy sense) the lower will be the activation energy. This is, in essence, the principle introduced by Evans.

Let us now adopt high resonance energy (i.e. low π-electron energy) as an indicator of aromaticity and see how it relates to the properties of chains and rings. Do six π electrons prefer to be in a ring (benzene) or a chain (hexatriene)? From §8.4 (with $N = 6$) we easily find total π energies

$$E_{\text{ring}} = 6\alpha + 8\beta, \qquad E_{\text{chain}} = 6\alpha + 6\cdot99\beta.$$

The ring is therefore favoured by $1\cdot01\beta$ (β being negative), and we say benzene is aromatic. The measure of aromaticity adopted here is not the same as the delocalization energy defined in (8.26), but allows for the fact that even in an open chain, where one classical structure might seem appropriate, there is appreciable delocalization of the π bonding. A similar comparison between cyclobutadiene and butadiene ($N = 4$) gives

$$E_{\text{ring}} = 4\alpha + 4\beta, \qquad E_{\text{chain}} = 4\alpha + 4\cdot47\beta$$

and shows that the *chain* is favoured by $0\cdot47\beta$. The cyclobutadiene ring is said to be *anti-aromatic*, and this anti-aromaticity is reflected in its extreme instability.

It is a simple matter to extend the comparison to other values of N using the results in §8.4. The conclusion (which may also be established analytically from equations (8.19) and (8.22)) is that if $N = 4n$ (n an integer) the chain will be more stable than the ring, i.e. the ring will be *anti-aromatic*, while if $N = 4n + 2$ the ring will be more stable than the chain, and hence *aromatic*. Dewar has pointed out that the same qualitative conclusion is reached (though the numerical values may be in error by 50 per cent or more) by the very simple argument used in establishing (10.17) which gives the π interaction energy

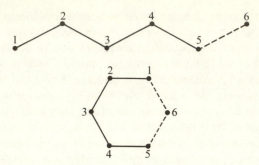

FIG. 10.18. Estimation of π electron energy, using the NBMO. Addition of a sixth conjugated carbon to the pentadienyl radical yields either hexatriene (one interaction) or benzene (two interactions).

when a single carbon atom is added to an odd chain. Hexatriene, for instance, may be formed by addition of a sixth carbon (methyl) to an open chain with $N = 5$ (pentadienyl radical) as in Fig. 10.18 (*a*). The coefficients in the NBMO are $c_1 = c_5 = c$, $c_3 = -c$, with $c = 1/\sqrt{3}$, and the interaction energy formula (10.17) gives, denoting the energy of the radical by E_0,

$$E_{chain} = E_0 + 2\beta c_1 = E_0 + 2\beta c.$$

However, if the methyl is joined to *both* terminal carbons to give benzene, we obtain

$$E_{ring} = E_0 + 2\beta[c_1 + c_5] = E_0 + 4\beta c.$$

Since $c = 1/\sqrt{3}$, the π energy of the ring should be *lower* than that of the chain, the difference being $2\beta/\sqrt{3}$. This value (1.15β) is about 15 per cent larger than the value obtained by properly solving the secular equations, but at least it has the correct sign! The same method suggests that for $N = 4$ the open chain has a lower π-electron energy than the ring, the difference being $2\beta/\sqrt{2}$ or 1.41β (cf. 0.47β). Although the numerical value is grossly in error, the sign is again correct and the butadiene ring is predicted to be anti-aromatic.

The '$4n+2$ rule' in this form, which is substantially equivalent to Hückel's rule (p. 243), follows at once from the NBMO approach because in an odd chain the AO coefficients in the NBMO alternate in sign. The end coefficients, such as c_1 and c_5 in the energy formula above, therefore exactly cancel when there are $3, 7, 11, \ldots$ or in general $4n - 1$ atoms in the radical, or $4n$ atoms in the ring, which is accordingly anti-aromatic. However, the coefficients reinforce to give greater binding energy and an aromatic ring when the radical contains $4n + 1$ and the ring $4n + 2$. It is remarkable that such a primitive method can give a qualitatively correct result of such generality and importance.

Let us now go back to Examples 1 and 2 of the last section and consider the activation energy in terms of the stability, and hence aromaticity, of the *transition complex.* In Example 1 it is clear that, for a distance of approach corresponding to the crossing point in Fig. 10.11, the distribution of energy levels resembles that in cyclobutadiene; there is a bonding MO (ψ_1), a degenerate pair (ψ_2, ψ_3 roughly non-bonding), and an antibonding MO (ψ_4). Figure 10.10 also shows that this is not merely a coincidence; we have a ring of four AO's and, at the appropriate distance, the bond integrals β_π and β_σ (which have the same sign since they each involve the overlap of lobes with the same sign) may take the same numerical value. If we only look at the secular equations we shall be unable to tell whether we are dealing with a σ–π transition complex or the π electrons of cyclobutadiene! Such systems are *isoconjugate* and necessarily yield the same pattern of energy levels and MO's with the same sets of AO coefficients; the only difference is in the *basis functions* $\phi_1, \phi_2, \ldots$ which in the present example lie *in the plane of the ring* and not normal to it as in cyclobutadiene. It is therefore clear that in the face-to-face approach of two ethylene molecules the system must pass through an *anti-aromatic transition state*. This observation was first made by Evans, who pointed out that dimerization was unlikely to take place by this mechanism owing to the relatively high energy of the transition state.

In Example 2, however, the distribution of orbital energies at the crossing point in Fig. 10.15 is exactly like that of the π-type MO's in benzene, a low-energy bonding MO, followed by a degenerate pair, and then their antibonding partners. Again the result is not coincidental. The corresponding transition complex is isoconjugate with the benzene π system, although two links in the ring are σ bonds instead of π, and consequently the transition state is *aromatic*, the energy *falling* as the conjugated ring forms, and electron delocalization occurs. As Evans pointed out, this Diels–Alder reaction should proceed easily by the postulated one-step process through the aromatic transition state. The generalization is obvious: cycloadditions of this type are thermally allowed if they proceed through an aromatic transition state, and are thermally forbidden if they proceed through an anti-aromatic transition state. From the $4n+2$ rule for aromaticity, this principle then leads again to the Woodward–Hoffmann rules for cycloaddition of any two polyenes.

Example 3 can be discussed in a similar way, but one new and interesting feature arises. In the disrotatory process (Fig. 10.16 (*a*)) the σ–π transition complex is roughly isoconjugate with the π-electron system of cyclobutadiene; β_{12} and β_{34} will be somewhat smaller than the normal π-bond integral (β_{23}) owing to the reduced overlap due to twisting, while β_{14} will become comparable with the others as twisting proceeds. If we gave all β's the same average value we should have secular equations exactly like those of cyclobutadiene and should conclude, rightly, that the reaction would not take place in disrotatory fashion because it would have to pass through an anti-aromatic transition state.

A novel situation arises, however, when we turn to the conrotatory process

(Fig. 10.16 (b)), for now, keeping the signs on the lobes of the 2p AO's as they are in butadiene, β_{12} and β_{34} will again be reduced in magnitude by twisting, but β_{14}, instead of growing to a comparable value, will *inescapably take the opposite sign*. Now the signs which we have attached to the lobes of the 2p AO's are clearly arbitrary; if we have an MO $\psi = c_1\phi_1 + c_2\phi_2 + \dots$ we can always rewrite it as $\psi = c_1\phi_1 + (-c_2)(-\phi_2) + \dots$, in which $-\phi_2$ has the reverse choice of signs on its lobes, merely by reversing the sign of the corresponding AO coefficient. In that sense we can reverse the sign of a particular β at will, and it is merely for convenience that in a conjugated system we line up the $2p_\pi$ AO's with their positive lobes on one side of the molecular plane and their negative lobes on the other. We may say that it is 'natural' to give them all the 'same phase' and then all β's for π bonds will have a common (negative) value corresponding to a positive overlap integral. We cannot, however, change one β value *in isolation* unless it connects a terminal atom to the rest of a chain; in the middle of a chain each atom has two neighbours and inverting one 2p AO will therefore change the signs of *two β's*. Our dilemma at the moment is that, by twisting a chain with all AO's chosen according to the 'in-phase' convention (giving β's the same sign) and bringing the ends together, we obtain *one β* which differs in sign from the rest. No matter how we choose the phases of the 2p AO's we cannot get rid of the one β of opposite sign to the rest, where a normal 2p AO meets an upside-down 2p AO; we say there is an essential 'phase dislocation'. As a result of this 'topological' difference, the secular equations for the conrotatory transition complex must differ in an essential way from those of the cyclobutadiene π system with the standard Hückel approximations (p. 238); to emphasize the difference, a system with an essential phase dislocation is said to be of 'anti-Hückel' type.

We may obtain the energy levels and MO's for an anti-Hückel system by algebraic methods, expanding the secular determinant along the row and down the column containing the 'odd' $-\beta$. However, as Dewar has noted, the rules for aromaticity or anti-aromaticity are suggested at once by the NBMO method. For a chain with an odd number of centres there is still a non-bonding MO, but if one of the β's connecting the ϕ_r to its neighbours ϕ_s and ϕ_t is negative the condition $(\alpha - \varepsilon)c_r + \beta_{rs}c_s + \beta_{rt}c_t = 0$, with $\varepsilon = \alpha$, namely

$$\beta_{rs}c_s + \beta_{rt}c_t = 0,$$

becomes $\beta c_s - \beta c_t = 0$; the *difference*, instead of the sum, of the coefficients on either side of the starred atom to which the phase dislocation is attached must now vanish. Whereas a chain containing $4n - 1$ atoms has an NBMO whose end orbitals appear with coefficients c and $-c$, the insertion of a dislocation thus results in a coefficient c at both ends. On attaching a methyl to one of the ends, or to both, to form a chain or a ring of $4n$ atoms, the argument used before (p. 320) shows that the $4n$-atom ring is *aromatic*. A similar argument shows that a $(4n + 2)$-atom ring is *anti-aromatic*.

We can now go back to the conrotatory mechanism in the electrocyclic cyclobutene $\leftrightarrow$ butadiene reaction. The transition state introduces an essential phase dislocation into a four-orbital ring, but the *anti*-Hückel ring with $4n$ centres is *aromatic* and the energy of the conrotatory transition state is therefore lowered by electron delocalization. Again the predictions of the Woodward–Hoffmann approach are confirmed.

The examples considered in this and the preceding section are typical of a vast number of reactions discussed in the recent literature. The fact that the two approaches lead to similar predictions is reassuring; the fact that they are usually in agreement with experiment is astonishing. Of the two formulations, that due to Evans and Dewar has the beauty of extreme simplicity and generality; it is topological rather then geometric and is therefore independent of any restrictive assumptions concerning the symmetry of the transition state—all that matters is the general nature of the orbital overlaps. The reactions we have considered are all of the type described by Woodward and Hoffmann as *pericyclic*; they involve a cyclic permutation of bonds around a ring (e.g. two π bonds breaking and two σ bonds forming in the dimerization of two ethylenes). The most economical statement of the results of this section is then

$$
\boxed{\text{Thermal pericyclic reactions take place preferentially via aromatic transition states}} \quad (10.43)
$$

which Dewar has called 'Evans's principle'.

Problems

10.1. In the polyene chain with $N = 6$ (hexatriene) the π bond orders are $p_{12} = 0.871$, $p_{23} = 0.483$, $p_{34} = 0.785$. Use the approximate formula (10.8) to predict which carbon atoms have the highest self-polarizability values. Obtain also the value of the free valence at each centre.

Which positions do you expect (on the basis of isolated-molecule theory) will be most prone to attack by (i) anions, (ii) cations, and (iii) neutral radicals?

10.2. Calculate the localization energies for positions 1, 2, and 3 in hexatriene (Problem 10.1). Do the results support or contradict the conclusions from isolated-molecule theory? (Hint: use the results from Problems 8.6 and 8.11 to get the energies of the fragments.)

10.3. Consider Problem 10.2 again and estimate the localization energy for position 1 (i) from formula (10.16), and (ii) from Dewar's formula (10.18). Compare the results with that obtained in Problem 10.2. Can you deal with the other positions in the same way?

10.4. Substitution reactions usually proceed more readily at the 1-position in naphthalene than at the 2-position. Try to account for this fact theoretically, using all the methods you can.

10.5. Which reaction do you expect will proceed more readily: 1-nitration of naphthalene or 4-nitration of quinoline? (Hints: obtain the NBMO for the residual molecule with the 1-position missing (Fig. 10.6). Decide how many π electrons will be localized. Fing the resultant π charges. How do the energies of the molecule and residual molecule, and hence the activation energies, depend on the presence of an oxygen at position 4 (which becomes 1, by convention, in quinoline)? Use equation (10.4) with $\delta\alpha$ referring to oxygen.)

10.6. Use Dewar's method (p. 320) to show that (i) a ring of 10 conjugated carbon atoms should be aromatic, and (ii) that whenever $N = 4n + 2$ (n being an integer) the N-atom ring should be aromatic. Make some numerical comparisons with the results from Problem 8.11.

10.7. Show that the cyclopentadienyl anion should be aromatic, the cation antiaromatic; but that the reverse is true for tropylium ions. (Hint: use equations (8.19) and (8.12) for chains and rings, filling the MO's appropriately. Why can you not use Dewar's method?)

10.8. Show that in a given alternant hydrocarbon a hetero atom such as nitrogen should accelerate nucleophilic substitution, and retard electrophilic substitution, most strongly at positions of opposite type (starred or unstarred) to its own.

10.9. Establish the $4n + 2$ rule of Problem 10.6 from the more accurate equations (8.19) and (8.22) instead of using Dewar's approximation. (Hint: evaluate the sums which gave the total energy, for chains and rings with appropriately occupied MO's, as in Problem 8.11 but for a general value of N. (Write $\cos k\theta = \mathrm{Re}\,.\,e^{ik\theta}$ and sum the geometric series.))

10.10. Use the model indicated in Fig. 10.10 (a), for cyclo-addition of two ethylene molecules, as the basis for a quantitative calculation of the correlation diagram shown schematically in Fig. 10.11. (Hints: assume $\beta_\sigma = k\beta_\pi$, with k increasing from 0 to, say, 1·2 as the ring is formed. Neglect overlap integrals and calculate the energies of the orbitals defined in (10.22), using $\varepsilon_1 = H_{11}$ etc. You may neglect changes of hybridization.)

10.11. Attempt a quantitative discussion of the conrotatory and disrotatory ring closing processes indicated in Fig. 10.16, to obtain a correlation diagram similar to that sketched in Fig. 10.17. (Hints: ignore changes in hybridization and neglect overlap integrals. Take β_{23} as a normal π-bond integral (β) and relate $\beta_{12}(=\beta_{34})$ and β_{14} to angle of twist (θ), using the vector property of p orbitals and assuming $\beta_\sigma = 1\cdot2\beta_\pi$. Use symmetry (cf. Problem 8.9) to reduce the secular equations and obtain orbital energies for several values of θ. For simplicity you might assume the carbon atoms form a square.)

10.12. Reconsider the conrotatory process in Problem 10.11, from the Evans–Dewar standpoint. Show that with the even simpler approximations indicated in the text (p. 320) similar conclusions follow. (Hint: you may either use symmetry or you may solve the secular equations by expanding the determinant. In each case take one bond integral as $-\beta$ and the others as β. Compare the levels with those which follow on replacing the $-\beta$ by 0 and β in turn.)

11

The solid state[†]

11.1. The four main types of solid

Our previous discussions have been concerned almost exclusively with molecules. Yet we may rightly expect that some at least of the principles applicable to separate molecules in the gas phase will also apply to solids. Before we attempt to make such application, however, we must distinguish four main types of solid. We shall see that in every case the theory of valency so far developed contributes to our understanding. The four types of solid are as follows.

(1) Metallic conductors and alloys.
(2) Molecular crystals (and liquids).
(3) Covalent crystals, with identical or similar atoms.
(4) Ionic crystals.

These four classes are not absolutely clear-cut. Some crystals are held together by hydrogen bonds (Chapter 12) and are intermediate between (2) and (4); similarly, graphite is intermediate between (1) and (3). Some of these will be referred to in our subsequent discussion, but the classification (1)–(4) will serve most conveniently as the basis of our account, and before continuing it will be convenient to give some examples of the four main types.

Metals have two very characteristic properties: (i) they are conductors of electricity, and (ii) they nearly always crystallize with a large co-ordination number. Thus the typical metallic structures are the body-centred cubic (b.c.c.), the face-centred cubic (f.c.c.), and the hexagonal close-packed (h.c.p.) shown in Fig. 11.1. In the first of these each atom is surrounded by 14 near neighbours, eight at the distance of closest approach R and six more at $2R/\sqrt{3}$ $= 1 \cdot 15R$; in the other two structures there are 12 equidistant neighbours.

From the first of the characteristic properties we infer that some electrons, at least, are relatively free to roam over the crystal, and even, under the influence of an applied electric field, to flow in any chosen direction.

From the second property we infer that the bonds cannot be of the familiar localized type discussed in earlier chapters. When there are 12 neighbours it is obvious that the octet rule has broken down completely. There is, in fact, no conceivable way in which an atom could form 12 simultaneous covalent bonds. It is perfectly clear that metals present us with an extreme case of the delocalization of bonds which we described in Chapter 8.

[†] A more detailed and general account is available in, for example, Kittel (1966).

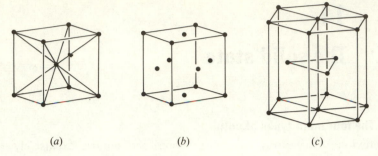

FIG. 11.1. Typical metallic structures: (*a*) body-centred cubic (e.g. lithium); (*b*) face-centred cubic (e.g. copper); (*c*) hexagonal close-packed (e.g. zinc).

In lithium for example (Fig. 11.1 (*a*)) each atom, with only a single 2s valence electron, must at least be bonded to its *eight* nearest neighbours, and the bonding must therefore be much more 'spread out' than in the molecule Li_2 where each atom has only one neighbour. The bond lengths of 0·303 nm in the metal and 0·267 nm in the molecule imply that the metallic 'bonds' are the weaker, but of course there are more of them, with the result that the total binding energy per atom increases from 0·564 eV in the molecule to 1·69 eV in the metal. In summary, then, the valence, or in this case 'conduction', electrons are more tightly bound in the metal than in the molecule, but their bonding effect must be distributed much more widely.

As an example of a molecular crystal we may take iodine. When a non-polar substance such as I_2 is cooled sufficiently, it crystallizes into a solid form. However, X-ray studies show that the unit of which the crystal is built is still the single diatomic molecule. It is true that, as Fig. 11.2 shows, there is a definite scheme in which the units are packed together, and it is equally true that the bond length is increased from 0·265 nm in the gas to 0·270 nm in the solid. However, this distance is still much less than the least distance 0·354 nm between atoms of adjacent molecules. It is quite evident that the forces responsible for the crystal formation are quite different from the valence forces

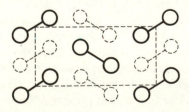

FIG. 11.2. Crystal of iodine, showing the presence of I_2 molecules (molecules shown by broken lines are in a plane beneath that of the others).

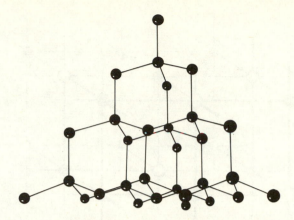

FIG. 11.3. Part of the diamond crystal. Each carbon atom is surrounded tetrahedrally by four others.

within each I_2 unit. They are 'van der Waals' forces and their weakness compared with valence forces is shown by the relatively low temperature at which melting takes place, and by the small energy changes involved in melting and evaporation. In Cl_2, for example, where the same situation obtains, the heat of dissociation of the molecule is 238 kJ mol^{-1} which may be compared with 17–20 kJ mol^{-1} for the heat of sublimation. The forces between the units in a molecular crystal tend to be non-directional, and the molecules pack together as closely as possible. Thus, for example, the rare gases Ne, Ar, Kr, ..., crystallize in the way in which spheres can be packed together most tightly.

In crystals of the third type the forces which operate to maintain the structure are almost identical with those occurring in the 'normal' covalent bond—hence the name 'covalent crystals'. A prime example is diamond (Fig. 11.3) in which the carbon atoms arrange themselves in an infinite tetrahedral pattern. This is not unexpected, knowing that a carbon atom with sp^3 hybridization is capable of forming strong tetrahedrally directed bonds as in methane and the paraffin chains. In covalent crystals the bonds are strongly localized between all pairs of neighbouring atoms, and the crystal is really nothing more than a huge saturated molecule held together by electron-pair bonds. Covalent crystals conform to an '$8-N$ rule', each atom having $8-N$ valence electrons and $8-N$ nearest neighbours, where N is the chemical group number (4 for C, 5 for N, etc.).

The fourth type of crystal occurs only when atoms of very different electronegativity are present. We then obtain ionic crystals, of which the alkali halides (Fig. 11.4) are the most outstanding examples. For sodium and chlorine, for example, using the electronegativities in Table 6.5 we have $x_{Cl} - x_{Na} = 2\cdot23$, and even for a diatomic molecule NaCl the ionic character of the bond would be large, about 53 per cent according to (6.35). In the NaCl *crystal*

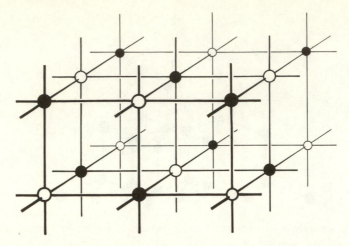

FIG. 11.4. The cubic lattice of NaCl. The open circles represent Na^+ ions, the solid circles Cl^- ions.

(Fig. 11.4) each sodium is under attraction from no less than six neighbouring chlorines, and each chlorine from six sodiums. In such circumstances it is hard to escape the conclusion that the outermost (3s) electron of each sodium will pass over to its neighbouring chlorine atoms to give a lattice consisting essentially of Na^+ and Cl^- ions. In such a lattice there may be almost no covalent character in the bonding; there are separate distinguishable units, but these are essentially *ions* rather than the electrically neutral units of a molecular crystal, and the crystal is held together basically by the electrostatic attractions between the oppositely charged ions.

11.2. Energy band theory. Metals

In §8.4 it was remarked that a long polyene could be regarded as a one-dimensional crystal† and that the MO method employed could be used equally well to study electron delocalization in metals. We noted also that when a large number of atoms were brought together to form a long chain, the energy levels of the resultant MO's formed a 'band', with a well-defined width which depended on the strength of the interaction between neighbours. It is now time to turn to two- and three-dimensional structures and to show how, as more and more atoms are added, MO theory turns into the 'energy band theory' of crystals. Other approaches have indeed been put forward, notably by Pauling (1938; 1960) who attempted a VB description, but it seems fair to say that the difficulties encountered have been found insuperable and that the band theory, the natural extension of MO theory, still provides the most satisfactory general method of dealing with crystals and in particular with metals.

† For an interesting review of one- and two-dimensional crystals see Yoffe (1976).

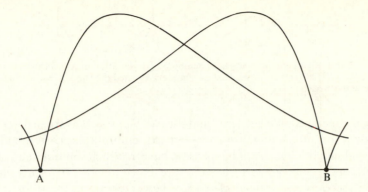

FIG. 11.5. The overlapping of two (Slater) 2s orbitals on adjacent atoms in metallic lithium.

Let us review the approach of §8.4, applying it to a hypothetical 'one-dimensional metal', a chain of N lithium atoms, instead of a polyene. There is a considerable overlap between the 2s valence orbitals of adjacent lithium atoms (Fig. 11.5), and we accordingly look for MO's of the form

$$\psi_\kappa = c_1^\kappa \phi_1 + c_2^\kappa \phi_2 + \ldots + c_N^\kappa \phi_N = \sum_m c_m^\kappa \phi_m \qquad (11.1)$$

which extend over the whole chain. The quantum number k is here replaced by κ, since in solid-state theory k is used for another purpose (§11.3). The equations to determine the AO coefficients are exactly as in §8.4, only the α's and β's being changed by the change from carbon 2p AO's to lithium 2s AO's, and the solutions are therefore given by (8.18a) and (8.19). In any given MO the coefficients follow a wave-like pattern and, since they determine the weights with which the AO's of Fig. 11.5 are superimposed, a plot of the resultant wavefunction ψ_κ would look something like Fig. 11.6. Wavefunctions of this general form are commonly called 'crystal orbitals'.

The coefficients given by (8.18a) were derived on the assumption that the crystal had ends at atom 1 and atom N. This is not the most useful assumption in solid-state theory because the corresponding 'standing-wave' solutions are not appropriate for describing current flow, with which much of solid-state

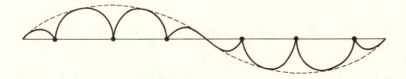

FIG. 11.6. Typical Bloch function for a chain of lithium atoms (for clarity, the spacing has been increased so that the 2s orbitals overlap less strongly).

FIG. 11.7. Illustrating periodic boundary conditions. The variation of AO coefficients (vertical lines) within a five-atom 'microcrystal' (shown bold) is repeated over and over again in adjacent microcrystals (broken lines).

theory is concerned. We need travelling-wave solutions, which can describe an electron whose momentum has a non-zero expectation value corresponding to motion along the chain. As it happens we have met such solutions already, for the ends of the chain are lost when they are joined together to form a ring and we can therefore use instead of (8.18*a*) the expression in (8.21) for the AO coefficients appropriate to the MO's of a cyclic polyene. The resultant MO's can be looked at in another way. The boundary condition we have used for a ring (8.20) merely states that, as we count the atoms along the chain, the wavefunction around the $(N+1)$th atom must be identical with that around the first atom. However, suppose instead of bending our 'metal' into a ring we had extended it indefinitely in both directions by adding further blocks of N atoms as in Fig. 11.7. The same boundary condition would then simply require that what happened in the original block was repeated over and over again in all the other blocks, e.g. that the mth coefficient in MO ψ_κ should be identical with the $(m+N)$th coefficient, the $(m+2N)$th coefficient, etc. This mathematical device for dealing with an 'infinite' crystal, by supposing it to be built up from 'microcrystals' (containing a finite but arbitrarily large number N of atoms) within which the wavefunction is repeated indefinitely, is referred to as 'using periodic boundary conditions'. It enables us to deal with crystals through which electrons may flow without suffering reflections at the boundaries, and use of the complex coefficients is also extremely convenient mathematically. A typical travelling-wave MO is thus, inserting (8.21) in (11.1),

$$\psi_\kappa = C_\kappa \sum_m \exp\left(\frac{2\pi i \kappa m}{N}\right)\phi_m \tag{11.2}$$

while the corresponding orbital energy is

$$\varepsilon_\kappa = \alpha + 2\beta \cos\left(\frac{2\pi\kappa}{N}\right). \tag{11.3}$$

The normalizing factor C_κ is usually chosen to normalize ψ_κ within the 'periodic volume' of N atoms and, neglecting overlap, would thus be given by $C_\kappa = 1/\sqrt{N}$. The real and imaginary parts of ψ_κ given by (11.2) are simply cosine and sine functions (corresponding to the alternative solutions used in (8.23)), and therefore both have the sinusoidal form shown in Fig. 11.6, one being merely shifted in phase relative to the other. The energy levels fall into a band of width 4β exactly as described in §8.4. Crystal orbitals built up in

LCAO approximation according to (11.2) are often called 'Bloch orbitals' after Bloch (1928) who first introduced them in solid-state theory. This work in fact antedated MO theory, and our approach therefore reverses the historical sequence. An important property of these orbitals is that they occur in degenerate pairs corresponding to $\kappa = 0, \pm 1, \pm 2, \ldots$, equal and opposite values of the quantum number describing the electron travelling through the crystal with the same energy but in opposite directions.

Similar considerations apply to other orbitals of the free atom: a lithium $2p_x$ AO, for example, will overlap with the equivalent $2p_x$ AO's on its neighbours and it will be possible to set up Bloch orbitals of the standard form (11.2), ϕ_m being a $2p_x$ AO on atom m instead of the 2s AO. With each AO energy we may therefore associate a Bloch-orbital energy *band*. If the various AO energies (e.g. 2s and 2p of lithium) are well separated, or if the distance between adjacent atoms is so great that there is no appreciable overlapping between neighbouring AO's (giving small β), the separate bands of energies will remain narrow and distinct. But if the original AO energies are close together, or if neighbouring atoms are sufficiently near to each other, then the bands will broaden and merge into one another. In rough terms this means that the MO's must not be regarded as composed solely of one, or of the other, type of AO, but that both types have to be included. More correctly, when band interaction is allowed, levels are 'repelled' from the overlap region (cf. the avoided crossing in Fig. 3.13) and a new energy gap will appear. If the starting AO's are degenerate, as with p or d orbitals, the resulting band may split into

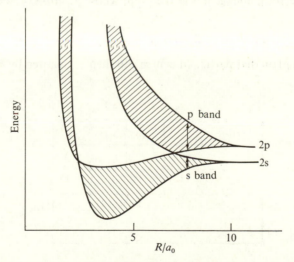

FIG. 11.8. Energy bands in metallic lithium, showing Bloch-orbital energies against nearest-neighbour distance (R). Where the two bands overlap, the Bloch orbitals contain both s and p character.

several parts (depending on the crystal symmetry) but the total number of states in all these bands will be the equivalent of three per atom for p-bands, and five per atom for d-bands.

We may therefore describe the energy levels for a metal in the following terms. When the atoms are at infinite separation, the energies are merely the energies of separate AO's. But as the lattice constant diminishes, interaction occurs, and each separate level splits into a band, whose width increases steadily. Ultimately the bands overlap, as is shown in Fig. 11.8 for the case of metallic lithium.

The electronic structure of the whole metal is determined by essentially the same principles as we used for atoms and molecules; using the *aufbau* approach electrons are added two at a time, with opposite spins, to the available Bloch orbitals in ascending energy order. The resultant wavefunction describes the electronic ground state of the crystal, from which the energy may be computed, in principle, exactly as for a molecule.

Now that we have a general understanding of the origin of the energy bands, and of the procedure to be used in obtaining the electronic structure of the crystal, we should verify that the mathematical analysis used for our one-dimensional model works equally well for real crystals in two or three dimensions. All the main features of energy band theory can in fact be illustrated using a *two*-dimensional crystal and it is therefore sufficient to confine attention to this case. Let us suppose the atoms are arranged on a regular lattice as in Fig. 11.9. The position of each atom is then specified by two integers (n_1, n_2) giving the numbers of steps, $\mathbf{a}_1$ and $\mathbf{a}_2$ respectively, required to reach it; in vector language it is at the point whose position vector, relative to the origin, is

$$\mathbf{R}_n = n_1 \mathbf{a}_1 + n_2 \mathbf{a}_2. \tag{11.4}$$

A given AO (ϕ) on that particular centre will then be denoted by $\phi_{n_1 n_2}$ and the

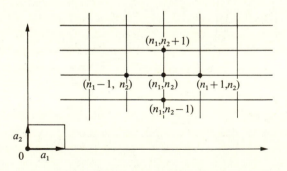

FIG. 11.9. Part of two-dimensional lattice, showing numbering of lattice points. Typical point (n_1, n_2) has four nearest neighbours, numbered as shown.

Bloch orbital constructed from the corresponding AO's on all centres will be of the form

$$\psi = \sum_{n_1 n_2} c_{n_1 n_2} \phi_{n_1 n_2}. \tag{11.5}$$

It is not difficult to show that, with the approximations of §8.4, the secular equations which connect $c_{n_1 n_2}$ with the coefficients of the AO's on its nearest neighbours are satisfied by taking $c_{n_1 n_2} = e^{in_1 \theta_1} \times e^{in_2 \theta_2}$ and determining θ_1 and θ_2 by the condition that the same values will recur on going N steps in *either* of the directions $\mathbf{a}_1$ or $\mathbf{a}_2$ (i.e. increasing either n_1 or n_2 by N). The resultant AO coefficients are then

$$c_{n_1 n_2}^{\kappa_1 \kappa_2} = \exp\{2\pi i(n_1 \kappa_1 + n_2 \kappa_2)\} \tag{11.6}$$

while the corresponding MO energy is

$$\varepsilon_{\kappa_1 \kappa_2} = \alpha + 2\beta_1 \cos(2\pi\kappa_1/N) + 2\beta_2 \cos(2\pi\kappa_2/N). \tag{11.7}$$

The κ_1 and κ_2 are again integer quantum numbers and the N^2 Bloch orbitals, constructed from the N^2 AO's of the periodic volume, arise from $\kappa_1, \kappa_2 = 0, \pm 1, \pm 2, \dots$. Since the cosines have extreme values ± 1 the orbital energies fall into a band of width $4\beta_1 + 4\beta_2$. Repetitions of the solution occur, as in the one-dimensional case, when κ_1 or κ_2 is changed by an integral multiple of N, and any convenient set of N^2 distinct pairs of values may therefore be used.

Exactly similar considerations apply to three-dimensional crystals, except that the Bloch orbitals and their energies depend on *three* quantum numbers $(\kappa_1, \kappa_2, \kappa_3)$. Summarizing, each AO of the atom (or atoms) in the unit cell of the crystal gives rise to its set of Bloch orbitals whose energies form a band with a width determined by the overlap of neighbouring AO's. As the AO overlap increases, the energy bands themselves may begin to overlap (i.e. Bloch orbitals arising from different AO's may correspond to the same energy); the bands may then merge, as in the case of metallic lithium (Fig. 11.8). This simply means that the appropriate orbitals in the crystal may not always be well represented as pure 2s-type, say, or pure p-type Bloch orbitals, but may correspond more closely to *mixtures* of Bloch orbitals formed from a variety of AO's. The determinations of such mixtures, for each choice of the quantum numbers $\kappa_1, \kappa_2, \kappa_3$, requires the solution of a secular problem whose dimensions depend on the number of Bloch orbitals to be mixed; this is one of the major technical problems of energy band theory.

The idea of energy bands separated by forbidden zones in which no energy levels fall permits a qualitative discussion of many properties of solids even without further detailed elaboration. For this purpose we need only one more concept, that of the 'density of states' which tells us how many levels per unit range of energy there are at a given point in a band. More precisely, we need the density-of-states function $N(\varepsilon)$ such that $N(\varepsilon)\,d\varepsilon$ is the number of energy

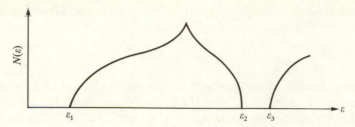

FIG. 11.10. Density of states for a typical discrete band.

levels in the range $(\varepsilon, \varepsilon + \mathrm{d}\varepsilon)$. A typical density-of-states curve is shown in Fig. 11.10, which corresponds to two energy bands with a gap (forbidden zone) between ε_2 and ε_3; in this example the states are densely packed near the middle of the band which extends from ε_1 (bottom) to ε_2 (top), but less so towards the top and bottom.

An immediate application of energy band theory is to the characterization of conductors and insulators. It is obvious from the way it was developed that our description of electron levels in terms of allowed and forbidden bands applies to all periodic systems, metallic or otherwise. Whether it is convenient to use it for non-metals, as well as metals, depends upon the phenomenon being discussed. There is no doubt, however, that in the matter of deciding whether a given solid is a conductor or an insulator, the concept of bands is extremely valuable. Let us consider an idealized substance in which the bands are all distinct and well separated (Fig. 11.11). As usual, the available electrons are supposed to occupy the allowed levels, working upwards from the bottom.

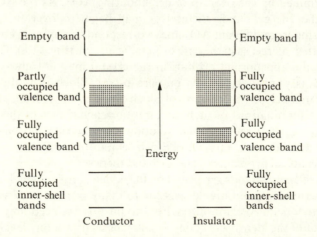

FIG. 11.11. Conductors and insulators. The conductor has a partly filled band; the insulator has only filled bands separated by gaps.

The shaded levels are those which are full: the unshaded ones are empty. In case (a) the top band is only partly filled. In case (b) it is completely full. Now as we have noted each energy level is at least doubly degenerate, the two states corresponding to the electron travelling in opposite directions through the crystal. Suppose that we apply an external electric field by joining the two ends of the metal to the poles of a battery. The immediate effect will be to try to make more electrons flow in the one direction than in the other. This may be achieved in case (a) by giving the electrons flowing in the direction of the field more energy and thereby raising them into some of the previously empty levels of the band. As a result a current flows, and the substance is a metallic conductor. Indeed the process would continue, with an indefinitely great current developing, were it not for collisions of the electrons with the positive nuclei, and this picture provides the basis for a theory of conductivity. In case (b), on the other hand, no such current can flow. For since the band is completely full, with equal numbers of electrons moving in all directions, we can never get any net flow in the direction of the field. The substance is therefore an insulator. Indeed the only way to get a current to flow is by applying so intense an electric field that enough energy can be given to a few of the electrons to raise them to the next band, previously quite empty, which then becomes a conduction band. This is the explanation of dielectric breakdown.

The argument above is easily generalized. If there are partly filled bands of electrons, the substance is a conductor: if there are only completely filled and completely empty bands, it is an insulator.

We may illustrate this in terms of lithium and diamond, of which the first is a metal and the second an insulator. The band structure of lithium has already been given in Fig. 11.8. The corresponding structure for diamond is reproduced in Fig. 11.12. In the case of lithium there is only one valence electron per atom, so that the lowest band is only half-full: the crystal should therefore be a conductor. In the case of diamond there are four valence electrons per atom. The shaded bands shown in the figure each contain N levels, assuming there are N atoms in the crystal, and there are two more very narrow bands which follow curves (a) and (b) and provide a further $2N$ levels. Thus at the equilibrium distance, the lower band (between a and c) is completely filled, and so is (a). This accounts for all $4N$ electrons and the higher bands are thus completely empty. There is a large energy gap between the filled and empty bands and this provides us with an immediate interpretation of the insulating property of diamond. Although this description of diamond is perfectly proper and correct, we shall presently find that for most purposes a description in terms of localized σ-bonds is simpler and equally valid.

Several other points are related to this distinction between metals and insulators. We must be content to list some of them and comment briefly upon them.

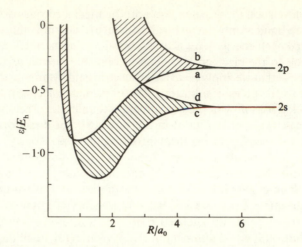

FIG. 11.12. Energy bands in diamond. The vertical line shows the equilibrium spacing.

(i) Our description of a metal leads us to regard each atom as ionized, and in a 'sea' of electrons. The binding is due to this approximately uniform electron cloud, and for that reason the internal forces in metals are different from those in, for example, polar crystals, where the individual units are more or less distinct positive and negative ions. Further, on account of the 'sea' of electrons, considerable movement of the positive ions is possible without great expenditure of energy. This observation leads to an explanation of the plasticity of metals, and of the effects of work-hardening.

(ii) In order that the 'electron sea' may be formed it is desirable that in the separate atoms there should be one or more easily ionized electrons, for electrons that are too tightly bound to their particular nucleus will not readily be shared with the rest of the crystal. This is one reason why the earlier groups of the periodic table give metallic structures, while later groups adhere to the 8–N rule (p. 327) characteristic of covalent crystals.

(iii) The importance of the electron concentration, i.e. number of valence electrons per atom, can hardly be exaggerated; it is perhaps the dominant factor in determining the properties of metals and alloys as has been stressed by Hume–Rothery and others (e.g. Hume–Rothery 1960). Indeed, the main purpose of adding a second type of atom to form an alloy is to vary the degree of filling of the energy bands by adding or subtracting electrons.

(iv) The band structure explains how an insulator may acquire a photo-conductivity if ultraviolet light shines upon it. All that is required is that the frequency v of the light shall be such that hv is at least equal to the energy gap between the top occupied band and the lowest unoccupied band. Absorption of such light carries electrons into the conduction band, where they may flow under the influence of an electric field.

(v) The important property of semi-conductivity depends jointly on band structure and electron concentration. In an 'intrinsic semi-conductor', there is a very narrow gap between a filled band and an empty band, across which electrons may be excited 'thermally' (i.e. by energy of the order kT); the excited electrons and the 'holes' left behind them act as negative and positive (n- and p-type) charge carriers and yield a temperature-dependent conductivity. By 'doping' such a crystal with electron-rich or electron-deficient impurities the number of n- and p-type carriers may be varied to produce predominantly 'n-type semi-conductors' or 'p-type semi-conductors'.

11.3. Brillouin zones†

In solid state theory it is usual to exhibit the dependence of energy levels on quantum numbers in a pictorial way. For this purpose the 'lattice vectors' (11.4) are introduced explicitly, and the quantum numbers are used to define a so-called 'k vector'. We continue to use the two-dimensional example of §11.2. In this case the **k** vector is defined as

$$\mathbf{k} = k_1(2\pi\mathbf{b}_1) + k_2(2\pi\mathbf{b}_2) \tag{11.8}$$

where $k_1 = \kappa_1/N$, $k_2 = \kappa_2/N$ and the vectors $\mathbf{b}_1$ and $\mathbf{b}_2$ are chosen in such a way that formula (11.6) reduces to

$$c_{n_1 n_2}^{\kappa_1 \kappa_2} = c_{\mathbf{R}_n}^{\mathbf{k}} = \exp(\mathrm{i}\mathbf{k} \cdot \mathbf{R}_n) \tag{11.9}$$

where $\mathbf{k} \cdot \mathbf{R}_n$ is the scalar product‡ of the two vectors **k** and $\mathbf{R}_n$. This is achieved simply by defining the vectors $\mathbf{b}_1$ and $\mathbf{b}_2$ such that $\mathbf{b}_1$ is perpendicular to $\mathbf{a}_2$ ($\mathbf{b}_1 \cdot \mathbf{a}_2 = 0$) and $\mathbf{b}_2$ is perpendicular to $\mathbf{a}_1$ ($\mathbf{b}_2 \cdot \mathbf{a}_1 = 0$), while the lengths of $\mathbf{b}_1$ and $\mathbf{b}_2$ are related to those of $\mathbf{a}_1$ and $\mathbf{a}_2$ so as to ensure that $\mathbf{b}_1 \cdot \mathbf{a}_1 = 1$, $\mathbf{b}_2 \cdot \mathbf{a}_2 = 1$; for when this is done

$$\mathrm{i}\mathbf{k} \cdot \mathbf{R}_n = 2\pi\mathrm{i}(k_1\mathbf{b}_1 + k_2\mathbf{b}_2) \cdot (n_1\mathbf{a}_1 + n_2\mathbf{a}_2) = 2\pi\mathrm{i}(\kappa_1 n_1 + \kappa_2 n_2)$$

and (11.9) then becomes identical with the original form (11.6). The vectors $\mathbf{b}_1$ and $\mathbf{b}_2$ define the 'reciprocal lattice' used by crystallographers, and each Bloch orbital and energy is completely specified by giving a **k** vector of the form (11.8) which has components of length $2\pi k_1$ and $2\pi k_2$ in the 'reciprocal space'; thus (11.5), with the coefficients written in the form (11.9), becomes

$$\psi_{\mathbf{k}} = \sum_n \exp(\mathrm{i}\mathbf{k} \cdot \mathbf{R}_n)\phi_n \tag{11.10}$$

while (11.7) reduces to

$$\varepsilon_{\mathbf{k}} = \alpha + 2\beta_1 \cos(2\pi k_1) + 2\beta_2 \cos(2\pi k_2). \tag{11.11}$$

† This section may be omitted on first reading without loss of continuity. Readers requiring more detail, at a semipopular level, should see Ziman (1963).

‡ That is the product of the lengths of the two vectors times the cosine of the angle between them (= 0 for perpendicular vectors).

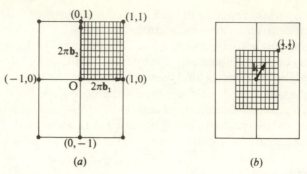

FIG. 11.13. Association between states and vectors in **k** space. (*a*) The vectors $2\pi\mathbf{b}_1$ and $2\pi\mathbf{b}_2$ define a cell in **k** space corresponding to the lattice in Fig. 11.9; every point defined by the cross hatching gives a pair of (k_1, k_2) values for a state. (*b*) Exactly the same states are defined by (k_1, k_2) values in the central zone which is symmetrical around the origin O.

Note that n in (11.10) runs over all lattice points, being short for the *pair* of integers (n_1, n_2). The part of '**k** space' corresponding to k_1 and k_2 values in the range 0 to 1 (i.e. κ_1 and κ_2 each running from 0 to N) for the lattice shown in Fig. 11.9 is indicated in Fig. 11.13 (*a*). The individual points shown in the cross-hatched area correspond to κ_1 and κ_2 increasing in unit steps, and to each point corresponds a state, i.e. a Bloch orbital. Since, as we have noted, points whose κ values differ by N (i.e. whose **k** vector components differ by unity) correspond to identically the same Bloch orbital, it is not necessary to represent the states by **k** vectors falling in the cross-hatched area in Fig. 11.13 (*a*); it is in fact more usual to use a zone which is symmetrical around the origin in **k** space, the central zone in Fig. 11.13 (*b*), which clearly contains exactly the same number of points, corresponding to all the distinct solutions. This 'Brillouin zone', which exhibits more clearly the implications of crystal symmetry, is bounded by the perpendicular bisectors of the lines joining the origin to the nearest neighbours with integer components, namely (Fig. 11.13 (*a*)) $(k_1, k_2) = (0, 1), (1, 0), (0, -1),$ $(-1, 0)$. For the two-dimensional crystal, with N^2 atoms in the periodic volume, there are just N^2 **k** vectors pointing from the origin to the lattice points within the Brillouin zone, and each point is thus associated with a permitted state of an electron moving through the lattice.

To show how the energy of a state depends on quantum numbers, i.e. on its **k** vector, the points may be labelled by corresponding ε values, and then points with the same energy may be connected to give a map showing *energy contours*. A typical contour map is shown in Fig. 11.14. Such maps have many applications, which we shall not take up in detail. We note only the following.

(i) When the available crystal orbitals are filled in ascending energy order, the highest occupied level defines the so-called 'Fermi energy' ε_F. There is in fact a complete contour of this energy, passing through all the

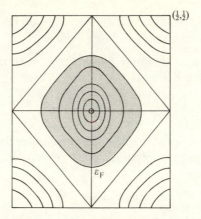

$(\frac{1}{2},\frac{1}{2})$

ε_F

FIG. 11.14. Typical energy contours within a Brillouin zone. Each contour connects points in **k** space corresponding to states of the same energy. If the states are filled (shaded area) up to the contour labelled ε_F, this defines the 'Fermi surface'.

points whose **k** vectors correspond to the same energy; this contour therefore marks the boundary of the occupied zone in **k** space. For a three-dimensional crystal the contour becomes a *surface*—the Fermi surface.

(ii) If the Fermi energy coincides with a discontinuity, so that the next contour lies a finite amount above ε_F and there are no states in between, we have a gap immediately above the filled band and consequently an insulator; if there are some regions, at least, in which no discontinuity occurs we shall have metallic properties. Clearly the nature of the energy contours within the Brillouin zone gives more detailed insight into the origin of the gaps between the energy bands.

(iii) By counting the points in **k** space corresponding to states lying between adjacent contours of energy ε and $\varepsilon + d\varepsilon$ we can directly determine the density of states $N(\varepsilon)$, which is needed in the detailed discussion of many properties.

(iv) By further analysis of the crystal orbitals it can be shown that the **k** vector is related formally to the momentum vector of the electron travelling through the lattice, but the electron does not behave as a *free* electron owing to its interaction with the periodic potential field produced by the nuclei. As an example, let us take a square lattice, with $\beta_1 = \beta_2 = \beta$ in (11.11) and $\mathbf{a}_1, \mathbf{a}_2$ perpendicular. Near the centre of the Brillouin zone, where the components of the **k** vector are small, we may then expand the cosines in (11.11) to obtain (taking $\beta_1 = \beta_2 = \beta$)

$$\varepsilon_{\mathbf{k}} = \alpha + 2\beta\{1 - \tfrac{1}{2}(2\pi k_1)^2 + \ldots\} + 2\beta\{1 - \tfrac{1}{2}(2\pi k_2)^2 + \ldots\}$$
$$= \text{constant} - 4\pi^2\beta k^2$$

where $k^2 = k_1^2 + k_2^2$. For a *free* electron, with momentum $\mathbf{p} = \hbar\mathbf{k}$, the corresponding energy is

$$\varepsilon_{\mathbf{k}} = \frac{\hbar^2}{2m}k^2 .$$

It follows that, for states near the zone centre the electron energy levels are rather like those of a free electron except that they correspond to an 'effective mass'

$$m_{\text{eff}} = -\hbar^2/8\pi^2\beta \tag{11.12}$$

and are shifted by a constant term which is unimportant. If β is small, as it would be for inner-shell electrons of the constituent atoms, mobility would thus be low, but for valence electrons with large negative values of β the effective mass would become very small. Similar considerations may be extended to the behaviour of electrons in states elsewhere in the zone, e.g. at points on the Fermi surface and at zone corners.

The study of Brillouin zones clearly provides the key to a detailed theory of the electronic structure and properties of crystals and, in particular, of metals, where the need to use delocalized travelling-wave solutions of the Schrödinger equation is essential in order to obtain an understanding of electrical conduction. We now turn to other types of solid, where the electronic structure may be described rather more simply in terms of localized orbitals.

11.4. Molecular crystals

In molecular crystals, as we have characterized them in §11.1, the individual units of the structure are molecules, not atoms, and the cohesive forces which bind them together are relatively weak and non-directional. The arrangement and orientation of the molecules is then determined largely by packing considerations and is dependent on their size and shape; the fact that each molecule behaves rather like a solid object means that as soon as their separate electron distributions begin to overlap appreciably, or interpenetrate, very strong repulsive forces are set up; the molecules are then 'in contact' and no further penetration occurs, the long-range cohesive forces being balanced by the short-range repulsions. Both types of force are correctly predicted by quantum mechanics (though their accurate calculation is extremely difficult), but both are weak compared with the valence forces which bind the atoms together within each molecule, and this may be regarded as the characteristic feature of molecular crystals.

The most efficient form of packing depends mainly on the shape of the

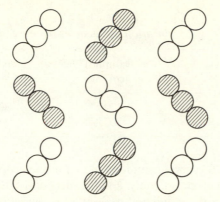

FIG. 11.15. Structure of crystalline benzene. Each circle represents a CH group, the plane of the hexagon being perpendicular to the paper. Molecules shown shaded are displaced above and and below the plane through one-half the unit cell height.

molecule. Molecules which are roughly spherical, such as HCl,† HBr, H_2S, and CH_4, adopt the close-packed cubic structure, in which each molecule has 12 nearest neighbours as in the rare-gas crystals. At temperatures high enough to produce thermal motion, but some way below the melting point, the individual molecules in such crystals actually rotate, as if each were in a cavity provided by the surrounding molecules. If, however, the shape of the molecule is very far from being spherical, as in benzene, the packing will still be such as to allow as many near neighbours as possible to each separate molecule. Fig. 11.15 shows what actually happens in benzene. The planes of the molecules are perpendicular to the plane of the paper, and any one molecule (say the centre one of the figure) is surrounded by 12 neighbours, four unshaded ones at the same horizontal level and four shaded ones above and below. By virtue of the two directions in which the planes of the molecules lie, a closer packing is possible than if all were parallel to the same plane. It is not surprising that molecules of this kind show almost the same internal vibrations in the solid, liquid, and gas phases, for the individuality of the molecules is always preserved and the internal binding is only very slightly weakened by the regular association in the solid or the irregular association in the liquid. Table 11.1 shows how little change there is in two of the benzene frequencies. A similar situation holds for all crystals of this class.

It is obvious that the intermolecular forces only exert a minor influence on the structure of such crystals; the important factor is essentially geometrical. Further, the chief distinction between the liquid and the solid is the regular and permanent association in the solid, to be contrasted with the largely random and constantly changing association in the liquid.

† Here, for example, the hydrogen atom produces only a small bump on the electron density contours of the halogen.

TABLE 11.1
Typical vibration frequencies in benzene

ν (gas) (cm^{-1})	ν (liquid) (cm^{-1})	ν (solid) (cm^{-1})
3099	3090	3089
3045	3035	3034

During the last 20 years our understanding of the properties of molecular crystals has expanded considerably, and the effect of the intermolecular forces on the properties of the individual molecules is now quite well understood. This is not the place for a detailed review but two aspects of the theory must be mentioned. The interested reader may refer to Craig and Walmsley (1968).

First we consider just two molecules, A and B, of the crystal and recall the expression already used (p. 194) for the energy of a system made up of weakly interacting parts. This will be a good first approximation so long as the wavefunctions Ψ_A and Ψ_B of the separate systems do not overlap appreciably; the ground-state energy in this approximation is, adding the energy of repulsion between the nuclei of A and those of B,

$$E_0 = E_A + E_B + J_{AB} + E_{AB}^{\text{nuc}} \tag{11.13}$$

where E_A represents the energy of the electrons of A in the field of *all* the nuclei (of both A and B), and similarly for E_B, while (cf. (7.19) et seq.) J_{AB} is the repulsion energy of two charge clouds, of density $-eP_A$ and $-eP_B$:

$$J_{AB} = \frac{e^2}{\kappa_0} \int \frac{P_A(1)P_B(2)}{r_{12}} \, d\tau_1 \, d\tau_2 \tag{11.14}$$

Now E_A consists of two parts: the energy E_A^0 (kinetic plus potential) of the electrons of molecule A *by itself*, together with their energy (V_{AB}) in the field of the nuclei of B. The same is true for E_B and thus

$$E_A = E_A^0 + V_{AB} \qquad E_B = E_B^0 + V_{BA} \tag{11.15}$$

Consequently (11.13) may be rewritten as

$$E_0 = E_A^0 + E_B^0 + (V_{AB} + V_{BA} + J_{AB} + E_{AB}^{\text{nuc}}) \tag{11.16}$$

The meaning of the terms in parentheses is indicated schematically in Fig. 11.16; altogether they yield the classical electrostatic interaction between the two molecules, each regarded as a set of nuclei surrounded by its charge cloud, of density $-eP_A$ or $-eP_B$. The interaction between the molecules, in this approximation, may thus be written

$$E_0 = E_A^0 + E_B^0 + E_{\text{elec}} \tag{11.17}$$

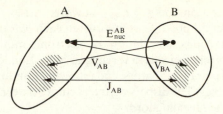

FIG. 11.16. Schematic representation of interaction between two molecules (for simplicity only one nucleus is shown in each molecule). E_{nuc}^{AB} is the repulsion energy between nuclei of A and nuclei of B; J_{AB} is the repulsion energy between charge clouds; V_{AB} is the attraction between the charge cloud of A and the nuclei of B, and *vice versa* for V_{BA}.

where E_{elec} is the classically computed electrostatic interaction. A similar result is valid, no matter how many molecules there may be, and the leading term in the cohesive energy of a molecular crystal is thus E_{elec}, calculated as a sum over all pairs of molecules.

When the interacting units carry a net charge, as in ionic crystals where they are positive and negative ions, the electrostatic interaction may dominate and can account for virtually the whole of the cohesive energy. For molecular crystals, however, especially when the molecules have high symmetry and therefore possess neither net charge nor even a dipole moment, E_{elec} may be much too small to account for the cohesive energy. In such cases it is necessary to go to a higher approximation. We know how to do this, using the variation method; if Φ_{AB} is the wavefunction used so far we may improve it by adding on other terms such as Φ_{A^*B}, Φ_{AB^*}, and $\Phi_{A^*B^*}$ where Φ_{A^*B}, for example, indicates that molecule A is in an excited state. The coefficients in the variation function

$$\Psi = \Phi_{AB} + c_{A^*B}\Phi_{A^*B} + c_{AB^*}\Phi_{AB^*} + c_{A^*B^*}\Phi_{A^*B^*} \tag{11.18}$$

may then be obtained, as usual, by solving the secular equations. Since Φ_{AB} itself would be an accurate wavefunction in the absence of interactions between the molecules, the coefficients are expected to be small quantities; the interaction is a small "perturbation" and its effect is to distort Φ_{AB} slightly, contaminating it with small amounts of the other terms which formally describe local excitations. An approximate solution may be obtained essentially as in §4.7; each term in (11.18) makes an additive contribution to the further lowering of the energy and the final result is to replace (11.17) by

$$E = E_A^0 + E_B^0 + E_{elec} + E_{pol} + E_{disp} \tag{11.19}$$

where the new terms E_{pol} and E_{disp} are called 'polarization' and 'dispersion' energies, respectively. E_{pol} arises from admixture of the single-excitation terms such as Φ_{A^*B} (summed over all possible choices of excitation A*) and is associated with a slight change of the electron density in each molecule in response to the field produced by the surrounding molecules, hence the designation 'polarization energy'. E_{disp}, however, arises from terms such as $\Phi_{A^*B^*}$ (summed over all possible excitations in each *pair* of molecules) and is due to the instantaneous 'correlation' of electronic motions in all pairs of molecules; such correlation was first recognized by London, who considered the interaction of a pair of oscillators, and following London it is still described as a 'dispersion' energy. The dispersion interaction between molecules A and B can be written in the revealing form

$$E_{disp}^{AB} = - \sum_{A^*B^*} \frac{J_{AA^*,BB^*}}{\Delta E(A \rightarrow A^*, B \rightarrow B^*)}$$ (11.20)

where the denominator is the excitation energy required to excite both molecule A and molecule B, the summation is over all possible excitations of each molecule, and

$$J_{AA^*,BB^*} = \frac{e^2}{\kappa_0} \int \frac{P_{AA^*}(1)P_{BB^*}(2)}{r_{12}} d\tau_1 \, d\tau_2.$$ (11.21)

This latter quantity has an electrostatic interpretation exactly similar to (11.14), except that the density functions involved are not ground-state electron densities but rather 'transition densities'. Such transition densities, which connect two different states, are also encountered in spectroscopy; for example, the x component of the electric moment of the density P_{AA^*} determines the probability of the transition $A \rightarrow A^*$ under the influence of radiation polarized in the x direction. It therefore seems likely that molecules which possess strongly allowed transitions (and hence transition densities with strong electric moments) to low-lying excited states (and hence small denominators in (11.20)) will exhibit strong dispersion interactions; a wide variety of organic molecules, particularly those with aromatic character, conform to this expectation. Such systems have been extensively studied theoretically (see, for example, Craig and Walmsley 1968).

Although the field of spectroscopy lies outside the scope of this book, it should be noted that the ground-state theory sketched above can be extended to deal with excited states. The resultant 'exciton' theory of crystal spectra has made great advances during recent years and deserves at least a passing reference.

In a crystal containing a large number of identical molecules it is possible to define single-excitation functions, $\Phi_n(M^*)$ say, in which the nth molecule (at lattice point R_n) is in excited state M^* while all the rest are in the ground state (M). This "locally excited state" cannot represent a true excited state of the crystal; for another equally eligible function, with the same energy, would be $\Phi_m(M^*)$—in which the excitation had 'jumped' to a neighbouring molecule at lattice point R_m. To describe an excited state we should take all such functions, allow them to mix, and solve secular equations to obtain the appropriate mixing coefficients. It will be recognised that this procedure is formally identical with that by which we arrived at the Bloch orbitals; but instead of getting an orbital which describes a delocalized *electron*, moving through the metal, we now get an *exciton* which describes a delocalized *excitation* moving from molecule to molecule . The coefficients with which the locally excited functions must be combined again turn out to be the Bloch factors $e^{i\mathbf{k}\cdot\mathbf{R}_n}$ and the wave vector $\mathbf{k}$ determines the direction in which the excitation travels; and the energy of the molecular excited state M^* is replaced by a *band* of excited states in the crystal. The whole theory runs parallel to the band theory of metals but the band width is determined by an interaction integral of the type (11.21) (for a pair of neighbouring molecules A and B) instead of the β integral for overlapping AO's. For a full account, reference may be made to Craig & Walmsley (1968); see also Davydov (1962) or Knox (1963).

11.5. Covalent crystals

The diamond crystal has been cited (p. 327) as an example of a covalent crystal. The bonds are essentially localized between nearest neighbours and are very nearly identical with similar bonds in a covalently bonded molecule. The difference in electronegativity of the bonded atoms is normally very small in such crystals and the bonds are strong and not significantly polar. Such crystals are typically hard, with a very high melting point and large latent heat. Silicon, germanium, and grey tin are all of the same type as diamond. Carborundum (SiC) also has a tetrahedral arrangement around each atom, and again we can speak of effectively localized bonds, alternate atoms being C and Si.

Tetrahedral structures of this kind are very open, and correspond to a large atomic volume (i.e. volume per atom). This can only be so when directional forces play a considerable part, thus providing further evidence that the bonds have an essentially molecular and valence character. In fact we may infer this type of binding whenever we find tetrahedral angles, even if, as in zinc sulphide ZnS, this may not be obvious on other grounds. In the case of ZnS it seems probable that each 'bond' is formed from sp^3 tetrahedral hybrids at both atoms. If such bonds were purely covalent, we should have a situation represented by Zn^-S^+; if they were purely ionic, we should have Zn^+S^-, though of course in this latter case it would not be easy to understand the open tetrahedral structure. It seems reasonable, therefore, to suppose that the bonds are partly covalent and partly ionic. Detailed calculations (Coulson, Redei, and Stocker 1962) suggest that the ionic character in this crystal, and in most other similar tetrahedral crystals, is such that the more electronegative element (here sulphur) carries a small excess of electrons, of the order of 0·3. However, the degree of covalency is sufficiently large for directional valence forces to determine the stereochemistry, and we therefore still class them as valence crystals. They are often referred to in terms of the group number (see §7.9) of the atoms involved. Thus diamond is IV–IV, and so is carborundum, but ZnS is II–VI, and the exceedingly important intermetallic semiconductors such as GaAs or InSb are III–V. A related system, ZnO, which is of hexagonal close-packed type, shows (as was first pointed out by James and Johnson 1939) concentration of electrons along some of the Zn—O bonds just as would be expected for a model of covalent–ionic resonance.

There are other solids of this same general type, but with different geometrical shapes. For example, sulphur, which is by nature divalent, forms long chains with the characteristic valence angle of 105° when it is heated to about 200°C (plastic sulphur). These chains contain essentially localized covalent S—S bonds, and the chains are held together partly by van der Waals polarization forces and partly by the fact that when they are sufficiently long they may thread in and out of one another to give a disorderly 'matted' effect. Similarly (Fig. 11.17) crystalline Se and Te form parallel zigzag chains in which each atom has two neighbours, and SiS_2 consists of infinite chains of SiS_4 tetrahedra with weak van der Waals forces between the chains. There does not seem to by any simple reason, however, why CS_2 remains molecular and does not form chains similar to SiS_2. This example reminds us that in solids various structures often differ by only a very small energy. It is because of this that polymorphism is so common, a given substance existing in two or more forms according to the external conditions of temperature and pressure. For the same reason it is much easier to understand a given structure *a posteriori* than to predict it in advance.

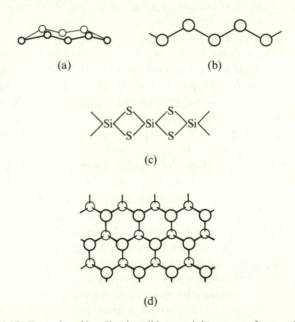

FIG. 11.17. Examples of bonding in solids containing atoms of groups IV–VII.

Arsenic is interesting in this connection because it crystallizes in double layers (see Fig. 11.17) in which each As atom has three neighbours at a distance of 0·251 nm. The distance between the layers is rather greater, 0·315 nm, showing that, though the bonds are almost as localized as in a molecule such as AsH_3, there is still a small degree of delocalization associated with interlayer bonds. This particular solid is therefore intermediate between types (1) and (3) of §11.1. Antimony and bismuth show a similar effect, but the layers become closer together, indicating an increasing shift from localized bonds towards a metallic structure.

There are plenty of other examples of this covalent type of binding. They all illustrate the $8 - N$ rule (p. 327) which, it will be recognized, is merely another version of the octet rule. The applicability of this rule to a crystalline solid may be regarded as strong evidence that the bonds are localized and of molecular-valence character.

We have seen that the crystal structure of diamond is associated essentially with the tetrahedral hybridization of its sp^3 orbitals; confirmation is obtained from the fact that the interatomic distance 0·154 nm is almost exactly the same as occurs in ethane and the heavier paraffins. It is a very natural step from here to associate the trigonal hybridization discussed for aromatic molecules in Chapter 8 with graphite, and indeed the very structure of graphite (Fig. 7.10 (*a*)), in which each layer is built in hexagonal fashion, reminds one strongly of a set of huge aromatic compounds. The distance between successive planes is 0·335 nm, a value so large that it can only arise from van der Waals forces. Within any one plane, however, there is a basic set of trigonal localized sp^2 σ bonds; the remaining electrons, which in the terminology of Chapter 8 we might call π electrons, must occupy MO's extending over the whole plane. We have here reached a situation half-way towards a metal, and in fact graphite does show a small electrical conductivity in its basal planes but not across them. Explicit calculations show the π bond order of the C—C bonds to be 0·53. Fig. 8.13 then predicts a bond length of 0·142 nm, in complete agreement with experiment. Just as diamond represents the extension to an infinite solid of the type of binding found in saturated carbon compounds, so graphite represents the extension of the type of binding found in aromatic molecules.

We have seen above that ZnS possesses a diamond structure but is held together by both covalent and ionic forces. In the same way boron nitride (BN) forms a layer lattice, exactly like graphite but with boron and nitrogen atoms alternating round each ring. In this lattice the σ bonds will be partly covalent and partly ionic, just as in ZnS, except that the hybrids are sp^2 trigonal rather than sp^3 tetrahedral. Again there will be mobile electrons which occupy π-type MO's extending over the whole plane, just as in graphite. Here we have a solid showing features of types (1), (3), and (4) of §11.1. It is interesting to note that, just like carbon, boron nitride forms not only a layer lattice, but also a tetrahedral diamond-like lattice of great hardness (Wentorf 1957). All the electrons form localized σ-type bonds, as in a typical group III–V compound.

It should by now be evident that no essentially new theoretical concepts are needed in interpreting the bonding in covalent crystals; they can for the most part be quite well understood in terms of the localized electron-pair bonds discussed in earlier chapters, and any 'left-over' electrons may provide a degree of non-localized bonding or metallic character just as in the case of the benzene π electrons. It is true that such crystals can be described entirely in terms of Bloch orbitals, as in §§11.2 and 11.3, but the valence crystal is characterized by the possibility of transformation of the occupied Bloch orbitals to an equivalent set of localized bond orbitals, and when such a possibility exists we can describe the bonds in familiar chemical terms, making use of our knowledge of similar bonds in small molecules.

11.6. Ionic crystals

In a formal sense an ionic crystal is analogous to a molecular crystal; it consists of separately identifiable units which attract each other at large and moderate distances but which repel strongly once their electron distributions begin to interpenetrate. The discussion of §11.4 is, with reservations, still applicable, the main difference arising because the individual units in the crystal are now *charged* (positively and negatively charged ions) and the attractions between them are therefore very much stronger and of longer range.

Let us deal first with the attractive part of the energy, using the semiclassical picture for which our earlier discussions provide a basis. The NaCl crystal (Fig. 11.4) is a good example; it contains positive and negative ions with charges Z^+e and $-Z^-e$ where the charge numbers are here $Z^+ = Z^- = 1$. In general, for a crystal of NaCl structure, with interionic distance R, a cation with charge Z^+e is surrounded by six anions with charge Z^-e and then by 12 more cations (Z^+e) at a distance $\sqrt{2}\,R$, and so on. The potential energy of the chosen cation in the field of all its neighbours is thus, taking out a common factor,

$$-\frac{Z^+Z^-e^2}{\kappa_0 R}\left\{6 - \frac{12}{\sqrt{2}}\left(\frac{Z^+}{Z^-}\right) + \frac{8}{\sqrt{3}} - \frac{6}{2}\left(\frac{Z^+}{Z^-}\right) + \ldots\right\}. \quad (11.22)$$

The ratio Z^+/Z^- is a constant, determined by the crystal type, and the quantity in braces is thus a structure-dependent sum, common to all crystals of given type. A similar sum arises from the interactions between each anion and all the rest. Now if we add together these two sums we shall in effect count every interaction energy twice. The energy per ion pair is thus half the sum of two expressions of the form (11.22). Various mathematical methods have been devised for making the summations for this and other possible geometrical arrangements. The results may always be expressed in the form

$$E_{\text{ion}} = -\frac{AZ^+Z^-e^2}{\kappa_0 R} \tag{11.23}$$

where A has a value, here about 1·75, dependent on the crystal structure. Values of A for a few important structures are given in Table 11.2. In every case R is the closest anion–cation distance and the energy (11.23) refers to one stoichiometric molecule. The constant A is called the Madelung constant for a particular structure.†

TABLE 11.2
Values of the Madelung constant

Structure	Example	A
Simple cubic	NaCl	1·748
Body-centred cubic	CsCl	1·763
Fluorite	CaF_2	2·519
Wurtzite	ZnS	1·641
Rutile	TiO_2	2·385

It is obvious from this table that any tendency towards ionic bonding will be considerably enhanced in the crystal, with the result that there is almost complete charge transfer. This transfer cannot, however, be quite complete, for we may regard the wavefunction for the whole crystal as a linear combination of functions representing purely ionic and purely covalent situations; the energy function will be stationary when the latter appear with small, but non-vanishing, coefficients.

There is considerable experimental evidence from X-ray diffraction for almost complete charge transfer in strongly ionic crystals. From electron-density maps (cf. Fig. 8.4) it is possible to calculate, by numerical integration, the number of electrons associated with each nucleus (there is of course some slight uncertainty depending on the precise definition of 'size' of the different ions); in this way it is found that about 17·9 electrons 'belong' to the chlorine atom in NaCl, i.e. that it is essentially a Cl^- ion. With the more electronegative fluorine, the degree of ionicity in the bonding is even higher. Further support for the belief that covalent character is negligible comes from calculation of the overlap integrals at the observed internuclear distance; for NaCl the greatest value of such an integral is about 0·06, indicating almost complete independence of the two ions as far as covalent bonding is concerned.

We must now consider the nature of the *repulsive* forces which arise when the charge clouds of the ions come into 'contact', and without which the crystal

† Sometimes the ionic charges which appear in (11.23) are included in the definition of Madelung constants; for a recent discussion see Quane (1970).

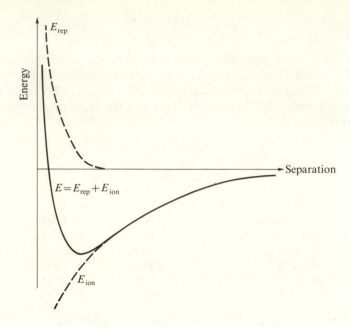

FIG. 11.18. Attractive and repulsive components of the lattice energy of an ionic crystal.

would collapse. The above calculation of the energy of attraction is based on a point-charge model of the ions and is valid (from classical electrostatics) as long as the spherical charge clouds do not overlap appreciably, but when they begin to interpenetrate there is a deformation of the charge clouds and new energy terms arise.

The classical approach to the inclusion of the repulsion energy is to assume some functional form such as†

$$E_{rep} = B/R^n \qquad (11.24)$$

where B is a positive constant and n is usually taken to be an integer; when n is large (appropriate values are normally in the range 5–12), E_{rep} rises very sharply for small R values as indicated in Fig. 11.18. The energy per stoichiometric molecule, multiplied by the Avogadro constant L, then gives the molar lattice energy of the crystal in the form

$$U = L\left(-\frac{AZ^+Z^-e^2}{\kappa_0 R} + \frac{B}{R^n} \right). \qquad (11.25)$$

The parameter B may then be determined empirically so as to make U a minimum at the observed internuclear distance $R = R_e$. The corresponding

† Note that this is the form used in Problem 1.4, to which the present discussion is clearly related.

equilibrium lattice energy is easily found to be

$$U_0 = -\frac{ALZ^+Z^-e^2}{\kappa_0 R_e}\left(1 - \frac{1}{n}\right) \tag{11.26}$$

and comparison with E_{ion} (given by (11.23)) shows at once that the repulsive contribution to the lattice energy arises from the $(1/n)$-term and is thus relatively small compared with the Madelung term. This approach is of course purely empirical and does not in any way 'explain' the origin of the repulsions, but it leads to a fairly good account of the properties of many ionic crystals.

The quantum-mechanical theory of the repulsive potential is difficult; the first calculations were made by Born and Mayer (1932), and accounts of more recent work appear in various books and reviews.[†] The repulsion arises, pictorially, from a deformation of the initially spherical charge clouds of the closed-shell ions, which arises when they are crushed together. The effect of this deformation may be related, in an orbital approximation, to overlap and similar integrals which show an exponential dependence on internuclear distance, or more precisely a combination of exponential times polynomials, in which exponential behaviour dominates. It is not surprising therefore that the overall repulsive effect is better represented by a term $E_{rep} = C\exp(-aR)$ than by (11.24). Such a form has been used widely, ever since the early work of Born and Mayer, and is usually recast as

$$E_{rep} = c\exp\left(-\frac{R - R_A - R_B}{\rho}\right) \tag{11.27}$$

where R_A and R_B are ionic radii while c and ρ are adjustable parameters. The 'Born–Mayer' radii for the alkali halides are shown in Table 11.3 and are somewhat smaller than the conventional ionic radii based on 'packing' considerations (to which we turn presently). The choice

$$c = 4{\cdot}0 \times 10^{-19}\,\text{J}, \qquad \rho = 0{\cdot}0345\,\text{nm}$$

leads to a good account of the properties of all the alkali halides; in particular, the minimum of the lattice energy $U = L(E_{ion} + E_{rep})$ occurs when R has the observed value, and the correct compressibility is also obtained.

It appears from the above discussion that the bonding in ionic crystals can be well understood in terms of the long-range electrostatic attractions and much stronger short-range repulsions (of quantum-mechanical origin), but we should also recognize the following extra terms.

    (i) The energy of attraction may contain a small covalent component (the charge transfer being incomplete).

    (ii) The resultant ions (e.g. Na^+, Cl^-) are polarizable, the large 'soft' anion

[†] See, for example, Margenau and Kestner (1969), Murrell (1974), or for a different approach McWeeny (1970).

TABLE 11.3

Born–Mayer radii (nm) for alkali halides

Li$^+$	0·0475	F$^-$	0·1110
Na$^+$	0·0875	C$^-$	0·1475
K$^+$	0·1185	Br$^-$	0·1600
Rb$^+$	0·1320	I$^-$	0·1785
Cs$^+$	0·1455		

much more so than the 'hard' compact cation, and a polarization energy should therefore be admitted.

(iii) There is a finite vibrational energy (the 'zero-point energy') even when the temperature approaches zero and the thermal motion is reduced to a minimum.

Term (i) is difficult to define precisely but is likely to be small and may be regarded as absorbed into the other, empirically fitted, non-classical term, the repulsion energy. Term (ii) may be estimated from polarizability data for free ions, but is small owing to the rather high symmetry of the electric field around each ion. Term (iii) amounts to $\frac{1}{2}h\nu$ per mode of vibration of classical frequency ν; with L ion pairs there are $2 \times (3L - 6)$ vibrational modes and the resultant energy, though a small fraction of the total, is significant. Table 11.4 gives an idea of the relative sizes of the various energy terms.

TABLE 11.4

Energy terms (in eV) in the NaCl crystal

Electrostatic energy	8·92
Polarization energy	0·13
Repulsion energy	1·03
Zero-point vibrational energy	0·08
Total lattice energy	7·94 (per NaCl unit)
Observed lattice energy	7·86

It should be noted that the dispersion interactions (p. 344), so important in providing the weak binding between the electrically neutral units in a molecular crystal, have been disregarded; in an ionic crystal they are in fact (although still present) almost completely negligible compared with the powerful electrostatic interactions.

Finally, we must ask why different crystals adopt different structures. Since the energy difference between one arrangement of the ions and another is an exceedingly small fraction of the total electronic energy, there is no reason to

TABLE 11.5
Univalent crystal radii (nm)

Li^+	0·060	H^-	0·208
Na^+	0·095	F^-	0·136
K^+	0·133	Cl^-	0·181
Rb^+	0·148	Br^-	0·195
Cs^+	0·169	I^-	0·216

expect a simple answer, but elementary packing considerations, based on the idea that every ion has a characteristic 'size', prove to be extremely useful. As in our discussion of covalent radii (§7.10) we must try to find a set of *ionic radii* such that the equilibrium distance between adjacent ions in any ionic crystal is approximately equal to the sum of the appropriate radii. Clearly, measurement of R_e for one given crystal tells us the *sum* of two such radii, but since we cannot say where the anion ends and the cation begins we cannot measure the individual radii. To overcome this difficulty, Pauling (1960) assumed that the sizes of ions with a given rare-gas structure should be inversely proportional to the screened nuclear charge $Z_e = Z - s$; this is theoretically justified by (2.23)† and there is no difficulty in estimating the screening constants (s) from Slater's rules. Let us consider, for example, K^+ and Cl^-, both with an argon-like structure. In each case $s = 11·25$ so that $Z - s$ has the values 7·75 and 5·75 respectively. The experimental KCl distance in the crystal is 0·314 nm; if we divide this in the ratio of the values of $3/(Z - s)$ we find $R(K^+) = 0·133$ nm and $R(Cl^-) = 0·181$ nm. Proceeding in this way we obtain the set of univalent radii shown in Table 11.5. With a few exceptions, to which we shall return in a moment, the interionic distance in the crystal is almost exactly equal to the sum of the appropriate entries in this table.

For more highly charged ions it is necessary to decrease the univalent radii, calculated as above, to correct for 'compression' of the ions by the greatly increased Coulomb attraction between them. We shall be content to list a few important crystal radii in Table 11.6. For further particulars see Pauling (1960). Improved and more comprehensive tables, particularly for metal ions, have been given by Shannon and Prewitt (1969). It will be noticed that the positive ions are always much smaller than the corresponding negative ions, and that the higher the positive charge, the smaller is the ionic radius. It will also be seen that for given net charge the ionic radius increases with atomic number; the only exception to this rule occurs when, as in $Ca^{2+} \rightarrow Zn^{2+}$ or $Sc^{3+} \rightarrow Ga^{3+}$, the difference in atomic number is associated with the filling-up of inner d-shell orbitals.

† It should be noted that the quantum number n also appears, in a way which depends on the definition of 'size'; a proportionality to n is commonly assumed but the dominant factor is the effective nuclear charge.

TABLE 11.6
Polyvalent crystal radii (nm)

Be^{2+}	0·031	B^{3+}	0·020		
Mg^{2+}	0·065	Al^{3+}	0·050	O^{2-}	0·140
Ca^{2+}	0·099	Sc^{3+}	0·081	S^{2-}	0·184
Zn^{2+}	0·074	Ga^{3+}	0·062		

The ionic radii given in Table 11.5 enable us to understand one important factor which determines the type of crystal structure with given atoms. This is the radius ratio of the ions. Suppose that, as in Li^{+}I^{-}, this ratio is much less than $\sqrt{2} - 1 = 0.414$. (In Li^{+}I^{-} the ratio is actually $0.60/2.16 = 0.28$, but in Li^{+}F^{-} it is $0.60/1.36 = 0.44$). Then, as Fig. 11.19 shows, it is not possible for the positive and negative ions on two adjacent corners of the fundamental unit cube to touch. Such contact as there is must be between the much larger anions. Consequently the interatomic distance is greater than the sum of the crystal radii (in LiI this is 0·302 nm instead of the expected 0·276), and the positive cations fit into the interstices of the closely packed negative anions.

Another point arises in this connection. So long as the ionic radii differ considerably the sodium chloride arrangement is the most stable for alkali halide crystals. When the radii become more nearly comparable, however, as in CsCl (0·169 and 0·181 nm respectively) there is no interstitial room available for the cations, and the structure changes to the body-centred form shown in Fig. 11.1 (*a*) where each ion is surrounded by as many oppositely charged ions as possible. In the caesium chloride lattice this is eight, instead of six as in the sodium chloride lattice. This situation applies to the chlorides, bromides, and iodides of caesium and rubidium.

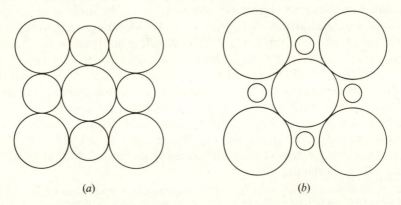

(*a*)   (*b*)

FIG. 11.19. Crystal structures of (*a*) LiF and (*b*) LiI.

We could summarize this discussion by saying that the crystal structure assumed by any given substance is chiefly determined by (a) the relative numbers of ions of each type, (b) the desire of ions of one kind to surround themselves with as many ions of different charge as possible, (c) the large repulsions that set in when the ions are closer together than the sum of their appropriate radii, and (d) the radius ratio of the ions, which governs the most satisfactory way in which they can be packed together. Rules of this kind were developed first experimentally by Goldschmidt.

In conclusion, it must be mentioned that there are many types of crystal which fit neither into the ionic category, nor into any of the other main categories distinguished in this chapter. In ice, for example, the units are neutral H_2O molecules, but fairly strong electrostatic forces arise from the polar nature of the individual molecules. Even when a crystal is essentially ionic, the individual groups (e.g. in ferricyanides and cobaltammines) may be of such complexity that the simple considerations of this section do not apply directly. Such structures can, however, be understood using the same basic ideas. One or two special cases will be discussed in Chapter 12.

Problems

11.1. Consider a quasi-infinite polyene of 10^8 carbon atoms with a spacing of 0.15 nm and a bond integral $\beta = -2.5$ eV, and plot ε_k (corresponding to cyclic boundary conditions) against the magnitude k of the wavevector. Calculate (i) the width of the π band, (ii) the spacing between adjacent levels at the top, the middle, and the bottom of the band, and (iii) the densities of states at the same three points. With one π electron from each atom would you expect the system to be a conductor or an insulator.† (Hint: here $k = \kappa/N$ with $\kappa = 0, \pm 1, \pm 2, \ldots$ and you may use (11.3). The density of states is simply the number per unit range of energy.)

11.2. Consider a particle of mass m moving freely between boundaries 1.5 cm apart (the length of the polyene in Problem 11.1), assuming periodic boundary conditions. Solve the free-particle Schrödinger equation to obtain the energy levels in the form $\varepsilon_k = (\hbar^2/2m)k^2$, and wavefunctions $\psi_k = \exp(2\pi ikx)$, and show how the values of k are quantized as a result of the periodic boundary conditions. Choose an effective mass m which will give the same level spacing for small k values as you obtained in Problem 11.1 and compare the two ε_k curves on the same diagram. (Hint: refer to Problem 3.9, but impose the condition that ψ_k must start repeating when x increases by L.)

11.3. Show that the Bloch orbital for a polyene π electron (Problem 11.1), namely

$$\psi_k = C_k \sum_n \exp(2\pi ikn)\phi_n \qquad (k = \kappa/N)$$

is normalized, with neglect of overlap, when $C_k = 1/\sqrt{N}$. Then obtain the expression (using a nearest neighbour approximation)

$$\langle p_x \rangle_k = 2\hbar d \sin 2k$$

† For large N the system is unstable. In a real chain the bond lengths alternate. How would your conclusions be affected? (Think of the extreme situation where alternate bonds become independent.)

for the expectation value of the electron momentum along the chain (x direction) in state ψ_k, d being defined by

$$d = \int \phi_1 (\partial/\partial x) \phi_2 \, d\tau$$

where ϕ_2 is the first neighbour of ϕ_1 in the positive x direction. Which states in the band correspond to the greatest momentum? Note how the momentum falls towards the top of the band, i.e. as an energy gap is approached. (Hint: evaluate the expectation value using (3.54), remembering that your wavefunction is complex. With real AO's

$$D_{mn} = \int \phi_m (h/i) \, (\partial \phi_n/\partial x) \, d\tau = D_{nm}^*$$

by Hermitian symmetry (p. 64), and you need only consider D_{12} for a typical pair of neighbours (C–C link).)

11.4. Explain how you would perform an LCAO MO calculation on a 1-dimensional chain of N beryllium atoms, admitting 2s and $2p_\sigma$ orbitals and applying periodic boundary conditions. Would the system be a conductor or an insulator (*a*) for a lattice spacing at which the bands are separate, and (*b*) for a spacing at which they overlap? Would the s- and p-type Bloch functions mix when the bands overlap, or not? If so, how could you determine the mixtures and how would the band structure be affected? (Hints: if ψ_{sk} and ψ_{pk} are typical Bloch functions both must have the same k if mixing is to be possible. The mixing (if any) will then be determined, for each k value, by a matrix element $\int \psi_{sk}^* \hat{h} \psi_{pk} \, d\tau$. Obtain an expression for this, in nearest neighbour approximation, along the same lines as in Problem 11.3. Put in some plausible values of the coulomb and bond integrals α_s, β_{ss}, β_{sp}, and β_{pp}, and try to discover what happens for a few typical values of k.)

11.5. Show that a Bloch orbital of the form (11.10) satisfies the secular equations for *any* lattice whose atoms lie at the points $\mathbf{R}_n$ given by (11.4), provided only one AO per atom is admitted. Show also that the corresponding energy (neglecting overlap integrals) is

$$\varepsilon_k = \alpha + \sum_\rho e^{ik\cdot\rho} \beta(\rho)$$

where $\beta(\rho)$ is the 'bond integral' for atoms separated by the vector ρ and the summation is over neighbours of *all* orders. (Hints: write the secular equations in the form

$$(\alpha - \varepsilon)c_n + \sum_m \beta(\mathbf{R}_n, \mathbf{R}_m)c_m = 0$$

where c_n and c_m are coefficients of the orbitals ϕ_n and ϕ_m centered on points $\mathbf{R}_n$ and $\mathbf{R}_m$. Note that $\beta(\mathbf{R}_n, \mathbf{R}_m)$ depends only on the vector from $\mathbf{R}_n$ to $\mathbf{R}_m$ (i.e. $\rho = \mathbf{R}_m - \mathbf{R}_n$) and substitute $c_n = e^{ik\cdot\mathbf{R}_n}$ corresponding to (11.10)).

11.6. Use the result of Problem 11.5 to generalize (11.11) to a square lattice with the admission of second- and third-neighbour interactions. Sketch the analogue of Fig. 11.13 (*b*) and shade the zone corresponding to the occupied orbitals, assuming each atom supplies one electron.

11.7. Set up an LCAO calculation for the benzene σ electrons, using as a basis the three sp^2 AO's (a_n, b_n, c_n, say) on each carbon (n) and the 1s AO's ($d_1, \ldots d_6$) on the hydrogens, and making approximations as suggested in Problem 8.14.

Show that, with neglect of intra-atomic β's there are two filled energy bands and that by suitable mixing of the corresponding doubly occupied Bloch-type MO's the system may be described in terms of doubly occupied bond orbitals localized in the C–C and C–H regions. What relevance do your results have for the interpretation of the electronic structures of (i) graphite and (ii) diamond? (Hints: the AO's may be combined into symmetry orbitals of the form $\phi^a = \Sigma_n \exp(2\pi i\kappa n/N)a_n$ etc.; these are essentially Bloch orbitals for a one-dimensional lattice of $N\,(=6)$ unit cells, and there are four such Bloch functions for each κ value ($\kappa = 0, \pm 1, \pm 2, 3$) corresponding to the four AO's in each unit cell. Set up and solve the secular equations for these Bloch functions.)

11.8. Show that, when two systems A and B with electron configurations $A[\phi_A^2]$ and $B[\phi_B^2]$ begin to overlap, the electron density is no longer of the form $2\phi_A^2 + 2\phi_B^2$ but becomes

$$P = 2(\phi_A^2 + \phi_B^2 - 2S\phi_A\phi_B)/(1 - S^2)$$

where S is the overlap integral $\int \phi_A \phi_B \, d\tau$. Hence show that charge amounting to $4S^2/(1 - S^2)$ electrons is pushed out of the overlap region. This is the basis for the interpretation of closed-shell repulsions on p. 351. (Hint: the whole four-electron wavefunction is a determinant corresponding to the electron configuration $[\phi_A^2 \phi_B^2]$; this is equivalent to the determinant based on $[\psi_+^2 \psi_-^2]$ where $\psi_{\pm} = N_{\pm}(\phi_A \pm \phi_B)$ are *orthogonal* linear combinations. Work in terms of ψ_+, determining the normalizing factors, and noting that for the orthogonal orbitals P *is* a sum of orbital contributions. Note how your discussion relates to that of the hypothetical molecule He_2 (p. 98).)

12
Weak interactions and unusual bonds

12.1. Unusual types of bonding

In preceding chapters we have considered all the more common types of valence interaction. Most molecules are held together by localized electron-pair bonds, of roughly σ type, occupying the regions indicated by the links (—) in the chemical formula: sometimes the resulting framework is strengthened or stiffened by delocalized π bonding in which only a fraction of an electron pair is associated with each link (giving a 'partial' double bond); sometimes, as in transition-metal complexes and the solid state (particularly when d orbitals are involved), the delocalization is of a more complicated nature and the bonding is more widely spread. We also know something of the non-valence interactions which play their part in determining both molecular shape (§7.8) and intermolecular forces (§11.4). However, there are many types of bond, with quite specific and characteristic properties, which do not fit easily into any of these categories: some of them are weak by comparison with the more usual electron-pair bonds (hydrogen bonding is an example); others, like the bonds in boron compounds, are quite strong but are odd in some other respect; some are at first sight totally unexpected, like the bonds involving the rare or 'inert' gases. In this chapter we discuss some of these new varieties of bonding, and also some of the other effects of interest which arise from interactions between different chemical groups (e.g. the 'hindered rotation' of groups connected by a σ bond which get in each other's way and cannot rotate freely, or the bending deformations which occur in 'overcrowded' molecules).

12.2. The hydrogen bond†

Most of the bonds with which we have so far been concerned have an energy lying in the range $200-500\,\text{kJ}\,\text{mol}^{-1}$, but there is one very common bond, whose energy is very much less than this, which we must now describe. In rough terms we can say that in this bond, usually called the hydrogen bond, a single hydrogen atom appears to be bonded to two distinct atoms, one of which is most commonly oxygen. The hydrogen bond is particularly important because its energy is so small, of the order of $25\,\text{kJ}\,\text{mol}^{-1}$, and also because the hydroxyl group occurs so frequently in most biological systems.

† For a more detailed discussion see Pimentel and McClellan (1960), Hamilton and Ibers (1968), and Schuster et al. (1976). A useful brief review is by Coulson (1957).

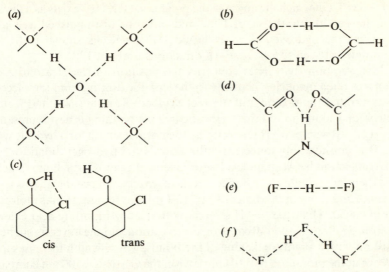

FIG. 12.1. Some types of hydrogen bond: (*a*) water, showing tetrahedral coordination around each oxygen atom; (*b*) the dimer of formic acid; (*c*) cis- and trans-orthochlorophenol (only the cis form shows hydrogen bonding); (*d*) bifurcated hydrogen bond in certain diacyl-diphenylamines; (*e*) the HF_2^- unit in the ionic crystal $K^+ (HF_2)^-$; (*f*) polymeric HF.

The most obvious example of this bond is in ice, where the binding of the water molecules to each other may be represented by the broken lines in Fig. 12.1 (*a*). At sufficiently low temperatures there is practically complete co-ordination of this kind, giving rise to a relatively open tetrahedral structure around each oxygen atom. If it were not for the directional character of these dotted hydrogen bonds, it is hard to see how such an open structure could be stable. The process of melting may be thought of largely as the breaking of a certain number of these bonds, and at room temperature only about one-half of the maximum possible number of such bonds still remain unbroken. The breaking of all these latter is associated with transition from the liquid to the gas phase (heat absorbed on melting, $6{\cdot}02\,kJ\,mol^{-1}$; heat absorbed on sublimation, $51{\cdot}2\,kJ\,mol^{-1}$). This argument suggests, as is indeed true, that highly associated liquids, with high dielectric constants, very often make use of this type of bond.

We can distinguish at least five distinct types of hydrogen bond. These are shown in Fig. 12.1 (*a*)–(*e*), and may be described as follows.

(*a*) Intermolecular, extending over many molecules.

(*b*) Intermolecular, extending over two molecules, which form a dimer.

(*c*) Intramolecular, in which the hydrogen is bonded to two atoms of the same molecule but there is only one hydrogen bond.

(*d*) Intramolecular, in which the hydrogen is held by two equal hydrogen bonds (bifurcated hydrogen bonds).

(e) (FHF)⁻, in which the anion of the polar crystal KHF_2 exists as a distinct charged unit in the solid. The bonds must be distinguished (see later) from the hydrogen bond in liquid HF where, as shown in (f), the angle between successive FHF directions is about 120°.

The first question that arises concerns the position of the H atom. X-ray evidence is often inconclusive owing to the small scattering from low electron density around the proton, but indirect evidence from infrared and Raman spectroscopy soon showed that in most cases the bonding is best represented as A—H .... B where A—H is of roughly the normal length for a hydride while H .... B is considerably longer. In other words the hydrogen bond forms a weak connection between molecules (or chemical groups) which nevertheless retain their individuality; the bonded units in ice are still recognizable as water molecules and those in the dimer of Fig. 12.1 (b) are recognizable as molecules of formic acid. Thus, the O—H stretching frequency in formic acid changes only from $3570\,cm^{-1}$ to $3110\,cm^{-1}$ on dimerization and the lower frequency is consistent with a slight weakening of the bonds, whose length increases from 0·098 nm in the monomer to 0·104 nm. In ice, the O—H is 0·100 nm compared with 0·096 in the gas phase; the O .... O distance, however, is about 0·276 nm and the length of the hydrogen bond H .... O is thus almost twice that of the normal O—H bond. In general, however, the position of the hydrogen in A—H .... B is dependent on the overall A .... B distance, and if the geometry of the compound requires A and B to be abnormally close the H atom may adopt a central position. Thus in potassium hydrogen malonate (see Currie and Speakman 1970) the O .... O distance is 0·249 nm, and the hydrogen is exactly central. Such cases are the exception rather than the rule, however, and more typically the A—H resembles a normal covalent bond while the H .... B, the hydrogen bond, is much longer and much weaker.

Owing to the variability of the hydrogen bond length for the same elements appearing in different compounds, it is not possible to tabulate unique bond energies, but Table 12.1 summarizes the orders of magnitude to be expected. It should be noted that the corresponding bond energy in the symmetrical ion (FHF)⁻ is about $220\,kJ\,mol^{-1}$, and it may be argued that the bonds in this system are delocalized over all three centres but essentially covalent in character—an interpretation which we develop in a later section.

From a theoretical standpoint the hydrogen bond can be understood most readily in terms of the discussion which led to eqn (11.19) for the energy of two interacting molecules. If the hydrogen bond is weak the interaction between the systems it connects is in first order electrostatic (represented in Fig. 11.16); the bond energy may then be estimated classically from a knowledge of the forms of the charge clouds of the two systems. That this picture is substantially correct is supported by the fact that hydrogen bonds only occur between atoms of high electronegativity, which accordingly produce strong dipoles, and by the observation that in such bonds an electron-deficient region (e.g.

TABLE 12.1

Energies of hydrogen bonds $(A—H \ldots . B)$

A \ B	N	O	F	S
C	12[a]–20	8–12		
N	12–50	12–16	~20	
O	16–30	12–30		
F		~45	25–35[b]	
S				~20

[a] Values of $-\Delta H$ in kJ mol^{-1}. The entry at the junction of row A and column B refers to the hydrogen bond in $A—H \ldots . B$.

[b] These values are for $(HF)_n$ polymers, not for the ion $(FHF)^-$.

around the proton) comes close to an electron-rich region such as a lone pair (e.g. an oxygen lone pair in H_2O (Fig. 7.18)). Indeed, a very primitive calculation on ice by Bernal and Fowler (1933), who used a model with excess negative charge centred on the oxygen and compensating positive charges on the hydrogens, yielded a remarkably good estimate of the cohesive energy. A more realistic model (Lennard–Jones and Pople 1951; Pople 1951) recognizes the fact that the various parts of the H_2O charge cloud do not have their centroids exactly on the nuclei, but the conclusions are similar and it does not seem to matter much exactly how the positive and negative charges are located, as long as they correctly reproduce the molecular dipole moment.

We can soon see why only hydrogen (or deuterium) will serve best as the middle atom of the bond. In order that the electrostatic energy of interaction shall be greatest, it is desirable that the units should approach as close together as possible. In this respect hydrogen possesses two favourable properties. Its atomic radius (0·03 nm) is extremely small and it possesses no inner-shell electrons. As a result the adjacent molecule can approach very closely without the introduction of large repulsive energy terms. It is, of course, necessary to have an electropositive atom in the centre, but if, for example, we tried to replace H by Na the larger size of the Na atom and the presence of a complete inner L shell would open out the structure to such an extent that the energy of binding would be inadequate. The desirability of having a rather electronegative atom at each end of the system is explained in a similar way, for here again the relatively small size of such atoms (Table 7.6 on p. 222) readily permits a close approach of the two molecules.

In a higher-order approximation, of course, the electrostatic binding would be modified by polarization effects (E_{pol} in (11.19)), each charge cloud being slightly perturbed by the presence of the other, and the van der Waals contribution (E_{disp} in (11.19)) might also be appreciable. However, such refinements do not affect the general picture of two distinct interacting systems, which is embodied in the choice of a wavefunction of *product* form (§5.2). A wavefunction of this form does not allow for any sharing of electrons between the two systems, however, and so we cannot expect it to remain appropriate if their charge clouds interpenetrate to any extent, in which case the link between them may assume some covalent character.

There are two main ways of allowing for covalent character in such situations: (i) the whole system† —A—H .... B— may be regarded as one 'super-molecule' to which we may apply standard MO methods (the MO's extending over at least all three centres), or (ii) we may proceed further with the formulation used so far but admitting wavefunctions to describe situations such as $(-A-H)^- (B-)^+$ in which an electron has been transferred from one system to the other. The super-molecule approach has been widely used in *ab initio* calculations (for example, in calculations on water dimers, see Del Bene and Pople (1970), Diercksen (1971)) and clearly needs no new theory. The alternative, which retains the idea of two molecules but with the admission of 'charge-transfer states' as components of the wavefunction, has been explored by Fujimoto and Fukui (1972) in another context. It has much in common with the VB approach, as may be seen by considering the system —O—H .... O—. By transferring an electron to give $(-O-H)^- O^+$ we can achieve a situation with two *singly* occupied and overlapping orbitals (one in an oxygen lone-pair orbital, the other on the O—H group), and on coupling the spins antiparallel the corresponding wavefunction will describe covalent bonding between the two systems in the usual way. Although *ab initio* calculations along such lines have seldom been attempted, early semi-empirical calculations by VB methods suggested that for normal hydrogen bonds covalent character is small and the predominant interactions are electrostatic. The most significant exception to this rule seems to be the ion $(FHF)^-$ where recent work (Kollman and Allen 1970; Noble and Kortzeborn 1970) shows that the system is essentially a linear triatomic species with MO's extending over all three centres. In VB language, we may say that there is considerable mixing of the structures represented by

$$F^- \ H-F, \quad F-H \ F^-, \quad F^- \ H^+ \ F^-$$

the symmetry of the system implying that the first two occur with equal weight to give two largely covalent H—F bonds. In fact the observed bond length (0·113 nm) is not very much longer than in the free molecule (0·092 nm). If we discount such exceptions, hydrogen bonding may be quite well understood at a qualitative level using simply electrostatic models.

It is hard to exaggerate the importance of hydrogen bonding. The

---

† We must remember that A and B are attached to other atoms, and that in referring to the A and B ends of the bond we are really referring to two molecules.

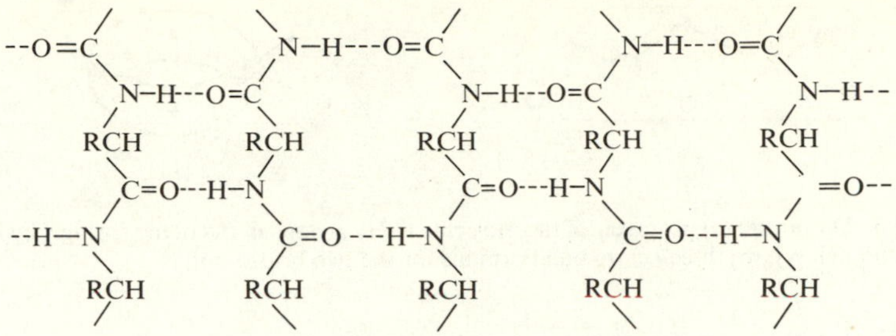

$$
\begin{array}{c}
\text{--O=C} \qquad \text{N--H---O=C} \qquad \text{N--H---O=C} \\
\text{N--H--O=C} \qquad \text{N--H--O=C} \qquad \text{N--H--} \\
\text{RCH} \quad \text{RCH} \qquad \text{RCH} \quad \text{RCH} \qquad \text{RCH} \\
\text{C=O---H--N} \qquad \text{C=O---H--N} \qquad \text{=O--} \\
\text{--H--N} \qquad \text{C=O ---H--N} \qquad \text{C=O---H--N} \\
\text{RCH} \quad \text{RCH} \qquad \text{RCH} \quad \text{RCH} \qquad \text{RCH}
\end{array}
$$

FIG. 12.2. Hydrogen bonds in a protein system.

abnormally high dielectric constants of liquids such as CH_3OH, H_2O and HCN compared with other liquids where the individual dipole moments are of the same order of magnitude, the association which frequently occurs in liquids, the mutual orientation of the molecules in many organic crystals such as the purines and pyrimidines, the process of attaching of ordinary 'dirt' to the skin of the human body, the regular arrangement of polypeptide chains in a protein structure in some such manner as that illustrated in Fig. 12.2, and the cross-linking in the double helix of a nucleic acid and its significance for the duplication of gene structure all depend on hydrogen bonding. For the importance of hydrogen bonding in such systems see, for example, Stryer (1975). In fact, nearly all biological processes seem to involve the hydrogen bond at some stage, an interesting example being anaesthesia (Paolo and Sandorfy 1974) in which anaesthetic potency often appears to depend on the capacity of a compound to *break* hydrogen bonds in the biological system.

12.3. Bonds in electron-deficient molecules

Another type of bonding which does not fit into the categories dealt with in earlier chapters arises in 'electron-deficient' molecules. In such molecules neighbouring atoms all appear to be joined by normal covalent bonds — except that there are not enough electron pairs to go round! In other words, the number of valence electrons is less than twice the number of bonds. The first molecules of this type to claim attention were the boron hydrides[†], but many other similar compounds, involving group 2 and group 3 elements such as beryllium and aluminium, have since been prepared and investigated.

The fact that only group 2 and group 3 elements form such compounds is associated with the electron deficiency of the atoms themselves, in the sense

---

[†] A review of the early work was given by H. C. Longuet-Higgins, *Qt. Rev.* **11**, 121 (1957). For a comprehensive account of hydrides and related compounds see W. N. Lipscomb, *Boron hydrides*, W. A. Benjamin, New York (1963).

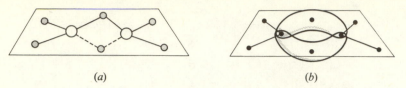

(a)   (b)

FIG. 12.3. Diborane: (*a*) geometry of the molecule (BH_2 groups in the plane, 'bridge' hydrogens above and below); (*b*) three-centre bond orbitals for the two bridge bonds.

that there are 'empty spaces' in their valence shells†. Thus, with s–p hybridization, Be and B have four valence-shell orbitals available for bonding but only two and three electrons respectively. In ionic crystals the deficiency is often made good by electron transfer; thus the complex ion BeF_4^{2-} is tetrahedral, and this is consistent with the idea that the two extra electrons have been given to the metal whose four singly occupied sp^3 hybrids may then be used in covalent bonding with the fluorines. In the so-called electron-deficient molecules, however, the bonding is of a more unusual type.

We shall begin by discussing diborane B_2H_6. At first it was believed that diborane had an ethane-like structure and it was often written $BH_3 . BH_3$, in spite of the fact that a borine radical BH_3 is unknown. There are, however, only 12 valence electrons, not enough to account for seven electron-pair bonds. Moreover, it was soon established that the geometry of the molecule was that shown in Fig. 12.3 (*a*) in which the BH_2 groups lie in a plane (like the CH_2 groups in ethylene); the boron atoms are joined by two 'bridge bonds', one above the plane and one below, each containing a proton. This led Pitzer (1945) to suggest that the correct analogy was not with ethane but with ethylene (with which $B_2H_4^{2-}$ would be isoelectronic) with two protons embedded in the double bond. In its simplest form, this description is at variance with the observations (i) that the 'double-bond' length is 0·015 nm *longer* than the single-bond length (0·162) based on the covalent radius in Table 7.6, and (ii) that the chemistry of the molecule shows no trace of the acidic character which would be associated with two relatively bare protons.

The most satisfactory resolution of these difficulties was proposed by Longuet-Higgins (1949) who set up MO's containing both the boron AO's and the hydrogen 1s AO's in the bridges, thus allowing the bridge protons to 'clothe' themselves adequately with electrons; he also demonstrated that the full wavefunction, with its σ and π type MO's, could be written equally well in terms of localized MO's, two of them having the form shown in Fig. 12.3 (*b*). The resultant description of the bridge region, in terms of two three-centre

† Group 1 elements are also electron deficient in this sense, but with only one valence electron outside a closed shell the preferred form of bonding is either 'metallic' (the electrons being shared over a large number of neighbours) or 'ionic' (in which the electron is virtually removed). The tendency for formation of electron-deficient molecules is at its maximum in group 3.

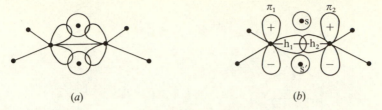

FIG. 12.4. Diborane: (*a*) formation of bond orbitals using roughly tetrahedral boron hybrids; (*b*) orbitals used in the σ-π description.

bond orbitals, won immediate acceptance and yielded an excellent account of the main properties of diborane (a useful review of recent studies is given by Lang and Lippard 1972). Subsequent work, largely by Lipscomb and his collaborators, rapidly led to a complete rationalization of the structures of the very large family of boron hydrides and their derivatives, including a wide range of compounds in which the boron atoms form a polyhedron or 'cage' (Lipscomb (1963)).

The direct way of determining the nature of the bonding in diborane is, of course, to make a full MO calculation using all the valence AO's, and then to examine the form of the charge cloud (e.g. by making a population analysis). We know, however, that instead of using the AO's of s and p type we may start from any alternative mixtures (e.g. hybrids) and that in either case solution of the secular equations should lead to the same MO's. Chemical intuition may thus be incorporated in the description; we may, for example, build up bent bonds by assuming roughly tetrahedral hybridization at each boron atom (Fig. 12.4 (*a*)), or we may emphasize the 'protonated double bond' picture by using trigonal hybrids in the molecular plane (i.e. σ orbitals) and setting up π-type combinations of the remaining orbitals (Fig. 12.4 (*b*)). The former approach was used first (Eberhart *et al.* 1954) but we shall adopt the latter (the BH$_2$ angle is in fact 121·5°, nearly trigonal) and shall show the equivalence of the two types of description.

Let us label the orbitals as in Fig. 12.4 (*b*), where the bridge hydrogen 1s orbitals (s and s′) are mirror images of each other across the molecular plane. The terminal B—H bonds appear to be 'normal' and thus utilize two of the sp^2 hybrids on each boron. From the remaining orbitals we may set up symmetry combinations in the usual way, using an additional label S or A for symmetry of antisymmetry across the plane which bisects the molecule at the bridge hydrogens. Thus, without normalizing, we obtain symmetry orbitals

$$\sigma S: \ (h_1 + h_2), \quad (s + s')$$
$$\sigma A: \ (h_1 - h_2)$$
$$\pi S: \ (\pi_1 + \pi_2), \quad (s - s')$$
$$\pi A: \ (\pi_1 - \pi_2).$$

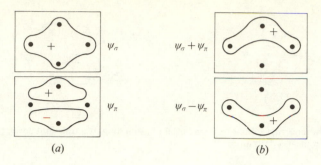

FIG. 12.5. Diborane: (a) σ- and π-type MO's; (b) bond orbitals formed by linear combination (sum and difference).

The resultant MO's for the description of the bridge region will thus be of the (un-normalized) form

$$\sigma S: \psi = (h_1 + h_2) + \lambda (s + s')$$
$$\pi S: \psi = (\pi_1 + \pi_2) + \lambda (s - s') \tag{12.1}$$

and similarly for the A-type MO's. With positive λ values the S-type MO's are of the form shown in Fig. 12.5 (a) and are both bonding in the bridge region. They can accommodate all four available electrons (one from each boron, one from each bridge hydrogen) to give a three-centre σ bond and a three-centre π bond. This is the delocalized MO description, and it shows that the protons are embedded in the σ bond as well as in the π bond.

It is a simple matter to pass to the bent-bond picture by forming more localized orbitals just as we did for H_2O in §7.3. Since B and H have similar electronegativities (Table 6.5) the λ values should be close to unity, and by taking the sum and difference of the σ and π MO's (12.1) we obtain

$$\psi \simeq (h_1 + \pi_1) + 2s + (h_2 + \pi_2)$$
$$\psi' \simeq (h_1 - \pi_1) + 2s' + (h_2 - \pi_2) \tag{12.2}$$

The closed-shell wavefunction for the electron configuration $(\psi_\sigma)^2(\psi_\pi)^2$ is identical with that for $(\psi)^2(\psi')^2$, but clearly (Fig. 12.5 (b)) the 'sum and difference' orbitals defined in (12.2) represent bent bonds above and below the molecular plane, respectively. Moreover $h_1 + \pi_1$ and $h_1 - \pi_1$ are hybrids pointing up and down, respectively, on centre 1, while $h_2 + \pi_2$ and $h_2 - \pi_2$ are similar hybrids on centre 2, and these bond orbitals might therefore have been set up using the prescription suggested by Fig. 12.4 (a).

Whichever description we employ it is clear that (a) there is no normal two-centre σ bond along the axis, and (b) the bridge hydrogens are very far from bare protons (indeed our simple calculation suggests they should have excess negative charge). These conclusions are substantiated by the many MO

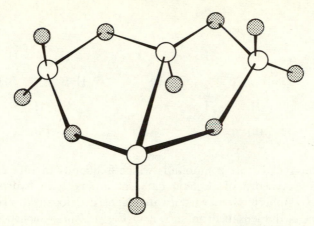

FIG. 12.6. Structure of B_4H_{10}, showing bridge bonds. Shaded circles represent the H atoms.

calculations, at various levels of sophistication, which have been performed in recent years (see, for instance, Laws *et al.* 1972).

Once the idea of bridge bonds has been accepted it is not difficult to understand a large number of otherwise mysterious structures. In addition to higher hydrides of boron, such as B_4H_{10} (Fig. 12.6), other group 3 elements such as Al and Ga also make use of hydrogen bridges. The aluminium compound

$$
\begin{array}{ccccccc}
 & \text{Al} & & & & \text{Al} & \\
 & & \text{H}_2 & \text{H}_2 & \text{H}_2 & \text{H}_2 & \\
\text{Al} & & & \text{Al} & & & \text{Al} \\
\text{H}_2 & & & \text{H}_2 & & & \text{H}_2 \\
\text{Al} & & & \text{Al} & & & \text{Al} \\
 & & \text{H}_2 & \text{H}_2 & \text{H}_2 & \text{H}_2 & \\
 & \text{Al} & & & \text{Al} & & \\
\end{array}
$$

(I)

which forms planes of indefinite extent, has two bridge bonds between each adjacent pair of Al atoms, one hydrogen above the plane and one below. Similarly, gallium forms a compound (II), with a geometry resembling that of diborane, and in some cases a group 2 atom may be involved as in (III) (Cook and Morgan 1969) where a BH_2 is replaced by BeH.

$$\begin{array}{ccc}
\text{Me} & \text{H} & \text{H} \\
\diagdown & & \diagup \\
\text{Ga} & \text{B} & \\
\diagup & & \diagdown \\
\text{Me} & \text{H} & \text{H}
\end{array}$$

(II)

$$\begin{array}{c}
\text{H} \\
| \\
\text{Be} \\
\diagup \quad \diagdown \\
\text{H} \qquad \text{H} \\
| \qquad\qquad | \\
\text{H}_2\text{B} \qquad \text{BH}_2 \\
\diagdown \quad \diagup \\
\text{H}
\end{array}$$

(III)

In the more elaborate compounds which frequently involve complete or incomplete polyhedra of electron deficient atoms, with hydrogen atoms around the periphery, a somewhat greater degree of delocalization occurs. The basic shape is the icosahedron shown in Fig. 12.7 (*a*) which occurs in the $B_{12}H_{12}{}^{2-}$ anion. This system has been fully discussed using simple MO theory (Longuet-Higgins and Roberts 1955) and the bonding has been interpreted in terms of population analysis by Hoffmann and Lipscomb (1962), who also suggest a simplified model based on 'semi-localized' MO's. Here the cage of boron atoms is much more like a metal than a molecule, but some degree of localization persists. The hydrides decaborane, octaborane, and hexaborane have structures which can be visualized by taking two, four, or six borons out of the $B_{12}H_{12}^{2-}$ icosahedron and adding a sufficient number of protons to bridge the 'exposed' boron atoms that remain. Decaborane ($B_{10}H_{14}$), formed by removing the boron atoms 1 and 6 in Fig. 12.7 (*a*) and adding two protons, is shown in Fig. 12.7 (*b*); the peripheral borons are bridged by the four starred hydrogens. For a discussion of a whole range of exotic compounds derived

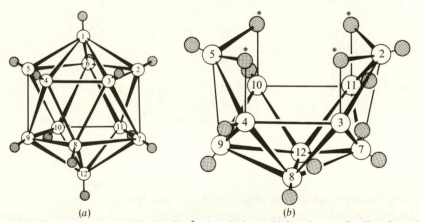

(a)   (b)

FIG. 12.7. Cage molecules. (*a*) The $B_{12}H_{12}{}^{2-}$ icosahedron. (*b*) Decaborane $B_{10}H_{10}$, formed from the icosahedron by removing two boron atoms and using hydrogens to make bridge bonds around the 'hole'. (From E. L. Muetterties (1967). The chemistry of boron and its compounds. Wiley, New York, with permission.)

from the polyhedral hydrides reference may be made to the books by Lipscomb (1963) and by Muetterties and Knoth (1968).

There are many other interesting compounds of boron in which the bonding is of the more conventional two-centre type. In some of these the electron deficiency in the boron valence shell is filled by the insertion of the lone pair of another chemical group. Thus BH_3CO exists and is tetrahedral around the boron atom; the strongly directed carbon lone pair of the CO may be regarded as overlapping a vacant sp^3 hybrid of a tetrahedral boron atom to give an electron-pair bond which is 'normal' except that both electrons are donated by the CO group. Another compound of the same kind is $BH_3 . NH_3$ in which the nitrogen lone pair has a similar function. In other situations the boron may be regarded as trigonally hybridized and the electron deficiency is then corrected by donation into the vacant π-type AO. The prime example of this type is borazine $(B_3N_3H_6)$ or 'inorganic benzene' (I). Each nitrogen, being in a trigonal valence state,

I

II

contributes two π electrons and the molecule is thus isoelectronic with benzene, the six π electrons giving some double-bond character to each B—N bond.[†] This is reflected in the observed bond length of 0·144 nm, which is considerably less than the single-bond value 0·158. An oxygen analogue of this system is also known (II) (Peters and Milberg 1964), the π-type lone pairs of the oxygen replacing those of the nitrogens, though in this case the conjugation will involve to some extent the attached OH groups.

The realization that two conjugated carbon atoms may be replaced by a boron and a nitrogen has led to the development of new types of hetero-aromatic molecules (cf. §8.7), of which the aza-derivative of phenanthrene (III) is one example (Dewar *et al.* 1958). The extreme case occurs in boron nitride (IV), already referred to in Chapter 11 (p. 347), a layer-lattice crystal in which each

[†] See, for example, Davies (1960) for simple MO calculations of bond orders. *Ab initio* calculations have been made by Armstrong and Clark (1970, 1972) and by Peyerimhoff and Buenker (1970).

III   IV

plane is structurally similar to a graphite layer. Many heteroaromatic systems containing boron, alongside nitrogen or oxygen atoms, are now known. In all such systems, however, the bonding may be pictured easily in terms of two-centre electron-pair bonds, sometimes with a degree of π bond character, and no further novel features appear.

12.4. Phosphonitriles and related ring systems

Other ring systems, besides borazine, are quite common in inorganic chemistry and in some of them the bonding is novel in the sense that, although essentially of π type, it involves the use of d orbitals. The best known of these are the phosphonitrilic halides, of which (I) is typical; it is essentially a trimer formed by bringing together three PCl_2N units and it is very nearly planar (Wilson and Carroll 1960), as also are the bromide (Zoer *et al.* 1969) and the fluoride (Dougall 1963). Other examples contain sulphur instead of phosphorus, as in (II), and tetramers such as III occur as well as trimers, though the tetramers are normally buckled into a 'tub' or 'crown' conformation. The halogens may also be replaced by NH_2 or CH_3 groups.

In all these systems, if we treat nitrogen as trivalent, sulphur as tetravalent, and phosphorus as pentavalent, there is an ambiguity in the placing of 'double bonds' and consequently some resemblance to the aromatic molecules considered in Chapter 8. There are important differences, however, and they cannot be regarded as 'inorganic aromatics'.

I   II   III

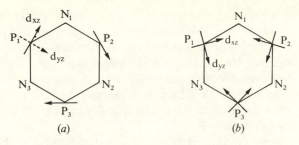

FIG. 12.8. Phosphonitrilic ring: (*a*) 'radial' and 'tangential' phosphorus d orbitals; (*b*) alternative orientation of the d orbitals.

Let us consider the trimer in Fig. 12.8 (*a*). If we form approximately tetrahedral sp^3 hybrids from the phosphorus 3s and 3p orbitals, we can establish a set of σ bonds of conventional type for the P—Cl and P—N bonds. We shall then be left with six unallocated electrons, one from each of the six atoms of the ring. The lowest remaining AO's from which we can form delocalized MO's are the 2p$_\pi$ orbitals on each N atom and the five 3d orbitals on each P atom. Now a suitable d orbital can enter into conjugation with adjacent p$_\pi$ orbitals as Fig. 12.9 indicates. With the directions of the axes as shown, the d$_{xz}$ orbital (call it B) overlaps equally both A and C and so we have the necessary condition for delocalized π-type bonding. The chief difference is that, as is shown by the signs in the various lobes of the orbitals, the overlap is positive for B and C but negative for A and B. This shows itself in the distinction between bonding and antibonding MO's. Thus, if all three orbitals had been p orbitals, a bonding MO would be similar in general form to $\phi_A + \phi_B + \phi_C$, but in Fig. 12.9 this would give antibonding in the region A–B and bonding in B–C. Bonding in both regions would come from $-\phi_A + \phi_B + \phi_C$. In the language of §10.6, the d orbital introduces a 'dislocation' in the conjugation path, two adjacent β's showing an unavoidable difference of sign.

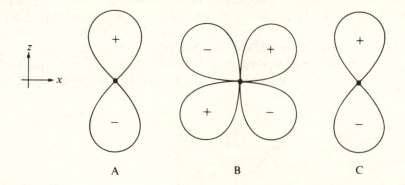

FIG. 12.9. Conjugation involving a d orbital and two adjacent p orbitals.

If we follow Craig and Paddock (1958; see also Craig 1959), the three 'tangential' d orbitals on the phosphorus atoms (Fig. 12.8 (*a*)) may be combined with 2p orbitals on the intervening nitrogens to give a six-centre conjugated path with one dislocation at each phosphorus. The system then resembles benzene except that the coefficients in the *anti*bonding benzene MO's yield the *bonding* MO's in the phosphonitrile. As there is a node at each phosphorus atom, however, the π charge density apparently breaks up into three allylic regions P—N—P.

In a more complete treatment, both tangential and 'radial' d orbitals would be admitted, or, instead we might employ any orthogonal mixtures. The set shown in Fig. 12.8 (*b*), formed by rotating each radial and tangential pair through 45°, is also well adapted for overlap with the nitrogen 2p AO's. This set provides the basis for the discussion given by Dewar *et al.* (1960). If we consider the region P_1—N_1—P_2 it is clear that we can set up a three-centre bond orbital of π type in the form

$$\psi_1 = c_1 (d_{xz}^{(1)} + d_{yz}^{(2)}) + c_2 \, p_N^{(1)}$$

Presumably, since N is slightly more electronegative than P the coefficient c_2 will be a little greater than c_1. The formation of this orbital is indicated schematically in Fig. 12.10. A similar bond orbital ψ_2 can be formed from $d_{xz}^{(2)}$, $p_N^{(2)}$, and $d_{yz}^{(3)}$, and a third ψ_3 from $d_{xz}^{(2)}$, $p_N^{(3)}$, and $d_{yz}^{(1)}$. These three orbitals are orthogonal, in good approximation, and the π-electron configuration $[\psi_1^2 \psi_2^2 \psi_3^2]$ thus gives an electron density which is the sum of three essentially distinct three-centre contributions, no two with any AO in common. The corresponding localization is thus rather more complete than in the Craig–Paddock description. Of course, if we allow free mixing of all nine AO's a certain amount of delocalization will result. As usual, the best description is somewhere between. A general comparison of the localized and de-localized descriptions of p_π–d_π bonding is given by Craig and Mitchell (1965).

Many other inorganic ring and cage molecules have been discovered during the last 15 years, involving not only sulphur and phosphorus but also silicon, selenium, tellurium and antimony. However, these do not necessarily introduce new types of bond. The simplest cage molecule, for example P_4 (white phosphorus), is a tetrahedron in which the bonds may be quite well represented by 'bent' electron pairs formed by the three singly occupied 3p

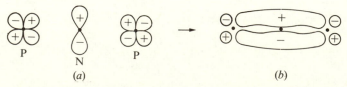

FIG. 12.10. Three-centre PNP bond: (*a*) AO's used; (*b*) resultant bond orbital.

AO's of each phosphorus atom. For more complicated structures reference may be made to recent textbooks of inorganic chemistry.

12.5. Bonds in rare-gas compounds†

In §12.3 we studied some electron-deficient molecules where the atomic valence shells contained too few electrons to fill two-centre orbitals connecting the bonded atoms. Clearly, the rare gases (if they are to form compounds at all) will lead to 'electron-rich' molecules, for their valence shells are completely filled. With helium the outer shell is $(1s)^2$, and with the other rare gases it is $(ns)^2(np)^6$ where $n = 2, 3, 4, 5, 6$ for Ne, Ar, Kr, Xe, and Rn respectively. At first sight, these atoms are expected to have zero valency, as they have no singly occupied orbitals, and for a long time it was believed that except perhaps in excited states (He_2 has a considerable electronic spectrum) these atoms formed no compounds. In 1962, however, Bartlett (1962) showed that a xenon atom could very easily play the same role as O_2 in the ionic species $O_2^+ (Pt\,F_6)^-$, and was able to isolate xenon hexafluoroplatinate $Xe^+(PtF_6)^-$. This is not altogether surprising, since the ionization potential of Xe is close to that of O_2 and their sizes are also similar. Very soon afterwards, and independently, Claassen *et al.* (1962) and Chernik *et al.* (1962) in America and Hoppe *et al.* (1962) in Germany showed that Xe and Rn could form covalent compounds with various numbers of F atoms. Before long a huge literature existed dealing with the synthesis, structure and properties of these rare-gas compounds. The early history is set out by Hoppe (1964), Chernik (1963), Hyman (1963) and Bartlett (1964). As a result of all this work, two outstanding facts emerged.

(*a*) Only the heavy inert gases (Kr, Xe, Rn) form compounds of this kind.

(*b*) The formation of stable compounds requires electronegative ligands. Some typical compounds are XeF_n ($n = 2, 4, 6$), $XeCl_2$, XeO_n ($n = 3, 4$), KrF_n ($n = 2, 4$), and $KrCl_2$, and there are also oxyfluorides such as $XeOF_4$. Xe—F bond energies in such molecules are of the order of $120\,kJ\,mol^{-1}$. Xe—Cl and Kr—F bond energies are distinctly smaller. The shapes of these molecules are also interesting. Molecules such as XeF_2 are linear, XeF_4 is square planar, XeF_6 is a distorted octahedron, XeO_3 is pyramidal, and XeO_4 is tetrahedral.

Despite the apparent breakdown of the octet rule it is not hard to understand the existence of these interesting molecules and to give an explanation of (*a*) and (*b*) above. We begin by noticing that the electronic structure of a rare gas is similar to that of a negatively charged halogen; for example I^- is isoelectronic with Xe. We are also familiar with interhalogen compounds (p. 220) and these include the ions ICl_2^-, which is linear, and IO_4^-, which is tetrahedral. These two ions should be isoelectronic with $XeCl_2$ and XeO_4, and the one set of molecules is thus apparently no more peculiar than the other.

---

† For a good review see Bartlett and Sladky (1973). Early theoretical papers are also well represented in Hyman (1963).

The simplest description proposed† makes use of VB theory. Let us illustrate it by reference to linear F—Xe—F. We start with two VB structures

$$\text{(i) } F^- \; Xe^+{-\!-}F, \qquad \text{(ii) } F{-\!-}Xe^+ \; F^-$$

in both of which the rare gas has lost one of its outermost p electrons (in Xe this is the 5p) and can therefore make one normal covalent σ bond by electron-pairing with a fluorine 2p electron. These two structures are more stable than might at first be thought on account of the large electron affinity of fluorine and the Coulomb attraction of the two unlike charges. Resonance between these two structures will lower the energy still further. Simple numerical estimates suggest that the molecule is likely to be stable. It will also be linear since it is the same p orbital in Xe which is used in both (i) and (ii), and the principle of maximum overlapping will then require that the two F atoms overlap symmetrically with this orbital. The Xe—F bonds in (i) and (ii) will not be strictly homopolar, but a rough description of the overall charge distribution would be $F^{-\delta} Xe^{+2\delta} F^{-\delta}$, where $\delta \simeq \frac{1}{2}$. Nuclear magnetic resonance measurements of the chemical shielding of the F atom when the ^{19}F isotope is used (cf. p. 174) are in agreement with this charge displacement.

The case of XeF_4 is dealt with similarly by noting that if in (i) and (ii) we have used the $5p_x$ orbital for bonding, there is then a lone pair of electrons $(5p_y)^2$ on the Xe atom. We may therefore repeat for the y direction just the same kind of argument as was used for the x direction. The molecule XeF_4 should therefore be square planar.

Similar arguments may also be applied to the oxides. Thus, in XeO_3 we may use the three separate 5p orbitals to give a pyramidal molecule, using VB structures with one $Xe^+{-}O^-$ link and two non-bonded oxygens. Resonance then delocalizes the bond over all three oxygens, but with one non-bonded atom in each structure the bond energy is somewhat lower ($\sim 80 \, \text{kJ mol}^{-1}$). In $XeCl_4$ we may use four sp^3 hybrids to obtain the observed tetrahedral arrangement, and the same considerations then apply.

It is also possible to discuss the rare gas compounds using MO theory.‡ Let us consider again XeF_2 and admit, for simplicity, only the $2p_\sigma$ AO on each fluorine and the xenon 5p AO. If we choose the phases as in Fig. 12.11 (a), so that both β's have the same sign, the secular equations which determine the MO's will be identical with those for the π-type MO's in the allyl radical. On using the corresponding AO coefficients we obtain the MO's shown in Fig. 12.11 (b). The bonding MO is a three-centre orbital giving two strong σ bonds, and there is an NBMO which shares a lone pair between the end atoms. It is interesting that just as three-centre MO's provided an explanation of the

---

† Coulson (1964). Other semi-empirical theories (some using xenon d orbitals) were put forward about the same time and are reviewed by Jortner and Rice (1965).

‡ Typical papers from a large literature are Jortner *et al.* (1963), Loht and Lipscomb (1963), and Coulson (1964).

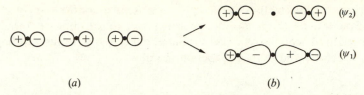

FIG. 12.11. The three-centre bond in xenon fluoride: (*a*) choice of 2p AO's to give positive overlap integrals; (*b*) forms of the bonding and non-bonding MO's.

bonding in electron-deficient compounds such as the boron hydrides (§12.3), so also they provide an acceptable picture of these electron-rich compounds.

We have still to give an explanation of the observations (*a*) and (*b*) at the beginning of this section and this follows most simply from the VB approach. The reason why only the heavier central atoms are found is apparently due to the variation in ionization potential along the series of rare gas atoms. The table below shows that this steadily falls as we move along the series. To set up

Atom	He	Ne	Ar	Kr	Xe	Rn
IP (eV)	24·7	21·6	15·7	14·0	12·1	10·6

the VB structures (i) and (ii) (p. 374) we must take an electron from the rare-gas atom, and if this requires too much energy we shall not be able to win back enough by establishing the electron-pair bond and the ± charge distribution to obtain a stable molecule. We should therefore expect that rare-gas compounds would be formed most easily with the later members of the series. Only a detailed numerical analysis could show that the dividing line between stable and non-stable molecules should occur after argon.

The requirement that the ligands should be electronegative depends on similar considerations, for evidently structures (i) and (ii) are energetically favoured as much by a high electron affinity of the ligand (acceptor) as by a low IP of the donor. In itself, however, this is not enough, as is shown by the much reduced stability of $XeCl_2$ as compared with XeF_2. The other important factor (which is in fact numerically dominant) is the Coulomb attraction between the ± charges. This is large if the separation of the charges is small, and, since this separation is approximately equal to the sum of the two atomic radii, it is clearly advantageous to choose as small a ligand as possible. It is therefore not surprising that many more compounds are formed with F and O as ligands than with Cl.

In retrospect, there is nothing very unexpected about this family of molecules. Although they were not discovered until 1962 the basic theoretical explanation of their existence and their structure could have been given 30 years earlier in terms of qualitative principles which were already well established at that time. More elaborate calculations (see, for example, the *ab initio* SCF calculations by Basch *et al.* (1971) are useful in confirming aspects of the simple picture (e.g. the drift of charge from xenon to halogens, or the lack of any real need for d orbital participation) but add little to our understanding of the basic structures and shapes of these molecules.

12.6. Stereochemical and related interactions

A large class of interactions may be considered under this one heading. Some of them are repulsive and others are attractive, but they normally involve different chemical groups within the same molecule. They determine, for example, the relative orientation of two CH_3 groups and the amount of energy required to twist one relative to the other (and hence the energy barriers for 'hindered rotation'), or the alignment of such a group with respect to a conjugated ring to which it is attached (widely attributed to a mechanism called 'hyperconjugation'), or the stereochemical distortions of molecules in which non-bonded pairs of atoms or chemical groups come closer together than their known 'sizes' would appear to permit (such molecules being referred to as 'overcrowded'). Since the interaction energies involved may be typically as small as 1/20 000 of the total energy it would seem unrealistic to expect that they could be computed with any confidence even by the most elaborate *ab initio* methods. We must therefore once again try to embody the results of calculation in simple pictorial models which can then provide a rational interpretation of structures and properties.

Hindered rotation†

Let us first consider the question of hindered rotation around a single bond, taking for illustration a series of molecules of the type CH_3X. As the CH_3 group is connected to X, whatever it may be, by what we have so far regarded as a single bond of σ type, it would appear that there should be little or no intrinsic resistance to rotation around the bond. In this respect there is a sharp distinction between σ and π bonds for, as we noted in §8.1, the reduced overlap of π-type AO's in a double bond when one group is rotated relative to the other certainly reduces the bonding substantially and thus produces an energy barrier (see Problem 8.2). The overlap of σ-type AO's, however, is not diminished by a similar twisting and we should therefore expect free rotation around a σ bond. In fact, there are frequently barriers of $1–20\,kJ\,mol^{-1}$ opposing such rotation. Table 12.2 lists some experimental barrier heights for CH_3X molecules. Such heights may be obtained with considerable accuracy by microwave spectroscopy; see, for instance, Wilson (1959).

   The explanation of these surprisingly high values presents some difficulty. Let us take the ethane molecule as an example. The barrier is high even though the polarity of the C—H bonds is low and we cannot therefore look for an

† For a good review see Pethrick and Wyn-Jones (1969).

TABLE 12.2
Barrier heights (in kJ mol^{-1}) for internal rotation in molecules CH$_3$X

X	CH$_3$	C$_2$H$_5$	CF$_3$	OH	CHO
Barrier height	11·8	13·9	6·3	4·5	4·8

interpretation purely in terms of electrostatic repulsions. Indeed, even if the six H atoms were completely stripped of electrons, leaving six bare protons, the barrier height for a 60° rotation would only be about 25 kJ mol^{-1}, and a more realistic net charge of $e/10$ on each hydrogen would lead only to one-hundredth of this value—about 2 per cent of the observed value. An interpretation in terms of electron-pair repulsions (cf. §7.8) meets similar difficulties† because here the electron pairs are not grouped around the same atom but are relatively remote.

Early attempts to obtain an explanation from MO or VB calculations were inconclusive, largely owing to the many approximations involved, but during the last 10 years many *ab initio* calculations have been made and it is now possible at least to identify the main factors involved.

First, in spite of the fact that the energy change involved is such a minute fraction of the total, good calculations of Hartree–Fock type have proved extraordinarily successful as a means of predicting barrier heights. In the case of ethane, calculations of varying accuracy and by several different groups have given surprisingly consistent results and barrier heights within 20 per cent of the experimental value. This is true also for many other molecules (Allen 1969) and the conclusion must be that most of the very large terms in the energy (e.g. the energies of inner-shell electrons) are quite unaffected by rotation and therefore cancel accurately (even if they are wrong!) in computing the barrier, which is of course an energy *difference*. Thus the errors of the Hartree–Fock approximation are not of crucial importance in such applications, and the geometry-dependent terms in the energy are apparently estimated with some degree of reliability. This is a reassuring situation and has encouraged even more ambitious calculations. Clementi and Popkie (1973), for example, have made *ab initio* calculations on a sugar–phosphate–sugar fragment of a polynucleotide chain, obtaining barrier heights which they consider accurate to within about 20 per cent. However, a computed barrier height does not aid our understanding and we must therefore look for an interpretation of such calculations. This may suggest simpler semi-empirical models which can be applied more widely and with less cost.

An 'energy-component analysis' is helpful. The total energy may be divided into terms whose physical meaning is clear and whose behaviour can be

†Qualitative discussions along such lines have nevertheless been given; see, for example, Lowe (1974).

followed during any proposed geometry variation. The dominant effects can then be distinguished and interpreted. Two main types of analysis have been proposed. One† depends on dividing the energy into terms which refer to one atom (A), two atoms (AB), three atoms (ABC), etc.; the other (Allen 1968) examines the balance between the repulsive terms in the energy and the attractive terms, and the way in which the balance changes as rotation proceeds. Here we shall use the latter method, which is very simple and does not depend on the type of wavefunction used.

There is only one attractive (i.e. negative) term in the energy, namely V_{en} the energy of the electron charge cloud in the field of the nuclei. The other terms are essentially positive and comprise V_{ee} (the energy of repulsion between different electrons), V_{nn} (a similar term for the nuclei), and T (the kinetic energy of the electrons). The attractive and repulsive components are thus

$$V_{att} = V_{en}, \qquad V_{rep} = T + V_{ee} + V_{nn}.$$

Allen has shown how these components vary as ethane is twisted around the C—C bond, starting from the reference conformation ($\theta = 0$) shown in Fig. 12.12(a). His results appear in Fig. 12.12(b). The energy falls to a minimum at

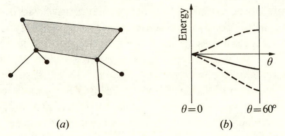

(a)   (b)

FIG. 12.12. Hindered rotation in the ethane molecule: (a) eclipsed conformation ($\theta = 0$); (b) energy variation against angle of twist, showing attractive and repulsive components.

$\theta = 60°$ because the energy of repulsion falls more than the energy of attraction rises; the equilibrium occurs at $\theta = 60°$ (staggered conformation) and the barrier may be described as 'repulsive dominant'. In other words the barrier is dominated by the urge of like charges to get as far apart as possible.

A similar analysis applied to hydroxylamine, starting from the reference conformation shown in Fig. 12.13 (a), indicates two types of barrier. As θ goes from 0 to 53·5° (the angle at which the OH is eclipsed by an NH bond) V_{rep} rises more than V_{att} falls, as shown in Fig. 12.14, and the first barrier is therefore repulsive dominant as in ethane. As the rotation continues, however, V_{att} begins to change more rapidly than V_{rep} and falls to such an extent that a minimum energy results at $\theta = 180°$, the conformation shown in Fig. 12.13 (b).

† Clementi (1967) for the first of a long series of papers—the ethane barrier is discussed in (1971) and (1972).

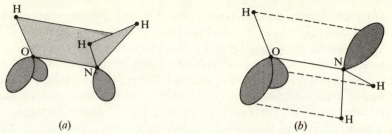

FIG. 12.13. Hindered rotation in hydroxylamine: (*a*) conformation corresponding to $\theta = 0$; (*b*) stable equilibrium conformation.

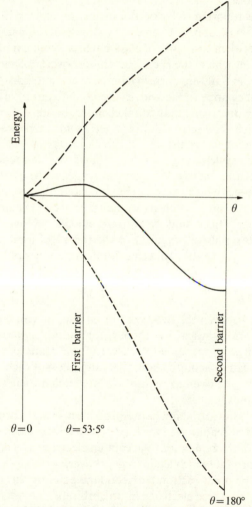

FIG. 12.14. Rotational barriers in hydroxylamine. The first barrier, at $\theta = 53{\cdot}5°$. occurs when the O–H bond is eclipsed by one N–H and is repulsive dominant; the second is attractive dominant.

The first barrier, that between the $\theta = 0$ and $\dot{\theta} = 53 \cdot 5°$ conformations, is thus repulsive dominant as in ethane; the energy maximum occurs when the nitrogen lone pair is eclipsed by an oxygen lone pair. As rotation proceeds, however, and the lone pairs become further apart, the effect of their repulsions diminishes and a quite strong attraction develops; this attraction reaches its maximum when each lone pair at one end of the molecule eclipses a bond pair at the other end. That the second barrier is *attractive* dominant suggests that an interpretation based on electron-pair repulsions alone (§7.8) is over-simplified, except perhaps when the pairs are grouped in close proximity around a single atom.

The calculations reported so far do not of course 'explain' how the barriers arise but they do lend support to two very simple ideas, namely (i) that the repulsive effects between two parts of a system arise from expulsion of charge from the region in which their charge clouds overlap, and (ii) that the attractions arise from build-up of charge in the overlap region. Apart from the context in which they appear, these ideas are no different from those we have used throughout in discussing individual chemical bonds—there are bonding interactions in which the electron charge cloud builds up between the positive nuclei, and 'non-bonded repulsions' in which charge is pushed out of the region between two systems (e.g. rare-gas atoms) with closed-shell electronic structure. A careful analysis (see, for example, Jorgensen and Allen 1971) of the charge density changes which occur during hindered rotation in ethane shows there is a mutual 'repulsion' of density in each pair of C—H pairs as they begin to eclipse each other. This is analogous to the closed-shell overlap interaction which leads to the repulsive potential in the theory of ionic crystals (see p. 351 and Problem 11.8). In the attractive dominant barrier of hydroxylamine, however, charge density in the overlap region of an NH bond and an oxygen lone pair *increases* as eclipse occurs, indicating a bonding interaction (Fig. 12.12 (*b*)).

The distinction between the two types of interaction can be accounted for qualitatively in various ways. VB theory, for example, would admit ionic structures in which an oxygen lone-pair electron was transferred to a nitrogen bond hybrid; it is not difficult to show that admission of such structures puts charge density into the overlap region. More detailed analysis suggests the following conclusions.

(i) There may be an attraction between a lone pair and a bond pair and this is favoured by readiness of the lone pair to lose an electron, readiness of one atom of the bond pair to accept an electron, and diffuseness of the lone pair orbital (which leads to good overlap).

(ii) There will be no attraction between lone pairs by this mechanism (for we cannot put three electrons in one orbital).

(iii) Attraction between tightly bound bond pairs will be insignificant, for electron transfer will be energetically expensive and secondly the overlap between relatively compact orbitals will be too small to give a substantial density change.

In short, for attraction to occur there must be a fairly weakly bound electron pair occupying an orbital which is diffuse enough to provide appreciable delocalization of the bonding over a second pair. Lone pairs, as Fig. 6.25(c) reminds us, are well adapted for this purpose. Of course, in a complicated molecule there may be a delicate balance between a number of attractive and repulsive terms but there is considerable support for a general interpretation along the lines indicated.

Hyperconjugation

The term hyperconjugation was introduced by Mulliken (1939) to describe the interaction between a π-electron system and an attached group such as CH_3

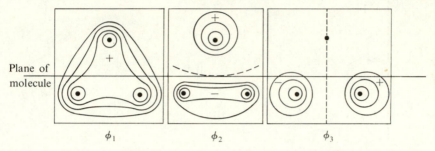

Plane of molecule

ϕ_1 ϕ_2 ϕ_3

FIG. 12.15. Hyperconjugation. Group orbitals in CH_3. Orbital ϕ_2 can conjugate with the π-electron system, resembling a π-type 2p AO normal to the molecular plane.

which, again, would apparently be free to rotate around the σ bond which connects it to the conjugated carbon atom. A large literature exists on this subject (see Dewar (1962) and references therein), but now that semi-empirical calculations on molecules of this type have become routine the original simple model is invoked rather less frequently. The idea is, however, still important enough to deserve a brief description.

Experimentally, quite apart from hindered rotation, a methyl substituent attached to a conjugated molecule often behaves as if it formed part of the region of conjugation; in the case of benzene, for example, charge appears to flow from the CH_3 group into the π-electron system of the ring. As a result, the group has ortho–para directing effects similar to those of other substituents, as discussed elsewhere (§8.7), and may also give rise to appreciable dipole moments in which the methyl group is positive relative to the benzene ring. This is true in general for alternant hydrocarbons, where in the absence of substituents the π charge distribution is uniform and the dipole moment is usually negligible. Other groups such as —$CH_2 . CH_3$, and CH_2 when it forms part of the ring, display similar behaviour though to a lesser degree.

Methylation, by feeding electrons into an alternant hydrocarbon, also reduces
π ionization potentials and excitation energies, and the resultant red shift of
electronic absorption bands is of some importance in the design of dye
molecules. Another effect of this hyperconjugation is a small but measurable
reduction of the C--CH_3 bond length below the value $0.154\,nm$ for a C--C
single bond; although some reduction would be expected because the sp^2
atomic radius is less than for sp^3 (p. 230), it appears that the link is indeed
strengthened by some kind of π bonding.

A simple qualitative interpretation of these effects can again be given in
terms of VB theory by introducing charge-transfer structures. Thus, in the case
of toluene, covalent structures such as (I) should be supplemented by
structures such as (II) and (III), in which an electron of the methyl group has
been given to the ring. If the methyl carbon is

(I)   (II)   (III)

sp^3 hybridized the 'double bond' connecting it to the ring will not be of the
conventional form, the overlap with a $2p_z$ AO being smaller than usual, but the
interpretation of some degree of π bonding and of the ortho–para directing
effects is remarkably simple.

For quantitative purposes the MO approach is more useful. We have in fact
already developed the method, first in the case of H_2O (§7.3) and secondly in
the treatment of transition-metal compounds (Chapter 9); in each case,
symmetry suggested the setting up of 'group orbitals', extending over two or
more ligand atoms, which could then interact with orbitals of similar
symmetry on the central atom. In the case of CH_3 attached to a ring we may set
up similar group orbitals which are *roughly* symmetric or antisymmetric
across the molecular plane, and then consider how they interact with the ring.
The MO's for a CH_3 group in fact have the forms shown in Fig. 12.15; when
they are occupied by six valence electrons (three from the carbon and three
from the hydrogens) they provide a charge distribution which is identical with
that arising from three localized C--H bond orbitals. Now, however, it is clear
that the two electrons in ϕ_2 occupy an orbital which is well adapted to overlap
with the $2p_z$ AO on the ring atom to which the CH_3 group is attached. The
whole group may be viewed as a pseudo-atom with two electrons in a (roughly)
π-type orbital, and toluene may then be treated formally in the same way as

aniline or nitrobenzene (p. 253); there is, in effect, a terminal hetero-atom X, with Coulomb integral α_X and a connecting link with resonance integral β_{CX}, which donates in this case two π electrons to a seven-centre conjugated system. Appropriate values of the parameters may be determined by semi-empirical procedures and a broadly satisfactory account of the effect of hyperconjugated groups can usually be obtained (see, for example, Streitwieser 1961). Certain general conclusions have been put forward by Mulliken (1959) who observed that hyperconjugation had a more marked effect on radicals and ions than on neutral molecules; this is in accord with the fact that the singly occupied highest orbital can readily accept electron density from the doubly occupied pseudo-π orbital of the hyperconjugating group, giving a bonding interaction.

Overcrowded molecules

In many molecules, particularly large organic molecules, one or more pairs of atoms may be forced into the same region of space as a result of their connection to a more or less rigid framework. Such atoms are commonly peripheral hydrogens of a planar conjugated hydrocarbon, an example being the pair shown below in phenanthrene

Since there is a repulsion between such non-bonded atoms, the molecule must experience some degree of distortion in order to accommodate this over-crowding. Since accurate quantum-mechanical calculation of the electronic energy as a function of all the atomic co-ordinates is not feasible,† the only realistic way of proceeding is to use an empirical model. The molecule is regarded as a mechanical framework, characterized by 'normal' bond lengths and angles and by force constants which determine the nuclear displacements produced by applied forces—in this case by the repulsion between the overcrowded atoms. The resultant molecular conformation will be the one in which the total energy, a function of the nuclear displacements, reaches its minimum value.

References to early work in this field may be found in a paper by Coulson and Haigh (1963) which gives an elegant treatment of planar molecules. Briefly, the energy of the molecule relative to a hypothetical system in which

† To investigate the variation of energy with displacement out of the plane, allowing two displacements up and two down for each atom, would require of the order 5^{16} complete energy calculations!

deformation is prevented and the atoms are in their 'normal' positions may be written

$$E = U(R) + V_{xy} + V_z$$

where $U(R)$ is the repulsion energy for the overcrowded pair at separation R, V_{xy} is the distortion energy for in-plane displacements, and V_z is that for out-of-plane displacements. A common form of repulsive potential (cf. p. 350) is

$$U(R) = -AR^{-6} + B\exp(-CR)$$

while V_{xy} and V_z are quadratic in the displacement co-ordinates with coefficients which are in principle obtainable empirically from vibrational spectroscopy. Since the distance between the overcrowded atoms depends on how the framework is distorted, R is also a function of the displacement co-ordinates (i.e. there is an internal relationship or 'constraint'). To obtain the equilibrium conformation it is then only necessary to find the minimum of E, subject to the constraint; this involves differentiating and setting the result equal to zero for each independent displacement and leads to a set of simultaneous equations.

It is remarkable that, with suitably chosen parameters in the repulsive potential $U(R)$, this primitive mechanical model can give molecular conformations in quite close agreement with those obtained experimentally by crystallographic methods. There have been many recent developments in this force-field approach to molecular conformations but as they are marginal to the theory of valence we shall not discuss them further (reference may be made to the reviews by Williams *et al.* (1968) and by Allinger (1976)). It is worth noting, however, that the same concept of 'natural' valence angles and bond lengths, from which deviations are caused by applied forces, is equally successful as a basis for the prediction of geometry changes when carbon atoms are replaced by heteroatoms such as nitrogen (Flory 1969). Again, it is assumed that there is a natural C—N—C valence angle ϕ_N and corresponding bending force constant k_N, and that empirical fitting of predicted and observed geometries is achieved when ϕ_N takes the eminently reasonable value of 109°.

Some of the most ambitious applications of this general type of approach, using empirical interaction potentials, have been to predictions of the statistical distribution of conformations of long-chain molecules such as polypeptides.

13
Self-consistent field theory

13.1. The need for improved theories

Simple MO theory, essentially of Hückel form, in which Coulomb and bond integrals are treated as disposable parameters and are fixed by appeal to experiment, has served us well in the last few chapters. We have been able to obtain, at this primitive level, a surprisingly satisfactory understanding of the structure and properties of both molecules and solids and even some insight into various types of chemical reaction, but we have also become aware of defects in the simple theory. Although we have been able to infer the general forms of the MO's we have had to treat all effects connected with *electron interaction* separately and in a much more qualitative way; the part played by electron–electron repulsions in determining the shapes of polyatomic molecules, for example, was discussed in purely pictorial terms (§7.8) *after* assigning the electrons to suitably localized MO's. The MO's themselves have always been determined as if all electrons moved *independently* in a common field. We know from Hartree's work on atoms (§2.5) that this independent-particle model can give excellent results provided the field acting on each electron is properly calculated and all electron interactions are included in calculating the energy, but in our semi-empirical discussions we have made no attempt to do so. The 'Hartree field' appears in the Hamiltonian for each electron and in semi-empirical theory, where the matrix elements of this Hamiltonian have been regarded as parameters, we have never needed to consider its explicit form.

Apart from its inability to provide a quantitatively satisfactory route to the total electronic energy, the simple MO theory also fails qualitatively in some situations. In §9.13 for example, we noted that a totally unrealistic charge distribution could easily result unless parameter values were chosen with extreme care. This was especially the case for transition-metal complexes because, with many neighbours around a central metal atom, even small charge shifts within each bond could lead to an excessive pile-up of charge on the metal, but it is generally true that when significant non-uniformities of electron density occur (e.g. in heterocyclic organic molecules) the theory becomes less satisfactory. The root of the difficulty, as we have remarked already, is that the parameters such as Coulomb and bond integrals are *not* constants which, once fixed, can be transferred unchanged from molecule to molecule; they depend on the prevailing charge distribution and should be adjusted until their values are consistent with the distribution to which they lead (via the secular equations). We have seen that simple methods of relating the parameter values (Coulomb integrals) to atomic valence shell populations as in (9.3) give a reasonable qualitative recognition of these ideas, but the only sure way of proceeding is to use some form of SCF theory in the Hartree–Fock sense.

Some results of *ab initio* SCF calculations have been presented in Chapter 6, but the implementation of SCF methods at the *ab initio* level is still unmanageable for many of the large molecules which are important in chemistry, and consequently a large number of approximate or semi-empirical variants have been developed. These go beyond Hückel-type MO theory in so far as they are based on a full quantum-mechanical treatment of the many-electron system, and therefore explicitly include electron interactions, but fall short of truly *ab initio* methods because they neglect large numbers of small terms and frequently estimate those which remain by appeal to experimental data; in other words they use the general mathematical framework of the *ab initio* calculations but make as many approximations as possible in order to achieve maximum simplicity. A detailed discussion of such developments is outside the scope of this book, but they are by now so well established that no introduction to valence theory would be complete without some explanation of them. We shall therefore indicate just one main line of development—the SCF theory for a system in a closed-shell ground state, which provides the underlying conceptual basis for the MO method as we have used it so far.

13.2. Ab initio SCF theory

The starting point for closed-shell SCF theory is the variational expression for the energy E associated with an antisymmetric wavefunction Ψ, formed (see p. 135) from spin-orbitals $\psi_1\alpha, \psi_1\beta, \psi_2\alpha, \ldots, \psi_n\beta$ (all orbitals doubly occupied). Provided the MO's are orthonormal the corresponding energy is[†]

$$E = 2\sum_K \langle \psi_K|\hat{h}|\psi_K\rangle + 2\sum_{K,L} [\langle \psi_K\psi_L|g|\psi_K\psi_L\rangle - \tfrac{1}{2}\langle \psi_K\psi_L|g|\psi_L\psi_K\rangle] \quad (13.1)$$

where in the summations K, L run over the (doubly) occupied MO's ($K, L = 1, 2, \ldots, n$) and we have used the convenient abbreviations

$$\langle \psi_K|\hat{h}|\psi_K\rangle = \int \psi_K^* h \psi_K \, d\tau \quad (13.2)$$

$$\langle \psi_K\psi_L|g|\psi_K\psi_L\rangle = \frac{e^2}{\kappa_0} \int \psi_K^*(1)\psi_L^*(2)\frac{1}{r_{12}}\psi_K(1)\psi_L(2) \, d\tau_1 \, d\tau_2 \quad (13.3)$$

and $\langle \psi_K\psi_L|g|\psi_L\psi_K\rangle$ is just like (13.3) except that the order of the K and L subscripts is reversed in the right-hand part of the integrand. The first integral

[†] This expression was first given by Slater (1929). For a sketch of the derivation and of its extension to non-orthogonal orbitals see McWeeny and Sutcliffe, Section 3.3 (1976).

is the expectation energy of an electron occupying orbital ψ_K and described by the one-electron Hamiltonian operator $\hat{h}$. The second integral is the electrostatic repulsion energy between 'electron 1' in ψ_K (charge cloud $|\psi_K|^2$) and 'electron 2' in ψ_L (charge cloud $|\psi_L|^2$), while the 'exchange integral' $\langle \psi_K \psi_L | g | \psi_L \psi_K \rangle$ is the electrostatic repulsion between two superimposed charge clouds of density $\psi_K \psi_L$, assuming the MO's real, otherwise the 'densities' are $\psi_K^* \psi_L$ and $\psi_K \psi_L^*$.

What we must now do, according to §3.6, is vary the forms of all the MO's until E reaches its minimum value; we shall then have the best wavefunction of one-determinant form. As it is not convenient to use purely numerical methods to vary the functions $\psi_1, \ldots, \psi_n$, we use the LCAO type of approximation, writing

$$\psi_K = \sum_r c_r^{(K)} \phi_r \qquad (13.4)$$

to express each MO in terms of a chosen basis of AO's $\phi_1, \phi_2, \ldots, \phi_m$. Instead of freely varying the *function* ψ_K we can then only vary the m numerical coefficients $c_r^{(K)}$ ($r = 1, 2, \ldots, m$), which, although more restrictive, is vastly simpler mathematically. On putting (13.4) in (13.1) and introducing the usual matrix whose elements P_{rs} (defined in (8.5)) correspond formally to charges and bond orders, we find after some rearrangement

$$E = \sum_{r,s} P_{sr} h_{rs} + \tfrac{1}{2} \sum_{r,s} P_{sr} G_{rs} \qquad (13.5)$$

where (cf. (13.2))

$$h_{rs} = \langle \phi_r | \hat{h} | \phi_s \rangle = \int \phi_r^* \hat{h} \phi_s \, d\tau \qquad (13.6)$$

is a matrix element of the operator $\hat{h}$ between the *atomic* orbitals, while G_{rs} arises from the two-electron integrals (13.3) and is defined by

$$G_{rs} = \sum_{t,u} P_{ut} [\langle \phi_r \phi_t | g | \phi_s \phi_u \rangle - \tfrac{1}{2} \langle \phi_r \phi_t | g | \phi_u \phi_s \rangle] . \qquad (13.7)$$

The two-electron integrals in this expression are analogous to (13.3) but again refer to the *atomic* orbitals chosen as the basis of the calculation; they are

$$\langle \phi_r \phi_t | g | \phi_s \phi_u \rangle = \frac{e^2}{\kappa_0} \int \phi_r^*(1) \phi_t^*(2) \frac{1}{r_{12}} \phi_s(1) \phi_u(2) \, d\tau_1 \, d\tau_2 \qquad (13.8)$$

and have a similar interpretation as electrostatic repulsion energies between little bits of the molecular charge cloud, $\phi_r \phi_s$ and $\phi_t \phi_u$ (assuming real functions), just as in (11.14). Such quantities are easy to visualize but exceedingly difficult to compute. Once we have chosen a basis of AO's, all the integrals (13.6) and (13.8) are in principle fixed and must be calculated before we can proceed. The number of distinct integrals which appear, when there are m AO's in the basis, is of the order $m^4/8$. So with, say, 20 AO's we should need to start by calculating about 20 000 very difficult integrals. The labour of integral evaluation was one of the 'bottlenecks' to progress before the advent of large computers.

The important thing about the 'electron interaction matrix', whose elements G_{rs} appear in the energy expression (13.5) which we are to minimize, is that it depends on the charges and bond orders (P_{rs}) which can only be calculated once we know the solution (i.e. the forms of the MO's in (13.4))! Let us ignore this difficulty for the moment. We must minimize E, given by (13.5), subject to the condition (assumed in Slater's derivation of the expression) that the MO's are varied only in such a way as to stay normalized and orthogonal to each other. This is simply a mathematical exercise; the AO coefficients in the MO's which minimize E under these conditions must, it turns out, then satisfy the equations

$$(h^{F}_{11} - \varepsilon)c_1 + (h^{F}_{12} - \varepsilon S_{12})c_2 + \ldots = 0$$
$$(h^{F}_{21} - S_{21}\varepsilon)c_1 + (h^{F}_{22} - \varepsilon)c_2 + \ldots = 0 \qquad (13.9)$$
$$\ldots\ldots\ldots\ldots\ldots\ldots\ldots\ldots\ldots\ldots\ldots\ldots\ldots\ldots$$

These are ordinary secular equations exactly like those used in simple MO theory (with overlap included), except that the matrix elements h^{F}_{rs} no longer need be regarded as parameters (Coulomb and bond integrals); they are given by

$$h^{F}_{rs} = h_{rs} + G_{rs}. \qquad (13.10)$$

The matrix elements of the 'Fock Hamiltonian', which describes an electron moving in the effective field provided by the nuclei *and all other electrons*, thus comprise two terms: the 'bare nuclear' term (h_{rs}) and the electron interaction term (G_{rs}).

The secular equations (13.9) are essentially the Hartree–Fock equations in LCAO form (see Roothaan 1951; Hall 1951). They must be solved by a self-consistent field (SCF) technique, akin to that invented by Hartree (§2.5), because h^{F}_{rs} requires a knowledge of G_{rs} and hence of the required MO's (which determine the charges and bond orders P_{rs}). The procedure is to 'guess' a set of MO's to start with (e.g. to take a set of Hückel-type MO's, calculated as in earlier sections); to calculate charges and bond orders using (8.5) and the electron interaction terms using (13.7); and then to solve the secular equations (13.9). The sets of coefficients defining the occupied (lowest-energy) MO's will in general differ somewhat from those used in the starting approximation; they enable us to calculate 'improved' electron interaction terms G_{rs} and thus to repeat the calculation of the MO's. When the MO's which 'come out' agree with those which went into the setting up of the secular equations, we say self-consistency has been achieved.

The *ab initio* calculations leading to the results discussed in Chapter 6 and elsewhere were mostly performed by the SCF method described above or by more sophisticated versions of it.

13.3. Semi-empirical SCF theory

It is clear from the last section that limits to the size of molecule on which we can make accurate calculations will be set mainly by the computation and processing of extremely large numbers of two-electron integrals, and that such calculations will be time consuming and costly even with the most powerful computers available.

In semi-empirical SCF theory, on the other hand, efforts are made to eliminate the need for calculations of vast numbers of integrals, only those of greatest numerical value being retained and even then frequently fixed by appeal to experimental data. It is possible to give some kind of theoretical rationale for such procedures. We have seen in §8.8 how the overlap integrals S_{rs} may be eliminated by orthogonalizing the AO's, and it is also possible to orthogonalize in such a way that the AO electron-repulsion integrals (13.8) take very small values whenever either of the charge densities, $\phi_r\phi_s$ and $\phi_t\phi_u$, refers to a pair of *different* AO's, i.e. when $r \neq s$ or $t \neq u$. The remaining integrals

$$\gamma_{rt} = \frac{e^2}{\kappa_0} \int \phi_r^2(1)\frac{1}{r_{12}}\phi_t^2(2)\,d\tau_1\,d\tau_2 \tag{13.11}$$

represent the Coulomb repulsion between AO-type charge clouds ϕ_r^2 and ϕ_t^2 and may be roughly estimated by classical electrostatics. This is commonly called a 'zero differential overlap' (ZDO) approximation, a term introduced by Parr, because it is equivalent to putting $\phi_r\phi_s = 0$ at all points in space, which is much more severe than putting $S_{rs} = \int\phi_r\phi_s\,d\tau = 0$; this means that all two-electron integrals containing an overlap density are neglected. The result is a dramatic simplification of the electron interaction terms in (13.7), for we now obtain, on omitting all terms in the summations for which $r \neq s$ and/or $t \neq u$,

$$G_{rr} = \tfrac{1}{2}P_{rr}\gamma_{rr} + \sum_s P_{ss}\gamma_{rs} \qquad \text{('diagonal' terms)} \tag{13.12a}$$
$$G_{rs} = -\tfrac{1}{2}P_{rs}\gamma_{rs} \qquad \text{('off-diagonal' terms, } r \neq s) \tag{13.12b}$$

in which (with m real AO's) only $\tfrac{1}{2}m\,(m+1)$ distinct γ integrals appear, i.e. for $m = 20$ about 200 integrals instead of 20 000.

Other obviously desirable simplifications are (i) to concentrate on the valence electrons alone (assuming the atomic 'cores' merely provide an effective field in which they move), and (ii) to approximate the bare nuclear integrals h_{rs} in terms of a suitable set of atom and bond parameters. The way in which the parameters are introduced depends somewhat on the type of system considered; we devote the next two sections to two important types of application.

13.4. π-Electron calculations

If we consider the π electrons alone, moving over a σ-bonded framework, we may assume the one-electron Hamiltonian $\hat{h}$ to have the form (using atomic units)

$$\hat{h} = -\tfrac{1}{2}\nabla^2 + (V_1 + V_2 + \ldots) \tag{13.13}$$

where the potential energy function V has been represented as a sum of parts referring to the constituent atoms of the framework; thus V_r is the potential energy of a π electron in the field provided by that bit of the σ-bonded framework which we imagine 'belongs' to atom r, or more precisely to the framework *ion* r without its π electrons. From (13.6) we obtain

$$\langle \phi_r | \hat{h} | \phi_r \rangle = W_r + \sum_{s(\neq r)} \langle \phi_r | V_s | \phi_r \rangle \tag{13.14}$$

where

$$W_r = \langle \phi_r | -\tfrac{1}{2}\nabla^2 + V_r | \phi_r \rangle \tag{13.15}$$

is essentially the orbital energy of a π electron in ϕ_r, in the presence of the σ-bonded framework atom r, with the rest of the framework removed. Thus $-W_r$ is often described as a 'valence state ionization potential', but this must not be taken too literally because V_r arises from part of a *bonded framework* (not a free atom) and W_r is not an observable quantity. What is important, however, is that W_r is characteristic of framework atom r, whether it appears in one conjugated molecule or another, and that the value of $-W_r$ is *roughly* that of an ionization potential.

The remaining terms in (13.14) may be estimated by noting that if atom s supplies Z_s π electrons, then

$$V_s + V(Z_s, \phi_s) = V_s^{\text{neut}} \tag{13.16}$$

where $V(Z_s, \phi_s)$ is the potential term for an electron in the field of Z_s π electrons in orbital ϕ_s, and V_s^{neut} is that for an electron in the field of a *neutral* framework atom s (i.e. the framework ion with its normal complement of π electrons restored). On making use of (13.16) and neglecting the field at ϕ_r due to the neutral atom s, we obtain from (13.14)

$$\alpha_r = \langle \phi_r | \hat{h} | \phi_r \rangle \simeq W_r - \sum_{s(\neq r)} Z_s \gamma_{rs} \tag{13.17a}$$

since γ_{rs}, defined in (13.11), is the energy of an electron in ϕ_r in the field due to one in ϕ_s. We use α_r for the Coulomb integral, noting that it refers to an electron in the field of the bare framework. The corresponding bond integral is

$$\beta_{rs} = \langle \phi_r | \hat{h} | \phi_s \rangle \simeq \langle \phi_r | -\tfrac{1}{2}\nabla^2 + V_r + V_s | \phi_s \rangle \tag{13.17b}$$

where the terms from V_t ($t \neq r, s$) can reasonably be neglected when ϕ_r and ϕ_s

are localized over the region r—s and are orthogonal; β_{rs} is adopted as a parameter characteristic of link r—s.

On adding to (13.17a) and (13.17b) the electron interaction terms (13.12a) and (13.12b), according to (13.10), the diagonal and off-diagonal matrix elements of the Fock Hamiltonian are found to be

$$\alpha_{rr}^{F} = W_r + \tfrac{1}{2}P_{rr}\gamma_{rr} + \sum_{s(\neq r)} (P_{ss} - Z_s)\gamma_{rs} \tag{13.18a}$$

$$\beta_{rs}^{F} = \beta_{rs} - \tfrac{1}{2}P_{rs}\gamma_{rs} \tag{13.18b}$$

These equations (Pople 1953) have formed the basis of most semi-empirical work in π-electron theory (beyond the Hückel approximation) during the last 20 years, though many minor variations have been proposed.† It is gratifying to find that the SCF treatment of conjugated systems runs closely parallel to that based on Hückel's approximations and that, although there are numerical differences, many of the concepts introduced in the simple theory are not substantially altered when electron interaction is admitted.

13.5. All-valence-electron calculations

In a saturated molecule, or a conjugated system with inclusion of the framework, or a general polyatomic species such as a transition-metal complex, it is usually necessary to invoke several valence AO's on each atom and the situation is therefore somewhat more complicated. Approximations which were good for π-electron systems may now be less good, but reductions similar to those made in §13.4 may still be applied and yield a variety of semi-empirical theories (CNDO, NDDO, INDO, etc.) which have been widely used in recent years.

It is convenient to use the notation of Fig. 13.1 in which ϕ_a, $\phi_{a'}$, $\phi_{a''}$ ... are the valence AO's on atom A, ϕ_b, $\phi_{b'}$, $\phi_{b''}$, ... those on atom B etc.; when we do not wish to specify which atom an orbital belongs to we shall use ϕ_r, ϕ_s,... as before. We shall consider only the simplest theory (Pople and Segal 1965), that of 'complete neglect of differential overlap' (CNDO), with approximations similar to those used by its authors. The ZDO approximation is assumed valid whether two orbitals, r and s, belong to the same or to different atoms; thus we may write (cf. (13.12a)

$$G_{aa} = \tfrac{1}{2}P_{aa}\gamma_{aa} + \sum_{B}\sum_{b} \gamma_{ab}P_{bb} \tag{13.19}$$

for $r = s = a$, an orbital on atom A. The summations are over all (valence) orbitals (b) on atom B, and over all atoms $B(\neq A)$.

In theories of CNDO type it is usual to give each electron repulsion integral (γ_{aa}, γ_{ab}) an 'average value' appropriate to the atom or pair of atoms concerned. If these average values are denoted by γ_{AA} and γ_{AB}, the summation over all orbitals on b introduces $\Sigma_b P_{bb} = P_B$, the total electron population of the

† For further discussion of semi-empirical theory see Streitwieser (1961) and Parr (1963).

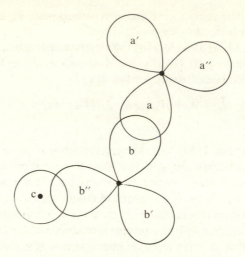

FIG. 13.1. Labelling of valence orbitals.

valence AO's on B. The three types of term (G_{rs}) then become

$$G_{aa} = \tfrac{1}{2}P_{aa}\gamma_{AA} + \sum_{a'(\neq a)} P_{a'a'}\gamma_{AA} + \sum_B P_B\gamma_{AB} \qquad (13.20a)$$

$$G_{aa'} = -\tfrac{1}{2}P_{aa'}\gamma_{AA} \qquad (a' \neq a) \qquad (13.20b)$$

$$G_{ab} = -\tfrac{1}{2}P_{ab}\gamma_{AB} \qquad (13.20c)$$

where the final sum in G_{aa} represents the Coulomb repulsion energy between an electron in ϕ_a and the valence electrons on *other* atoms (P_B electrons on atom B), the earlier terms being *intra*-atomic.

Further approximations are introduced in dealing with the quantities h_{rs}. In particular, it is often assumed explicitly that $h_{rs} = k_{rs}S_{rs}(r \neq s)$, where the proportionality constant depends only on the nature of the atoms to which ϕ_r and ϕ_s belong. When, on the other hand, $r = s = a$ (an orbital of A), the reduction is parallel to that which led to (13.17a). To summarize,†

$$h_{aa} = \alpha_a = U_{aa} - \sum_B \gamma_{AB}Z_B \qquad (13.21a)$$

$$h_{aa'} = \beta_{aa'} \simeq 0 \qquad (13.21b)$$

$$h_{ab} = \beta_{ab} = \beta^0_{AB}S_{ab} \qquad (13.21c)$$

Here, although (13.21a) looks alike (13.17a), it should be noted that U_{aa} represents the energy of an electron in ϕ_a in the field of the *core* of atom A alone (stripped of *all* valence electrons) and Z_B is the *total* number of valence electrons provided by atom B. The constant β^0_{AB} determines the bond integral

---

† The approximation $\beta_{aa'} \simeq 0$ is good only for orbitals with small overlap (e.g. hybrids), but in other cases (e.g. for s and p AO's on the same atom) may easily be relaxed.

for a given pair of atoms A, B, and a dependence on electronegativities, along the lines of (9.1), may be incorporated if desired.

On putting (13.20) and (13.21) together and noting that

$$P_{aa} + \sum_{a'(\neq a)} P_{a'a'} = P_A$$

we obtain the diagonal and off-diagonal elements of the Fock Hamiltonian in the form (cf. (13.18))

$$\alpha_{aa}^{F} = U_{aa} + (P_A - \tfrac{1}{2}P_{aa}) \; \gamma_{AA} + \sum_{B} (P_B - Z_B)\gamma_{AB} \qquad (13.22a)$$

$$\beta_{aa'}^{F} = -\tfrac{1}{2}P_{aa'}\gamma_{AA} \qquad (13.22b)$$

$$\beta_{ab}^{F} = \beta_{AB}^{0} S_{ab} - \tfrac{1}{2}P_{ab}\gamma_{AB}. \qquad (13.22c)$$

These equations summarize the so-called CNDO/2 approximation, which provides one of the most commonly used semi-empirical schemes for performing approximate SCF calculations.

It should be noted that the nature of the valence orbitals has been left unspecified. This is one of the big advantages of semi-empirical procedures, for instead of having to define explicit orbital forms and to calculate accurately corresponding matrix elements and repulsion integrals, we merely have to fix the numerical values of a few parameters (such as U_{aa}, β_{AB}^{0}) in order to get agreement with experiment for a few simple molecules. On the other hand, if we wish to discuss the actual electron distribution, or any properties that depend on it, we must clearly know what the orbitals really look like. The *interpretation* of the various semi-empirical methods therefore raises new and complicated issues. Should we suppose the AO's on each atom to be ordinary s, p, d orbitals? And if so how should we choose their orientation? Or should we imagine them to be hybrids, chosen by some kind of maximum overlap criterion? Or, since we have assumed them to be orthogonal in setting up our simplified equations, should we take then to be *orthogonalized* AO's of the type used in §8.8.

It is now widely recognized[†] that internal consistency of such theories demands the interpretation of the basic ϕ's not as free-atom orbitals but as linear combinations of AO's chosen so that each is as nearly 'atomic' as possible, consistent with orthogonality requirements. The question of whether the orbitals should be regarded as s, p, d, ... type or as hybrids raises other important issues, and of course the choice will be reflected in the way we set up our MO's and estimate suitable values of the parameters. Usually, in CNDO-type theories, the orbitals are assumed to be of s, p, d form and overlap integrals are calculated accordingly from standard formulae. The repulsion integrals γ_{AB} in (13.22) are usually estimated from the (Mataga-Nishimoto)

[†] For the early work, on π-electron systems, see McWeeny (1955, 1956), Parr (1960), Fischer-Hjalmars (1965); and, more generally, McWeeny and Ohno (1960), Klessinger and McWeeny (1965), Pople and Segal (1965), Cook *et al.* (1967).

formula

$$\gamma_{_{AB}} = (e^2/4\pi\varepsilon_0)(R_{_{AB}}+a_{_{AB}})^{-1} \tag{13.23a}$$

where $R_{_{AB}}$ is the separation of atoms A and B and

$$a_{_{AB}} = 2(e^2/4\pi\varepsilon_0)(\gamma_{_{AA}}+\gamma_{_{BB}})^{-1}, \tag{13.23b}$$

while $\beta^0_{_{AB}}$ is also interpolated in the form

$$\beta^0_{_{AB}} = \tfrac{1}{2}(\beta^0_{_A}+\beta^0_{_B}). \tag{13.24}$$

Suitable values of the basic parameters for a number of first-row atoms are given in Table 13.1; these provide a bisis for the theoretical study of a large number of molecules.

TABLE 13.1

Some parameter values for CNDO *calculations†*

Atom (A)	H	Li	Be	B	C	N	O	F
U_{aa} (s)	$-13\cdot60$†	$-5\cdot00$	$-15\cdot54$	$-30\cdot37$	$-50\cdot69$	$-70\cdot09$	$-101\cdot31$	$-129\cdot54$
U_{aa} (p)		$-3\cdot67$	$-12\cdot28$	$-24\cdot70$	$-41\cdot53$	$-57\cdot85$	$-84\cdot28$	$-108\cdot93$
β^0_A	$-9\cdot0$	$-9\cdot0$	$-13\cdot0$	$-17\cdot0$	$-21\cdot0$	$-25\cdot0$	$-31\cdot0$	$-39\cdot0$
γ_{AA}	$12\cdot85$	$3\cdot46$	$5\cdot95$	$8\cdot05$	$10\cdot33$	$11\cdot31$	$13\cdot91$	$15\cdot23$

† All table entries in eV. The values are those recommended by J. M. Sichel and M. A. Whitehead, *Theoret. Chim. Acta* 7, 32(1967), except for the β^0_A values which are due to J. A. Pople and G. A. Segal, *J.chem. Phys.* 43, S136 (1965).

The values of all the integrals (e.g. (13.6) and (13.8)) in a *complete* SCF calculation are of course dependent on the choice of orientation of the AO's on the various atoms, but the values relating to different choices of basis are related by 'transformation equations' which ensure that the *results* of the SCF calculation are independent of this choice, or *invariant*. An interesting feature of the CNDO approximations is that if we assume all integrals vanish except those of type $\langle \phi_a\phi_b|g|\phi_a\phi_b\rangle$ and then use the transformation equations to calculate the two-electron integrals for orbitals ϕ_a, ϕ_b, ... with a different orientation, then we again find all integrals vanish except those of type $\langle \bar\phi_a\bar\phi_b|g|\bar\phi_a\bar\phi_b\rangle$. In other words the CNDO *approximations* are invariant to choice of orientation and we need not worry too much about our choice of the orientation of the AO's. This invariance, however, although apparently giving a certain kind of internal consistency to the approximations, is *unnecessary* and is bought at the expense of requiring (p. 391) that $\gamma_{aa'}$ and γ_{ab} be given *average values* (γ_{AA}, γ_{AB}) irrespective of the orientation of the orbitals concerned—an approximation which is demonstrably poor. The alternative procedure, of choosing the ϕ's to be (orthogonalized) hybrids, also has considerable attractions; it enables us to retain close contact throughout with elementary concepts (e.g. the MO for a localized bond can be well represented as a combination of just two such hybrids) and, because the hybrids tend to be localized in different regions of space (cf. Fig. 7.16), the approximate validity of the zero differential overlap approximation is enhanced. In calculations (see, for example, Cook, Hollis, and McWeeny 1967) based explicitly on the use of hybrids, the parameter values computed *non-empirically* have been found to reproduce quite closely those based on the fitting of experimental data; this lends general support to the CNDO type of formulation and suggests a more rigorous basis for the interpretation of results. Much work remains to be done in this interesting area.

13.6. Comments on applications and future prospects

The development of semi-empirical methods has been a major preoccupation of quantum chemists for almost half a century. Alongside the fully *ab initio* calculations, which have achieved heights unimaginable even 20 years ago, there is now a jungle of more approximate variants. A vast literature has grown up even around the different procedures for choosing parameter values to reproduce, for example, spectroscopic data or calorimetric data. It is not the purpose of this book to lead the way through the jungle (there are other admirable books designed specifically for this purpose, for example Murrell and Harget (1972), Pople and Beveridge (1970)), but to anyone who has read so far the undergrowth should be neither fearsome nor unfamiliar. Most current methods of calculation are built up from some kind of SCF theory, to determine a basic set of MO's. This is followed, if necessary, by CI (along the lines indicated in §5.4) to overcome the defects of a one-configuration MO wavefunction and to obtain excited states and spectroscopic information. Moreover, these methods have been embodied in computer programmes which are now freely available.

It remains only to comment on the value of such calculations, for so far we appear to have had little need for them. For rather small molecules it is possible, at great computational expense, to calculate by direct approximate solution of the Schrödinger equation a few of the quantities we should like to know—perhaps a dissociation energy, a bond length, or a bond angle. We have made little use of such calculations, and as a means of obtaining such information they are, at best, merely competitive with experimental techniques. They have been more valuable to us as a means of introducing concepts and of confirming the superiority of one simplified theoretical model over another. Our simple ideas about orbital forms, built up in Chapter 4 and elsewhere from pictorial superposition of AO's, are not only confirmed but are greatly enriched by accurately computed contour diagrams such as those in Figs. 6.1 and 6.25, and the generality of the idea of hybridization emerged very clearly in §6.2 from an analysis of computed MO's. The history of the subject also contains, however, just as many theories and concepts which have been discarded; they have been tested, when the necessary computations became feasible, and found to be untenable. One very important function of more sophisticated and rigorous theory is thus to help us to discriminate between 'good' and 'bad' theories of a more primitive nature in order to distinguish a 'physically sensible' model from one which will lead to inconsistencies and contradictions.

To illustrate these points it is worth turning to a few examples, first in carbon chemistry, then in transition-metal chemistry, and finally in solid state theory.

A good review of early work in the first area has been given by Parr (1963), particularly that in the field of conjugated molecules. Hückel theory, as used in earlier chapters of the present book, was very successfully applied to molecules such as alternant hydrocarbons where the charge distribution turned out to be uniform ($q_r = 1$ at every conjugated centre) and the initial premise, $\alpha_r = \alpha$ at every centre, was therefore self-consistent. Discrepancies appeared when this was not the case and these could not be resolved until SCF theory was developed along the lines described in §13.4. The resultant Pariser–Parr–Pople theory enjoyed a considerable period of success, but was still a π-electron theory. It was already known, however, that the π electrons did not move 'outside' a σ-bonded framework, for simple calculations had shown the π-type MO's to be more or less embedded in the σ-electron distribution; any degree of independence between π and σ electrons could therefore only be due to orthogonality between their MO's and not to any kind of physical separation. It was therefore not surprising that all-electron *ab initio* calculations performed towards the mid-1960s by Clementi and others[†] showed a remarkable interdependence of the σ and π electrons. To take pyrrole (§8.7) as an example, calculation shows that although the two highest occupied MO's are indeed of π type, the next three (going down) are of σ type and only then do we come to the lowest occupied π MO. There is a similar merging of the charge clouds. Finally, although the more electronegative nitrogen atom does succeed in gaining electrons it does so by a two-way mechanism; the population values show a gain of 0.75 σ electrons but a *loss* of 0.34 π electrons. The nitrogen is thus a σ acceptor but a π donor and the extent of the two shifts can only be reliably determined by an *all*-valence-electron SCF procedure. For such reasons present-day semi-empirical calculations are usually based on theories of the type outlined in §13.5. The simpler calculations may quite adequately recognize the physical mechanism involved and may even reproduce quite well the results of *ab initio* calculation.[‡] When this has been established the semi-empirical method may be used more widely with some confidence. Since both approaches are approximate it would be foolish, and extravagant, to use an *ab initio* method on every occasion.

In the field of transition-metal complexes, the need for all-valence-electron SCF calculations—either *ab initio* or semi-empirical—is rather more acute, as

[†] See for example Schaefer (1972) and for computational progress Clementi (1972) and Steiner (1976).

[‡] See, for example, the calculation on formaldehyde by Cook *et al.* (1967) where similar donation and back donation occurs at the oxygen atom. A systematic discussion of the coupling between the σ and π distributions has been given by McWeeny (1970).

we saw in §9.13. Even now, good *ab initio* SCF calculations on such systems are at the limits of computational feasibility, but the rapid expansion of the field is clear from recent reviews.† Such calculations in the main confirm the qualitative principles adopted in Chapter 9; they can give a fairly satisfactory account of features such as d-orbital contraction (p. 209), electron populations (and hence oxidation states') of the metal ion, and the sequence of orbital energies, as revealed experimentally by photoelectron spectroscopy. Semi-empirical models are now in widespread use in inorganic chemistry and have been well reviewed by Dahl and Ballhausen (1968).

Ab initio calculations of the band structure and Brillouin zones for crystals present even greater difficulty, largely owing to the vast numbers of one- and two-electron integrals to be calculated. However, there is no intrinsic difficulty in formulating the SCF problem for a periodic system described in terms of Bloch orbitals. A graphite layer, for example, may be regarded as a gigantic polycyclic hydrocarbon and may consequently be treated by the theoretical methods presented in §13.4. In an early calculation of the graphite π band, Peacock and McWeeny (1959) formulated the SCF method in the context of solid state theory; they used the Pariser–Parr–Pople approximations (§13.3) with the same parameter values as for aromatic hydrocarbons, making a self-consistent calculation of the Brillouin zone structure and obtaining a computed work function in good agreement with experiment. This is a convincing demonstration of the general applicability of LCAO SCF methods for systems ranging from small molecules to solids. The method has since been developed by Del Re *et al.* (1967) and others, and a number of interesting applications to polymers have appeared. For example, André *et al.* (1971) have carefully compared various approximations in calculations on the 'one-dimensional crystal' of polyethylene. Work in this field is expanding rapidly and there is no doubt that some of the next major advances of MO theory will be in applications to biopolymers ('helical crystals') and to the many curious crystals produced by solid state chemistry.

†A large bibliography has been prepared by Hammet, Cox, and Orchard (1972). This complements the references to complete *ab initio* calculations given by Richards, Walker, and Hinkley (1971, 1974, 1978).

APPENDIX 1

Probabilities

Definitions and properties

The term 'probability' is used throughout this book with a perfectly precise meaning, so we ought to be familiar with the definition and elementary properties of probabilities even though we shall not need much in the way of mathematical theory.

Suppose we are making observations of some quantity, let us call it A, capable of taking certain values $A_1, A_2, \ldots, A_k, \ldots$. How do we measure the likelihood, or *probability*, of getting a particular result A_k? We make a large number of observations, N say, and count the number of times (N_k) that we find $A = A_k$, which we call a 'favourable' result. We then notice that the relative frequency N_k/N of getting a favourable result approaches a definite limiting value p_k as N becomes larger and larger. This limit is the probability of finding $A = A_k$, under the given experimental conditions, and we write

$$p_k = \lim_{N \to \infty} \left\{ \frac{N_k \,(\text{number of favourable results})}{N \,(\text{number of observations})} \right\} \qquad (A1.1)$$

If, for example, A stands for the number which appears on the top face of a die after throwing, the observable values $(A_1, A_2, \ldots)$ are $1, 2, \ldots, 6$ and a series of 'experiments' might give the following results.

Frequency of occurrence (N_k/N)

Number of throws (N)	$A_k =$ 1	2	3	4	5	6
10	0·2	0·1	0·0	0·2	0·2	0·3
100	0·16	0·20	0·14	0·20	0·18	0·22
1000	0·155	0·175	0·165	0·171	0·155	0·179
10 000	0·1464	0·1740	0·1668	0·1721	0·1505	0·1902
$N \to \infty$	p_1	p_2	p_3	p_4	p_5	p_6

Once the probabilities have been established they may be used to *predict* future results; for example, with this particular die, the 6 is more likely to occur than any of the other numbers—in a long series of future throws it is expected to occur with the greatest relative frequency, approaching 0·1902. This is clearly not an *ideal* die, for which each face might be expected to appear uppermost 'in the long run' with the same frequency, and the one bearing the 6 would thus have a probability $\frac{1}{6}$. The 'frequency definition' in (A1.1) is based on experiment, not on *a priori* speculation, and makes no hypotheses about 'ideal' properties.

We can already formulate some of the most important properties of probabilities.

(i) The probability (of anything!) is a *positive number*, lying in the range 0 to 1.

(ii) The limits $p = 0$ and $p = 1$ correspond, respectively, to *impossibility* (the favourable result is *never* obtained) and *certainty* (the favourable result is *always* obtained).

(iii) The sum of the probabilities of all the possible results must be 1.

The last observation (iii) is a special case of another basic property. Suppose we are interested in the probability of obtaining (in throwing the die) *either* a 1 or a 2; this amounts to extending the class of results that we regard as favourable. If we obtain these two results N_1 and N_2 times respectively, then the probability of obtaining *either* (in N throws) is estimated as

$$p = \frac{\text{no. of favourable results}}{\text{no. of observations}} = \frac{N_1 + N_2}{N} = p_1 + p_2$$

where $p_1 = N_1/N$ gives the probability of obtaining 1 only, and $p_2 = N_2/N$ that of obtaining 2 only. This result is general and may be stated as follows.

(iv) The probability of obtaining a result belonging to *either* of two classes (no matter which) is the sum of the probabilities for the separate classes, $p = p_1 + p_2$.

Observation (iii) is what we obtain if we apply (iv) repeatedly. On extending the class until we count *any* result as favourable we must get $p_1 + p_2 + \ldots + p_6 = 1$ since the numbers in each of the six separate classes must add up to the total number of observations.

The only other results we need (in Chapter 5) relate to the making of *two* observations and the recording of *two* particular results. We might be measuring two quantities, A and B, capable of taking values $A_1, A_2, \ldots A_i, \ldots$ and $B_1, B_2, \ldots B_j, \ldots$ respectively, and might want the probability of obtaining a specific *pair* of results, $A = A_i$ and $B = B_j$. Let us denote this probability by p_{ij}^{AB}; thus

p_{ij}^{AB} = fractional number of experiments yielding $A = A_i$ and $B = B_j$, where each 'experiment' consists of a measurement of *both* A and B.

Now suppose we wish to know the probability of finding $A = A_i$ *without* measuring B. It would be a nuisance to have to do the experiment another 10 000 times, measuring A but *not* B! Instead, we may use result (iv) above. The probability of finding $A = A_i$, $B = B_1$ is p_{i1}^{AB}; the probability of $A = A_i$ but $B = B_2$ is p_{i2}^{AB}, and from (iv)

$$\begin{pmatrix} \text{probability of } A = A_i \\ \text{and } B = B_1 \text{ or } B_2. \end{pmatrix} = p_{i1}^{AB} + p_{i2}^{AB}$$

Clearly we can continue the argument, finding for the probability of $A = A_i$ but $B = B_1$ or B_2 or B_3 or ... the value $p_{i1}^{AB} + p_{i2}^{AB} + p_{i3}^{AB} + \ldots$. We usually write this sum with the Σ notation and state the result in the following form.

(v) If the probability of finding simultaneously $A = A_i$ and $B = B_j$ is p_{ij}^{AB}, then that of finding $A = A_i$ without reference to B is

$$p_i^A = \sum_j p_{ij}^{AB} \qquad \text{(A1.2)}$$

where j runs over all possible values, $1, 2, \ldots, j, \ldots$.

In other words, by summing all the fractions corresponding to $A = A_i$, but with B taking each of its permitted values in turn, we find the fractional number of results with $A = A_i$ and B *anything* (or, equivalently, not measured!).

The result (v) is true quite generally, even when the values recorded for A and B are interdependent or 'correlated'. It may be, for example, that large values of A and B are found together less frequently than pairs of values in which one is large, the other small. This introduces the idea of a 'conditional' probability, which may be defined as follows.

(vi) If we write

$$p_{ij}^{AB} = p_i^A \times p_j^B (A = A_i) \qquad \text{(A1.3)}$$

the second factor on the right is a conditional probability; it is the probability of finding $B = B_j$ *given* that $A = A_i$.

The interpretation of $p_j^B(A = A_i)$ follows by the same kind of argument used previously; briefly (A1.3) states that the fractional number of experiments (p_i^A) giving $A = A_i$ is reduced by the further factor $p_j^B(A = A_i)$ to obtain the fractional number with *both* $A = A_i$ *and* $B = B_j$.

An important special case of (A1.3) occurs when $p_j^B(A = A_i)$ has the value p_j^B and is completely independent of A; then $p_{ij}^{AB} = p_i^A \times p_j^B$, the product of the probabilities obtained by observing A and B separately. The values of A and B are then said to be 'uncorrelated' or 'statistically independent'.

The idea of correlation in quantum mechanics is an important one, barely touched on in this book, but the main problem of calculating accurate wavefunctions is to describe the correlation between the motions of different electrons. The probability of finding two electrons simultaneously with similar values of their co-ordinates is less than it would be if their motions were uncorrelated; the correlation arises from their mutual repulsion.

Probability densities

Suppose the quantity we are measuring can take values which vary *continuously*; for example, the x co-ordinate of a particle oscillating along the x axis may be found to have any value (within a certain range) not just the discrete values $x_1, x_2, ..., x_k, ...$. To describe the probability of finding it at a certain point, or more precisely in a certain small interval, we then introduce a probability *density* or probability *per unit range*. This probability density function, $P(x)$ say, may be constructed experimentally by marking off the x axis in 'unit intervals' or 'boxes' and making repeated observations of the position of the particle. Every time we find the particle at a certain position we put a dot in the corresponding box. By counting the dots we can obtain the fractional number in each box, i.e. the probability of finding the particle in the corresponding interval. On making the boxes smaller and smaller we find the relative probabilities are described by a continuous function $P(x)$.

For a particle described by a wavefunction $\psi(x)$, $P(x) = |\psi(x)|^2$ is a probability density such that $P(x)\,dx$ is the probability of finding the particle between x and $x + dx$, i.e. of finding its x co-ordinate with a value within the corresponding range dx.

The properties of probabilities are of course echoed in the properties of probability densities. We list them below.

(i) $P(x)$ is a positive quantity.†

(ii) The probability that x has a value in the range x to $x + dx$ is $P(x)\,dx$.

(iii) The *integral* of $P(x)$ over all possible values of x must be unity,

$$\int P(x)\,dx = 1 \qquad (A1.4)$$

in which case we say the probability density is normalized.‡

(iv) The probability of finding x with any value in the range x_1 to x_2 is

$$P = \int_{x_1}^{x_2} P(x)\,dx \qquad (A1.5)$$

---

† Not a dimensionless number—in our example, $P(x)$ has the dimensions $(\text{length})^{-1}$ since it must be multiplied by dx to obtain an actual probability).

‡ The integration limits are usually understood. In the example used they would be $-\infty$ to $+\infty$.

Again, (iii) is a special case of (iv); the probability of finding some result (no matter what) must be unity, corresponding to certainty.

If we are observing two quantities, the co-ordinates x and y say, of a particle moving in a plane, the preceding discussion can be repeated. We divide up the x–y plane into unit boxes and count the number of times we find the particle in each box. The probability density is then a probability *per unit area* and depends on the values of both x and y:

$$P(x, y)\, dx\, dy = \text{probability of finding the particle within the area } dx\, dy$$
$$\text{at point } x, y.$$

Again there are the analogues of our previous results, but in this book we need only the following.

(v) If the probability of finding x and y values simultaneously in the ranges x to $x+dx$ and y to $y+dy$ is $P(x, y)\, dx\, dy$, then the probability of finding x in the range x to $x+dx$ and y with *any* value is†

$$dx \int P(x, y)\, dy.$$

If the result is denoted by $P_x(x)\, dx$, then the *one*-variable probability function is

$$P_x(x) = \int P(x, y)\, dy \qquad (A1.6)$$

integration over the whole range of y values being understood.

All the preceding statements can be extended to any number of variables. Thus (see p. 20) $P(x, y, z) = |\psi(x, y, z)|^2$ for a particle moving in three dimensions is a probability *per unit volume*; $P(x, y, z)\, d\tau$ (with $d\tau = dx\, dy\, dz$) is the probability of finding the particle in volume element $d\tau$ at point x, y, z. Similarly, using for brevity $P(1, 2) = |\Psi(1, 2)|^2$, $P(1, 2)\, d\tau_1\, d\tau_2$ is the probability of finding particle 1 in $d\tau_1$ and particle 2 in $d\tau_2$, at points x_1, y_1, z_1 and x_2, y_2, z_2 respectively, and the probability per unit volume of finding electron 1 at point x_1, y_1, z_1 and electron 2 anywhere is

$$\int |\Psi(1, 2)|^2\, d\tau_2$$

a result we have used in Chapter 5 (p. 128).

---

† Note how the summation in (A1.2) is replaced by integration.

Angular momentum

Orbital and spin angular momentum

When we first introduced atomic orbitals in Chapter 2 we noted that the quantum number m served to distinguish among the $2l+1$ different degenerate states of given n and l; thus the five different 3d states, all with $n = 3$ and $l = 2$, could be labelled by $m = 0, \pm 1, \pm 2$. Subsequently, we chose to work mainly in terms of new linear combinations, obtained by taking the sum and the difference of orbitals with equal and opposite m values, because the resultant AO's were *real* and therefore easier to visualize than complex wavefunctions. We later discovered (§3.10) that the quantum numbers l and m had a fundamental significance, distinguishing states in which the square of the angular momentum L^2 and one of its components L_z had definite measurable values along with the energy E. Such states are important in spectroscopy and in dealing with many-electron atoms for reasons which must at least be mentioned.

The fact that, for an electron moving in a central field, L^2 and L_z may have definite values in a stationary state is often interpreted by means of a *vector model*. The angular momentum is represented by a vector of length l (even though the allowed magnitude in quantum mechanics is in fact $\{l(l+1)\}^{1/2}$, not l), and the orientation of the vector is quantized so that its component along the z axis takes only integer values (m_l) going from l down to $-l$ in unit steps (Fig. A2.1 (a)). The closest classical analogue of the quantum state with wavefunction ψ_{nlm} is then provided by a particle moving in an orbit, with angular momentum of fixed magnitude (arrow of length l along the axis of the orbit), whose orientation is changing by precession around the z axis as in Fig. A2.1 (b); the z component of the angular momentum is then a constant of the motion, but the x and y components vary as the vector sweeps out the conical surface shown. This vector model, in spite of its limitations, provides a simple pictorial basis for classifying the states of many-electron atoms.

Any two angular momenta may be 'coupled'. Thus an orbital angular momentum (components L_x, L_y, L_z) and a spin angular momentum (components S_x, S_y, S_z) may be coupled to give a *total* angular momentum, with components

$$J_x = L_x + S_x, \qquad J_y = L_y + S_y, \qquad J_z = L_z + S_z \qquad \text{(A2.1)}$$

When the free-atom Hamiltonian is of the form used in this book (spin terms

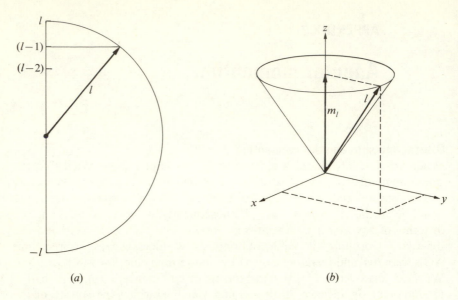

FIG. A2.1. Vector model for angular momentum. (*a*) Length of the arrow represents the magnitude of the angular momentum, each allowed projection indicating a definite component along the *z* axis. (*b*) When the vector precesses the *z* component remains a constant of the motion; the other components do not.

assumed negligible) the corresponding operators $\hat{J}^2$ and one component, $\hat{J}_z$ say, commute with $\hat{H}$. Consequently we may find one-electron states ψ_{jm_j} with

$$\hat{J}^2\psi_{jm_j} = j(j+1)\psi_{jm_j} \tag{A2.2}$$

$$\hat{J}_z\psi_{jm_j} = m_j\psi_{jm_j}. \tag{A2.3}$$

These are the analogues of (3.51) and (3.50) and ψ_{jm_j} is a state with definite 'spin–orbit coupling'. The maximum and minimum values of m_j are $(m_l+\frac{1}{2})$ and $-(m_l+\frac{1}{2})$, in the case of parallel coupling (Fig. A2.2 (*a*)), but $(m_l-\frac{1}{2})$ and $-(m_l-\frac{1}{2})$, in the case of antiparallel coupling (Fig. A2.2 (*b*)). There are then two corresponding families of states, corresponding to $j = l+\frac{1}{2}$ and $j = l-\frac{1}{2}$, respectively, as shown in the Fig. A2.2. Again, the coupling may be visualized in terms of a classical model (Fig. A2.3); the j vector precesses around the z axis, so that only one component (m_j) is constant, while the l and s vectors precess around j. The diagram suggests that no *components* of the l and s vectors are any longer constants of the motion, and this is borne out by the quantum-mechanical analysis, for $\hat{L}_z$ and $\hat{S}_z$ do *not* commute individually with $\hat{J}^2$ and the corresponding angular momenta cannot therefore be constants of the motion at the same time as the total angular momentum. Of course, the vector diagrams must not be taken too seriously but they are often useful as a pictorial aid.

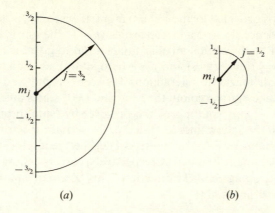

(a) (b)

FIG. A2.2. Alternative states with spin–orbit coupling. (a) With 'parallel coupling' and $l = 1$, $s = \frac{1}{2}$, the resultant states have $j = \frac{3}{2}$ and allowed components m_j (four states). (b) With 'antiparallel coupling' $j = \frac{1}{2}$ (two states).

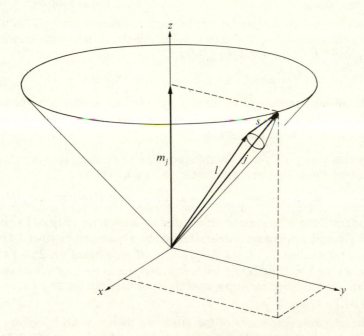

FIG. A2.3. Vector model with spin–orbit coupling. The vector labelled j precesses about the z axis, while l and s precess around j. The only constants of the motion are j, m_j, l, and s. (Note that the actual length of the vectors should be $\{l(l+1)\}^{1/2}$, instead of l, etc.)

An example is provided by the sodium atom. If we consider an excited state in which the valence electron is in a 3p orbital ($l = 1$), the spin may set 'parallel' or 'antiparallel' to the orbital angular momentum to give the two families of states, with $j = \frac{3}{2}$ (four states) and $j = \frac{1}{2}$ (two states), as shown in Fig. A2.2. When the small magnetic interactions between orbital and spin angular momenta are taken into account, the $j = \frac{3}{2}$ and $j = \frac{1}{2}$ states differ in energy by about $17 \cdot 2 \, \text{cm}^{-1}$; this splitting of the 3p level by 'spin–orbit' coupling is associated with the yellow lines (doublet) in the sodium spectrum. Moreover, by applying a magnetic field the two sets of degenerate levels ($j = \frac{3}{2}$ and $j = \frac{1}{2}$) are split into four and two branches, respectively, corresponding to J_z values of $\pm\frac{3}{2}, \pm\frac{1}{2}$, and $\pm\frac{1}{2}$ along the field direction. This 'Zeeman' splitting is also in accord with experiment.

Many-electron atoms

For a many-electron atom both the orbital angular momenta and the spin angular momenta of the different electrons may be coupled separately, and consequently the L and S in (A2.1) may each be a *resultant*; for example $\hat{L}_z = \hat{L}_z(1) + \hat{L}_z(2)$ is the operator for the z component of the resultant orbital angular moment of a two-electron system. To indicate the procedure it will be sufficient to consider a two-electron system. The steps are as follows.

(i) The electrons are assigned to orbitals with l quantum numbers l_1 and l_2, and the *resultant* orbital momentum is characterized by a quantum number† L which may take values

$$L = (l_1 + l_2), (l_1 + l_2 - 1), \ldots |l_1 - l_2|.$$

The allowed z components, for any given resultant, then correspond to $M_L = L, L - 1, \ldots, -L$. The states of many-electron atoms are classified as S, P, D, ... according as $L = 0, 1, 2, \ldots$.

(ii) With $s_1 = s_2 = \frac{1}{2}$ the resultant spin angular momentum is characterized by a quantum number S which may take values

$$S = s_1 + s_2, \qquad s_1 + s_2 - 1 \qquad \text{(i.e. 1 and 0)}.$$

(iii) The resultant orbital and spin angular momenta are coupled to give a total angular momentum characterized by a quantum number J which may take values $J = L + S, L + S - 1, \ldots, |L - S|$; there are $2S + 1$ such values for $S < L$, or $2L + 1$ for $L < S$, and this is the *multiplicity* of the state. The allowed z components then correspond, for given J, to $M_J = J, J - 1, \ldots -J$.

(iv) The spectroscopic states of the atom are then defined by giving the values of L, S, J, and M_J. A state is indicated by the letter S, P, D, ...

† Capital letters are used to denote quantum numbers for *resultant* angular momenta in a many-electron system; thus the analogue of (3.50) would be $\hat{L}^2 \Psi_{LM_L} = L(L + 1)\Psi_{LM_L}$.

with a raised prefix 1, 2, 3,... to show the spin multiplicity $2S + 1$. When necessary the value of J is added as a subscript.

In extending these rules to many-electron atoms, the presence of closed subshells of the type $(ns)^2$, $(np)^6$, $(nd)^{10}$,... leads to a great simplification; such filled shells give zero contributions to the resultant orbital and spin angular momenta and thus, in classifying the states, they may be ignored. The spectroscopic states are determined by the electrons outside the closed shells and are similar for atoms which lie in the same column of the periodic table.

It must be remembered that the above procedure is based on the use of a Hamiltonian in which spin terms are absent or very small; it is described as the L-S or Russell–Saunders coupling scheme and applies well for light atoms. For heavy atoms, the spin terms become more important, spin–orbit coupling dominates for each individual electron, and a *jj* coupling scheme must be adopted.

When there are only two electrons outside the closed shells, the rules above apply directly; in other cases they may be extended by coupling the vectors two at a time. The ground state of carbon provides a simple example. The carbon $(1s^2 2s^2)2p^2$ configuration has $l_1 = l_2 = 1$ for the two 2p electrons, giving possible L values of 2, 1, 0 and hence D, P, and S states. The spins may be coupled to give $S = 0$ or 1 (multiplicities $2S + 1 = 1$ or 3) and it might therefore be expected that ^{1}S, ^{1}P, ^{1}D, ^{3}S, ^{3}P, and ^{3}D states would arise. A moment's reflection shows, however, that some of these states are inadmissible. Thus a typical member of the ^{3}D set would require the two electrons to be in the same orbital ($m_l = 1$) with the same spin component ($m_s = \frac{1}{2}$), a situation incompatible with the Pauli principle. For carbon only ^{1}S, ^{3}P, and ^{1}D states occur, and in accordance with Hund's rules† the ground state is therefore ^{3}P. For nitrogen, on the other hand, with *three* singly occupied 2p orbitals, the ground state is expected to be *quartet* ($S = \frac{3}{2}$, $2S + 1 = 4$); this case illustrates the important fact that a *half-filled* subshell has *zero* orbital angular momentum ($L = 0$) and that accordingly the ground state is ^{4}S. For further applications any textbook on atomic spectroscopy may be consulted (for example, Kuhn 1962).

Angular momentum in linear molecules

The states of linear molecules are classified in a way parallel to, but simpler than, that for atoms. Instead of spherical symmetry the potential energy function has axial symmetry and only *one component* of angular momentum (that around the axis, taken as the z axis by convention) is a constant of the motion. In Chapter 4 we classified the MO's of diatomic molecules as $\sigma, \pi, \delta, \ldots$

---

† In their more complete form Hund's rules state that
  (i) the states of maximum spin multiplicity lie lowest,
  (ii) for given multiplicity, the states of maximum L lie lowest,
  (iii) for given S and L the states of maximum or minimum J lie lowest, according as the subshell is more or less than half-filled.

and remarked that they were associated with the possession of $0, \pm 1, \pm 2, \ldots$ units of angular momentum around the axis, and were thus in a limited sense analogous to the s, p, d, ... atomic orbitals. Again there was a choice between using *real* MO's (e.g. π_x and π_y) or *complex* MO's ($\pi_{+1} = \pi_x + i\pi_y$, $\pi_{-1} = \pi_x - i\pi_y$). The latter describe states of definite angular momentum around the axis and are most generally useful in spectroscopy. When the degenerate pairs of MO's are unequally occupied there is a resultant axial angular momentum (e.g. $\pi_{+1}^2 \pi_{-1}$ gives a state with $+1$ unit of angular momentum) and the many-electron states are then classified as Σ, Π, Δ, ... according to the resultant. The resultant *spin*, on the other hand, is still described by S and M_S, and the molecular axis provides the 'natural' axis for the spin–orbit coupling; thus a 'doublet pi' state $^2\Pi$ is one in which there is one unit of angular momentum around the axis (e.g. $M_L = +1$) and a half unit of spin angular momentum coupled parallel ($M_S = +\frac{1}{2}$) or antiparallel ($M_S = -\frac{1}{2}$). As in the case of atoms, different couplings between the spin and orbital angular momenta give rise to slightly different energies when magnetic effects are taken into account.

The electronic states of simple diatomic molecules shown in Table 4.3 have been classified according to the above principles, with the addition of a g or u subscript, together with a superscript $\pm$. The many-electron state is of u type only if the number of electrons in u-type MO's is odd (giving an odd number of sign changes under inversion); otherwise it is of g type. The superscript $\pm$ is often added to show how the orbital part of the wavefunction changes under reflection across a plane containing the molecular axis. For further details reference may be made to textbooks on molecular spectroscopy. See, for example, Herzberg (1950), or for a less detailed account Barrow (1962).

Group theory

Groups

On many occasions we have referred to molecular symmetry and its implications. The mathematical tool for dealing with symmetry is group theory, for a set of symmetry operations forms a *group* in the mathematical sense, but for the most part we have been able to manage without it. A certain amount of terminology has crept into several chapters, however, and to understand and use this terminology correctly some appreciation of the basic ideas of group theory is necessary.

The operations which bring a geometrical object into 'self-coincidence' (i.e. carry it into an 'image' which is quite indistinguishable from the original) form a *symmetry group*. The water molecule (Fig. A3.1 (*a*)) is brought into self coincidence by (i) rotation around the z axis through half a turn (180°), (ii) reflection across the molecular plane (xz plane), (iii) reflection across the bisecting plane (yz plane), and (iv) the operation of 'do nothing'—the so-called 'identity operation'. The last operation may seem trivial but the set forms a group only if it is included.

The naming of such operations is standard (Herzberg 1945):

E = the identity operation

C_n = rotation through $(2\pi/n)$ about the principal axis of symmetry (taken as the vertical axis). For the 'cubic groups' there is no unique principal axis; rotations about other axes are distinguished by a superscript.

σ_v = reflection through a plane containing the principal axis (v = vertical)

σ_h = reflection through a plane normal to principal axis (h = horizontal)

σ_d = reflection through a plane containing the principal axis and midway between a pair of twofold axes perpendicular to the principal axis (d = dihedral)

S_n = rotation through $(2\pi/n)$ combined with (i.e. followed by) reflection through a plane perpendicular to the axis

i = inversion, in which (p. 94) every point is sent into an image across the origin.

The σ, S, and i operations have the property of turning a right-handed corkscrew into a left-handed and are described as 'improper', though they are perfectly acceptable as symmetry operations.

The water molecule has a two-fold axis, so C_2 is a symmetry operation, and there are two reflection operations of σ_v type—let us call them σ_v and σ'_v for the

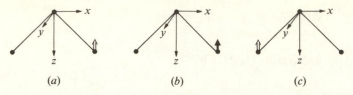

FIG. A3.1. Symmetry of the water molecule: (*a*) choice of co-ordinate axes; (*b*) effect of σ_v; (*c*) effect of σ_v followed by C_2, $C_2\sigma_v = \sigma_v'$. The 'white' arrow is added to reveal the symmetry operations; the black arrow represents its reflection beneath the molecular plane.

yz and *xz* planes, respectively. The group is called C_{2v}, containing C_2 and two σ_v-type reflection planes, and consists of the operations

$$\{E, \quad C_2, \quad \sigma_v, \quad \sigma_v'\}.$$

Any pair of operations may be combined to give a third. Thus, σ_v followed by C_2 is written $C_2\sigma_v$ and, as Fig. A3.1 shows, is exactly equivalent to the single operation σ_v'; this is indicated by writing $\sigma_v' = C_2\sigma_v$. Note the conventional order of the operations, that on the *right* being performed first. Similarly, C_2C_2 is equivalent to doing nothing and, with an obvious abbreviation, we write $E = C_2^2$. Thus all the operations in the group, the *group elements*, may be expressed in terms of C_2 and σ_v; the latter are then called *generators*. For many purposes it is sufficient to consider the generators alone and this leads to great economy; the symmetry group of the cube, for example, contains 48 operations but these may all be described by combination of only *three* generators. When only rotations and reflections are present (no translation operations) the group is called a *point group*, one point being unmoved by all the symmetry operations.

The naming of groups with a principal axis of symmetry (the axial groups) is straightforward. There are groups C_n, C_{nv}, C_{nh} with generators $\{C_n\}$, $\{C_n, \sigma_v\}$ and $\{C_n, \sigma_n\}$, respectively. There are groups D_n with generators $\{C_n, C_2'\}$, the C_2' operations being a rotation around an axis *normal* to the principal axis; repeated combination of these operations shows that there must be n such secondary axes. There are groups D_{nd}† with generators $\{C_n, C_2', \sigma_d\}$, where σ_d is a reflection through the plane midway between two adjacent two-fold axes; and there are groups D_{nh} with generators $\{C_n, C_2', \sigma_h\}$.

The remaining groups of most importance in chemistry are the cubic groups, which we encountered in Chapter 9; they describe the symmetries of a cube, the inscribed tetrahedron (edges joining four non-adjacent cube corners), and the inscribed octahedron (edges joining the centres of the eight cube faces). T and O are the groups of the tetrahedron and octahedron when improper operations are not admitted; the groups of the corresponding geometrical figures, which also contain reflection symmetries are T_d and O_h. The remaining

† For $n = 2$, 3. *For higher n values*, σ_h occurs among the operations and the groups are accordingly named D_{nh}.

group T_h follows on adding i to the generators of T; the tetrahedron itself does not possess an inversion symmetry and T_h is the symmetry group for a figure consisting of two interlocking tetrahedra, a di-dodecahedron.

Occasionally, higher symmetries are encountered (e.g. the icosahedron of boron atoms, Fig. 12.7(a)) but no new concepts appear. A more special case is provided by linear molecules, in which C_n becomes C_∞ (the generator being an infinitesimal rotation); but such groups are simple enough to be discussed from first principles.

Group representations

We have often made use of symmetry, in an intuitive way, in simplifying the secular equations which relate to the LCAO approximation; and we have borrowed from group theory, without attempting to understand it, the terminology used in classifying the resultant MO's according to symmetry. Such applications of group theory are now so commonplace that we ought at least to be familiar with the principles involved.

In the discussion of H_2O in §7.3 we were able to classify the MO's as symmetric or antisymmetric according to their behaviour on reflection across the plane (the yz plane in Fig. A3.1) which bisected the molecule, and in fact we constructed such MO's from linear combinations of the AO's which already exhibited such properties; for example the combination† $H_1 - H_2$, of hydrogen 1s AO's, and the oxygen $2p_x$ AO were both antisymmetric under the σ_v operation, changing sign under the reflection. We can now classify the symmetries rather more fully by making up a table (Table A3.1) to show whether a function is multiplied by $+1$ or -1 under each of the symmetry operations of the group C_{2v}. Inspection of the table shows that the first three functions have identical properties under the group operations; so do the next two; and the $2p_y$ AO has properties unlike either of the other sets. Each set contains orbitals of the same symmetry type or 'species', and when they are combined to form MO's we need mix only orbitals of the *same symmetry species*—hence the simplification of the secular equations.

The numbers in the table have a very fundamental property. We have noted that, for example, $C_2\sigma_v = \sigma_v'$, and it now appears that the numbers associated with these operations, in any row of the table, echo this property; for the first three rows we have $1 \times 1 = 1$; for the next two rows, $-1 \times -1 = 1$; and for the last row $-1 \times 1 = -1$. In other words the numbers associated with two operations, when multiplied together, give the number associated with the third. The set of numbers in any row is said to provide a *representation* of the group; the function whose behaviour they describe is said to 'carry' the representation, and functions with identical behaviour are said to 'belong' to the same representation.

† It is now obvious why in §7.3 we called this combination a 'symmetry orbital'.

TABLE A3.1

Behaviour of H_2O symmetry orbitals under operations of C_{2v}

	E	C_2	σ_v	σ'_v
s	1	1	1	1
p_z	1	1	1	1
$H_1 + H_2$	1	1	1	1
p_x	1	-1	-1	1
$H_1 - H_2$	1	-1	-1	1
p_y	1	-1	1	-1

The representations in Table A3.1 are said to be 'uni-dimensional', being carried by single functions. They are named in accordance with the following conventions:

(i) *A* or *B* indicates that $+1$ or -1, respectively, is associated with a generator C_n (or S_n) about the principal axis or, in the case of the cubic groups, about a cube diagonal. The principal axis is taken as the *z* axis.

(ii) Subscript 1 or 2 indicates that $+1$ or -1, respectively, is associated with the generator σ_v, in the case of the C_{nv} groups, or C'_2 in the D_n groups.† (The normal to the reflection plane for σ_v, or the secondary axis for C'_2, is used to define the *xz* plane).

(iii) Subscript g or u indicates that $+1$ or -1, respectively, is associated with the inversion *i*, whenever it occurs.

(iv) Any further classification is indicated by primes. Thus, when σ_h occurs, even or odd character ($+1$ or -1) may be indicated by a single or double prime.

Such conventions allow us to name unambiguously the symmetry types of all wavefunctions belonging to uni-dimensional representations such as those in Table A3.1. A useful extension of the convention is to use the capital letters when referring to complete many-electron wavefunctions, but the corresponding lower case letters for orbitals. Thus the water molecule MO's denoted by ψ_1 and ψ_2 in Fig. 7.5 are of a_1 and b_1 symmetry, respectively. Similarly, the ligand group orbitals $(A + B + G + D)$ and $(A + B - C - D)$ in Fig. 9.8 are of symmetry a_{1g} and b_{1g}, respectively, while the d_{xy} orbital (whose lobes fall *between* the co-ordinate axes) is of b_{2g} type because it is multiplied by -1 on reflection across the *xz*-plane.

In addition to the uni-dimensional representations of the type shown in Table A3.1, other types may occur. For a square pyramidal molecule, with symmetry group C_{4v}, the ligand group orbitals of Fig. 9.8 are again

---

† In the groups D_2 and D_{2h}, where there are three two-fold axes, B_1, B_2, or B_3 is used when $+1$ is associated with $C_2^{(z)}$, $C_2^{(y)}$, or $C_2^{(x)}$, in turn, -1 being associated with the other two. For a discussion of conventions see Mulliken (1955).

appropriate and it is evident that the combinations $(A-B)$ and $(C-D)$, although being multiplied by ± 1 under certain operations, are *interchanged* under other operations. Thus, under C_4, $A \rightarrow C$, $B \rightarrow D$, and consequently $(A-B) \rightarrow (C-D)$; and the functions p_x, p_y behave in exactly the same way, $p_x \rightarrow p'_x = p_y$ and $p_y \rightarrow p'_y = p_x$. In general, we should say that the 'images' p'_x and p'_y were expressible as linear combinations of the original functions. It is customary to describe the transformation mathematically by writing

$$(p'_x \, p'_y) = (p_x \, p_y) \begin{pmatrix} 0 & -1 \\ 1 & 0 \end{pmatrix}$$

where the square matrix on the right contains the coefficients which express the images in terms of the original functions. Thus, performing the matrix multiplication (e.g. first element = row × first column),

$$p'_x = C_4 p_x = p_x \times 0 + p_y \times 1 = p_y,$$
$$p'_y = C_4 p_y = p_x \times (-1) + p_y \times 0 = -p_x.$$

The set of matrices obtained in this way by considering each operation of C_{4v} in turn is shown in Table A3.2. This set forms a *two-dimensional representation* of the group: when, for example, C_2 and C_4 combine to give† $C_2 C_4 = \bar{C}_4$, the associated matrices are related in a similar way:

$$\begin{pmatrix} -1 & 0 \\ 0 & -1 \end{pmatrix} \begin{pmatrix} 0 & -1 \\ 1 & 0 \end{pmatrix} = \begin{pmatrix} 0 & 1 \\ -1 & 0 \end{pmatrix}$$

the rule for combination now being matrix multiplication. The last line of Table A3.2 shows the 'character system' of the representation, each entry being obtained as the sum of the diagonal elements of the corresponding matrix. The character systems play a basic role in group theory and have been tabulated

TABLE A3.2
Representation of C_{4v} carried by a pair of p orbitals (p_x, p_y)

	E	C_2	C_4	$\bar{C}_4$	$\sigma^{(x)}$	$\sigma^{(y)}$	$\sigma^{(x\bar{y})}$	$\sigma^{(xy)}$
Matrix	$\begin{pmatrix} 1 & 0 \\ 0 & 1 \end{pmatrix}$	$\begin{pmatrix} \bar{1} & 0 \\ 0 & \bar{1} \end{pmatrix}$	$\begin{pmatrix} 0 & \bar{1} \\ 1 & 0 \end{pmatrix}$	$\begin{pmatrix} 0 & 1 \\ \bar{1} & 0 \end{pmatrix}$	$\begin{pmatrix} 1 & 0 \\ 0 & \bar{1} \end{pmatrix}$	$\begin{pmatrix} \bar{1} & 0 \\ 0 & 1 \end{pmatrix}$	$\begin{pmatrix} 0 & \bar{1} \\ \bar{1} & 0 \end{pmatrix}$	$\begin{pmatrix} 0 & 1 \\ 1 & 0 \end{pmatrix}$
Character (χ)	2	-2	0	0	0	0	0	0

Notes: $\bar{C}_4$ denotes rotation through $(2\pi/4)$ in the *negative* sense and $\bar{1}$ is an abbreviation for -1. A superscript indicates the direction of the *normal* to a reflection plane; thus (x) indicates the x axis, $(x\bar{y})$ an axis midway between the (positive) x axis and the negative y axis.

† Note that in discussing symmetry positive rotation through three quarters of a turn is not distinguished from negative rotation through one quarter of a turn; what matters is the *resultant effect*.

for most of the groups of interest. For a convenient tabulation see Atkins *et al.* (1970).

In systems of higher symmetry, such as transition metal complexes (Chapter 9), *three*-dimensional representations are also important. Thus, the three p orbitals pointing along the axes in the octahedral complex of Fig. 9.9 would be mixed among themselves under the symmetry operations of the group O_h; just as the orbitals p_x and p_y carried a two-dimensional representation of C_{4v}, the set p_x, p_y, p_z would carry a *three*-dimensional represesesentation of O_h.

The nomenclature for two- and three-dimensional representations has been used in Chapter 9 and conforms to the following conventions:

(i) E and T indicate two- and three-dimensional representations, respectively. The T representations occur only for the *cubic* point groups; a typical representation is that carried by the set (p_x, p_y, p_z), pointing along the co-ordinate axes (Fig. 9.9). The E representations occur more widely; a typical representation is carried by a set (p_x, p_y) perpendicular to the principal axis (axial groups) or a three-fold axis (cubic groups).

(ii) If there are two distinct E representations, they are distinguished according to the behaviour of the functions which carry them as follows:

Both multiplied by $+1$ or by -1 under C_2: $E \rightarrow E_2$ or E_1, respectively.
Both multiplied by $+1$ or by -1 under σ_h: $E \rightarrow E'$ or E'', respectively.
Both multiplied by $+1$ or by -1 under i : $E \rightarrow E_g$ or E_u, respectively.
(For the cubic groups only the E_g and E_u types occur.)

(iii) Of two distinct T representations, the one carried by (p_x, p_y, p_z) is called T_1 in the case of the octahedral groups O and O_h (the second being T_2) but T_2 in the case of the tetrahedral group T_d (the second being T_1). A subscript g or u is added, as usual, to indicate even or odd behaviour under the inversion i.

With these rules it is evident that the representations to which the functions in Table 9.2 belong have been correctly named.

Irreducible representations and their significance

The representations we have defined owe their importance to a property of 'irreducibility' which may be quite easily understood with reference to an example. If in Fig. 9.8 the axes were to be tilted a little, so that the z axis was no longer normal to the plane of the ligands, all *three* p orbitals would be mixed among themselves by the symmetry operations, and we should therefore obtain a *three*-dimensional representation of the group C_{4v} instead of the two-dimensional form in Table A3.2. For any given choice of axes, the corresponding matrices could be worked out using the vector properties of the p orbitals (p. 185); in general they would have non-zero elements in all positions. However, on choosing the particular orientation (Fig. 9.8) with p_z normal to the plane, all the matrices take the special form

$$\begin{pmatrix} M_{11} & M_{12} & 0 \\ M_{21} & M_{22} & 0 \\ \hline 0 & 0 & M_{33} \end{pmatrix}$$

in which the 2×2 blocks (top left) describe the mixing of p_x and p_y, while the 1×1 block (bottom right) is 1 in every matrix—showing that p_z is sent into itself by all the symmetry operations. The process of choosing the orientation of the vectors or functions which carry a representation so that all the matrices take a common 'block diagonal' form is referred to as 'reduction of the representation'. When reduction can be taken no further the blocks define *irreducible representations* of the symmetry group; the functions which carry the representation then fall into sets (carrying the *irreducible* representations) which are mixed *only among themselves* by the symmetry operations, like (p_x, p_y) and p_z in the example.

When the functions which carry a representation are solutions of a Schrödinger equation, for a system of the given symmetry, the existence of irreducible representations has an immediate physical significance. A symmetry operation sends a wavefunction Ψ into a new function Ψ', but since the operation makes no observable difference to the molecule the Hamiltonian operator is left unchanged ($\hat{H}' = \hat{H}$). Thus if Ψ is a solution of $\hat{H}\Psi = E\Psi$, Ψ' must satisfy the same equation $\hat{H}\Psi' = E\Psi'$. If the energy level is non-degenerate, the only solution for that E value is Ψ or $c\Psi$ (c being an arbitrary constant). The image of Ψ can therefore only be a multiple of Ψ; $\Psi' = c\Psi$. But this is the property of a wavefunction which carries a *unidimensional* representation. As Table A3.1 shows, for a system of C_{2v} symmetry, some such functions will be multiplied by $+1$ under σ'_v but by -1 under C_2 and σ_v, and will thus be of B_2 type, and so on. If, however, an energy level is doubly degenerate, Ψ_1 and Ψ_2 being the two independent wavefunctions with energy E, Ψ'_1 and Ψ'_2 must be linear combinations of Ψ_1 and Ψ_2 (since $c_1\Psi_1 + c_2\Psi_2$ is the most general solution with energy E) and thus Ψ_1 and Ψ_2 carry a two-dimensional representation of the symmetry group.

If two solutions of the Schrödinger equation are mixed under the group of symmetry operations they must form a *degenerate* pair; if three functions are so mixed they must form a *triply degenerate* set; and so on. The mere existence of molecular symmetry—and hence of a point group with certain irreducible representations—thus enables us to predict the nature of the energy spectrum and the behaviour of the corresponding wavefunctions under the symmetry operations. On the other hand, functions which are *not* mixed under the symmetry operations need not correspond to the same eigenvalue. By looking for the *smallest* sets of functions which carry a representation (i.e. finding the *irreducible* representations) we thus find the *essential* degeneracies; there is no need for the functions in different sets to correspond to a common eigenvalue.

The resolution of degeneracies by crystal fields, as discussed in Chapter 9, is now easy to understand. Three orbitals p_x, p_y, p_z are degenerate because they carry an irreducible representation of the symmetry group (all rotations in three-dimensional space, the system being spherically symmetrical); on imposing a C_{4v} field (Fig. 9.8) the pair (p_x, p_y) and the single function p_z are no longer mixed by symmetry operations, belonging to different irreducible representations, and as the C_{4v} field builds up from zero their corresponding energies will diverge—the threefold degenerate level will be resolved into a non-degenerate plus a doubly degenerate level. A similar discussion applies to all the cases in Fig. 9.4, the five degenerate d functions falling into different irreducible representations as the symmetry is lowered by the crystal field.

There are many beautiful applications of group theory in the quantum mechanics of atoms and molecules, but for these we must refer the reader elsewhere. Short accounts, sufficient for most purposes, are available in Atkins (1970), McWeeny and Sutcliffe (1969), and McWeeny (1973).

References

ALLEN, L. C. (1968). *Chem. Phys. Lett.* **2**, 597.
—— (1969). *Ann. Rev. phys. Chem.* **20**, 315.
ALLINGER, W. L. (1976). *Adv. in phys. org. Chem.* **13**, 1.
ALLRED, A. L. and ROCHOW, E. G. (1958). *J. inorg. nucl. Chem.* **5**, 264.
ANDRÉ, J. M., KAPSOMENOS, G. S., and LEROY, G. (1971). *Chem. Phys. Lett.* **8**, 195.
ARMSTRONG, D. R. and CLARK, D. T. (1970 and 1972). *J. Chem. Soc. D.* **99** (1970);
     Theor. Chim. Acta **24**, 307 (1972).
ARRIGHINI, G. P. and GUIDOTTI, C. (1970). *Chem. Phys. Lett.* **6**, 435.
ATKINS, P. W. (1970). *Molecular quantum mechanics*. Clarendon Press, Oxford.
—— CHILD, M. S., and PHILLIPS, C. S. G. (1970). *Tables for group theory*. Clarendon
     Press, Oxford.
BADER, R. F. W. (1970). *An introduction to the electronic structure of atoms and
     molecules*. Clarke-Irwin, Toronto.
—— HENNEKER, W. H., and CADE, P. E. (1967). *J. chem. Phys.* **46**, 3341.
—— KEAVENY, I., and CADE, P. E. (1968). *J. chem. Phys.* **47**, 3381.
BAIRD, N. C. and WHITEHEAD, M. A. (1964). *Theor. chim. Acta* **2**, 259.
BALLHAUSEN, C. J. and GRAY, H. B. (1962 and 1965). *Inorg. Chem.* **1**, 111 (1962);
     Molecular orbital theory, W. A. Benjamin, New York (1965).
BAKER, A. D. (1970). *Acc. Chem. Res.* **3**, 17.
BARROW, G. M. (1962). *Introduction to molecular spectroscopy*. McGraw-Hill, New
     York.
BARTLETT, N. (1962). *Proc. Chem. Soc.* p. 218.
—— (1964). *Endeavour*, **23**, 3.
—— and SLADKY, F. P. (1973). In *Comprehensive inorganic chemistry*, p. 213. Pergamon
     Press, Oxford.
BASCH, H., MOSKOWITZ, J. W., HOLLISTER, C., and HANKIN, D. (1971). *J. chem. Phys.* **55**,
     1922.
BATES, D. R., LEDSHAM, K., and STEWART, A. L. (1953). *Phil. Trans. R. Soc.* A **246**, 215.
BAUER, S. H. (1970). In *Physical chemistry*, Vol. IV, (eds. H. Eyring, D. Henderson, and
     W. Jost). Academic Press, New York.
BERNAL, J. D. and FOWLER, R. H. (1933). *J. chem. Phys.* 1, 515.
BETHE, H. (1929). *Annln Phys.* **3**, 133.
BLOCH, F. (1928). *Z. Phys.* **52**, 555.
BONE, R. K. and HAALAND, A. (1966). *J. organomet. Chem.* **5**, 470.
BORN, M. and MAYER, J. E. (1932). *Z. Phys.* **75**, 1.
—— and OPPENHEIMER, J. R. (1927). *Annln. Phys.* **84**, 457.
—— and HUANG, K. (1954). *Dynamical theory of crystal lattices*. Clarendon Press,
     Oxford.
BOWEN, M. J. M., *et al.* (eds.) (1958). *Tables of interatomic distances and configurations in
     molecules and ions*, *Chem. Soc. Spec. Publ.* No. **11**.
BOYS, S. F. (1950). *Proc. R. Soc.* A **200**, 542
BURRAU, Ø. (1927). *K. Danske Vidensk. Selsk. Mat.-Fys. Medd.* **7**, 1.

CADE, P. F., SALES, K. D., and WAHL, A. C. (1966). *J. chem. Phys.* **44,** 1973.
CARRINGTON, A. and LONGUET-HIGGINS, H. C. (1960). *Quart. Rev.* **14,** 427.
—— and McLACHLAN, A. D. (1967). *Introduction to magnetic resonance.* Harper and Row, New York.
CARVER, J. C., GRAY, R. C. and HERCULES, D. M. (1974). *J. Am. chem. Soc.* **96,** 6851.
CHATT, J. (1953). *J. Chem. Soc.* 2939.
CHERNIK, C. L. (1963). *Rec. Chem. Prog.* **24,** 139.
CHURCHILL, M. R. and WORMALD, J. (1969). *Inorg. Chem.* **8,** 716.
CIPOLLINI, E., OWEN, J., THORNLEY, J. H. M., and WINDSOR, C. (1962). *Proc. Phys. Soc.* **79,** 1083.
CLAASSEN, H. H. *et al.* (a total of 17 co-authors) (1962). *Science* **138,** 136.
CLEMENTI, E. (1967). *J. chem. Phys.* **46,** 3842.
—— (1971). *J. chem. Phys.* **54,** 521.
—— (1972). *J. chem. Phys.* **57,** 4870.
—— (1972). *Physics of electronic and atomic collisions,* Vol. III. ICPEAC 1971; North Holland, Amsterdam.
—— *et al.* (1963). *IBM Tech. Rept.* R-J256.
—— (1964). *J. chem. Phys.* **40,** 1944.
—— (1967). *J. chem. Phys.* **47,** 1865.
—— and POPKIE, H. (1973). *Chem. phys. Lett.* **20,** 1.
—— and RAIMONDI, D. N. (1963). *J. chem. Phys.* **38,** 2686.
—— and ROETTI, C. (1974). *Atomic data and nuc. data tables* **14,** 177.
CLINTON, W. L. and RICE, B. (1959). *J. chem. Phys.* **30,** 542.
COATES, G. E., GREEN, M. L. H., POWELL, P., and WADE, K. (1968). *Principles of organometallic chemistry.* Methuen, London.
COOK, D. B., HOLLIS, P. C., and McWEENY, R. (1967). *Mol. Phys.* **13,** 553.
COOK, T. H. and MORGAN, G. L. (1969). *J. Am. Chem. Soc.* **91,** 774.
COTTON, F. A. (1969). *Acct. of Chem. Res.* **2,** 240.
COTTRELL, T. L. (1958). *The strengths of chemical bonds* (2nd edn.) Butterworths, London.
—— and SUTTON, L. E. (1947). *J. chem. Phys.* **15,** 685.
COULSON, C. A. (1937). *Proc. Camb. phil. Soc.* **33,** 111.
—— (1937). *Trans. Faraday Soc.* **33,** 1479.
—— (1938). *Proc. Camb. phil. Soc.* **34,** 204.
—— (1939). *Proc. R. Soc.* A **169,** 413.
—— (1941). *Waves.* Oliver and Boyd, Edinburgh.
—— (1942). *Proc. Camb. phil. Soc.* **38,** 210.
—— (1947). *Discuss. Faraday Soc.* **2,** 9.
—— (1948). *Contribution à l'etude de la structure moléculaire.* Desoer, Liége.
—— (1949). *J. Chim. phys.* **46,** 198.
—— (1957). *Research.* **10,** 159.
—— (1960). *Rev. mod. Phys.* **32,** 190.
—— (1963). *J. Chem. Soc.* 5893.
—— (1964). *J. Chem. Soc.* 1442.
—— (1970). *Pure appl. Chem.* **24,** 257.
—— and Fischer, I. (1949). *Phil. Mag.* **40,** 386.
—— and Gianturco, F. A. (1968). *J. Chem. Soc.* 1618.
—— and Haigh, C. W. (1963). *Tetrahedron* **19,** 527.
—— and Longuet-Higgins, H. C. (1947). *Proc. R. Soc.* A**192,** 16.
—— REDEI, L., and STOCKER, D. (1962). *Proc. R. Soc.* A**270,** 357.
CRAIG, D. P. (1959). *J. Chem. Soc.* 997.

CRAIG, D. P. and MAGNUSSON, E. A. (1956). *J. Chem. Soc.* 4895.
—— and MITCHELL, K. A. R. (1965). *J. Chem. Soc.* 4682.
—— and PADDOCK, N. L. (1958). *Nature, Lond.* **181**, 1052.
—— and WALMSLEY, S. H. (1968). *Excitons in molecular crystals.* Benjamin, New York.
—— and ZAULI, C. (1962). *J. chem. Phys.* **37**, 601, 609.
CRIEGEE, R. and SCHRÖDER, G. (1959). *Annln. Chem.* **623**, 1.
CURRIE, M. and SPEAKMAN, J. C. (1970). *J. Am. Chem. Soc.* A1923.
DAHL, J. P. and BALLHAUSEN, C. J. (1968). *Adv. in quantum Chem.* **4**, 170.
DARWENT, B. DE B. (1970). *Natl Bur. Stand. (U.S.) Publ.* NSRDS-NBS 31.
DASENT, W. E. (1970). *Inorganic energetics.* Penguin, Harmondsworth.
DAVIES, D. W. (1960). *Trans. Faraday Soc.* **56**, 1713.
DAVYDOV, A. S. (1962). *Theory of molecular excitons.* McGraw-Hill, New York.
DEL BENE, J. and POPLE, J. A. (1970). *J. chem. Phys.* **52**, 4858.
DELRE, G., LADIK, J., and BICZO, G. (1967). *Phys. Rev.* **155**, 997.
DEMORE, B. B., WILCOX, W. S., and GOLDSTEIN, J. H. (1952). *J. chem. Phys.* **22**, 876.
DEWAR, M. J. S. (1946). *J. Chem. Soc.* 406.
—— (1951). *Bull. Soc. Chim. Fr.* **18**, C71.
—— (1962). *Hyperconjugation.* Ronald Press, New York.
—— (1969). *The molecular orbital theory of organic chemistry.* McGraw-Hill, New York.
—— and DOUGHERTY, R. C. (1975). *The PMO theory of organic chemistry.* Plenum Press, New York.
—— KUBBA, V. P., and PETTIT, R. (1958). *J. Chem. Soc.* 3073.
—— LUCKEN, E. A. C., and WHITEHEAD, M. A. (1960). *J. Chem. Soc.* 223.
DIBELER, W. H., WALKER, J. A., and MCCULLOH, K. E. (1970). *J. chem. Phys.* **53**, 4414.
DICKENSON, B. N. (1933). *J. Chem. Phys.* **1**, 317.
DIERCKSEN, G. (1971). *Theor. chim. Acta* **21**, 335.
DOUGALL, M. W. (1963). *J. Chem. Soc.* 3211.
DUNITZ, J. D., ORGEL, L. E., and RICH, A. (1956). *Acta Crystallogr.* **9**, 373.
EBERHART, W. H. CRAWFORD, B. L., and LIPSCOMB, W. N. (1954). *J. chem. Phys.* **22**, 989.
ELLISON, I. O. and SHULL, H. (1955). *J. chem. Phys.* **23**, 2348.
EMERSON, G. F., WATTS, L., and PETTIT, R. (1965). *J. Am. chem. Soc.* **87**, 131.
EVANS, M. G. (1939). *Trans. Farad. Soc.* **35**, 824.
FIGGIS, B. N. (1966). *Introduction to ligand fields.* Wiley, New York.
—— and LEWIS, J. (1964). *Prog. inorg. Chem.* **6**, 37.
FISCHER-HJALMARS, I. (1965). *Adv. in quant. Chem.* **2**, 25.
FLORY, P. J. (1969). *Statistical mechanics of chain molecules.* Interscience, New York.
FOCK, V. (1930). *Z. Phys.* **61**, 126.
FUJIMOTO, H. and FUKUI, K. (1972). *Adv. in quant. Chem.* **6**, 177.
GIANTURCO, F. A., GUIDOTTI, C., LAMANNA, U., and MOCCIA, R. (1971). *Chem. phys. Lett.* **10**, 269.
GILLESPIE, R. J. (1963). *J. Chem. Ed.* **40**, 295.
—— (1970). *J. Chem. Ed.* **47**, 18.
—— (1972). *Molecular geometry.* Van Nostrand, Princeton, N.J.
—— and NYHOLM, R. S. (1957). *Q. Rev. chem. Soc.* **11**, 339.
GODDARD, W. A. III, DUNNING, T. H., HUNT, W. J., and HAY, P. J. (1973). *Acct. chem. Res.* **6**, 368.
GOLE, J. L., SIU, A. K. Q., and HAYES, E. F. (1973). *J. chem. Phys.* **58**, 857.
GORDY, W. (1955). *Discuss. Faraday Soc.* **19**, 14.
GREENWOOD, H. and MCWEENY, R. (1966). *Adv. in physical organic Chem.* **4**, 73.

GRIFFITH, J. S. (1961). *The theory of transition metal ions.* Cambridge University Press.
—— and ORGEL, L. E. (1957). *Q. Rev.* **11,** 381.
GRIFFITHS, J. H. E., OWEN, J., and WARD, I. M. (1953). *Proc. R. Soc.* A **219,** 526.
GUEST, M. F., HALL, M. B., and HILLIER, I. H. (1973). *J. Chem. Soc. Faraday II* **69,** 1829.
HAALAND, A. and NILSSON, J. E. (1968). *Chem. Commun.* No. **2,** 88.
HALL, G. G. (1951). *Proc. R. Soc.* A **208,** 328.
HAM, N. S. and RUEDENBERG, K. (1958). *J. chem. Phys.* **29,** 1215.
HAMILTON, W. C. and IBERS, J. A. (1968). *Hydrogen bonding in solids.* Benjamin, New York.
HAMMERSLEY, R. E. and RICHARDS, W. G. (1974). *Nature (Lond.)* **251,** 597.
HAMMET, A., COX, P. A., and ORCHARD, A. F. (1972). *Chem. Soc. Spec. Period. Rep.* **1,** 185.
HAMRIN *et al.* (1968). *Chem. phys. Lett.* **1,** 557.
HANNAY, N. B. and SMYTH, C. P. (1946). *J. Am. chem. Soc.* **68,** 171.
HARTMAN, A. and HIRSCHFELD, F. L. (1966). *Acta Crystallogr.* **20,** 80.
HARTREE, D. R. (1928). *Proc. Camb. Phil. Soc.* **24,** 89.
—— and HARTREE, W. R. (1938). *Proc. R. Soc. A.* **166,** 450.
HAYES, R. G. and EDELSTEIN, N. (1972). *J. Am. Chem. Soc.* **94,** 8688.
HEITLER, H. and LONDON, F. (1927). *Z. Phys.* **44,** 455.
HERZBERG, G. (1944). *Atomic spectra and atomic structure* (2nd edn). Dover, New York.
—— (1945). *Infra-red and Raman spectra of polyatomic molecules.* Van Nostrand, Princeton.
—— (1950). *Spectra of diatomic molecules.* Van Nostrand, Princeton.
—— (1970). *J. mol. Spec.* **33,** 147.
—— and MONFILS, A. (1960). *J. mol. Spectrosc.* **5,** 482.
HILLIER, I. H. and SAUNDERS, V. R. (1970). *Int. J. quant. Chem.* **4,** 203.
HINZE, J., WHITEHEAD, M. A., and JAFFÉ, H. H. (1963). *J. Am. Chem. Soc.* **85,** 148.
HOFFMANN, R. (1963). *J. chem. Phys.* **39,** 1397.
——and LIPSCOMB, W. N. (1962). *J. chem. Phys.* **36,** 2179 ; **37,** 3489.
HOLLANDER, J. M. and JOLLY, W. L. (1970). *Acct. of chem. Res.* **3,** 193.
HOPPE, R. (1964). *Angew. Chem. Int. Ed. Engl.* **3,** 538.
—— DAHNE, W., MATTAUCH, H., and RODDER, K. M. (1962). *Angew. Chem. Int. Ed. Engl.* **1,** 599.
HUHEEY, J. E. (1972). *Inorganic chemistry.* Harper and Row, New York.
—— and EVANS, R. S. (1970). *J. inorg. nucl. Chem.* **32,** 383.
HUME-ROTHERY, W. (1960). *Atomic theory for students of metallurgy.* Institute of Metals, London.
HUND, F. (1931). *Z. Phys.* **73,** 24, 565.
—— (1932). *Z. Phys.* **74,** 429.
HUO, W. M. (1965). *J. chem. Phys.* **43,** 624.
HURLEY, A. C., LENNARD-JONES, J. E., and POPLE, J. A. (1953). *Proc. R. Soc.* A **220,** 446.
HYMAN, H. H. (ed.) (1963). *Noble gas compounds.* University of Chicago Press.
ICZKOWSKI, R. P. and MARGRAVE, J. L. (1961). *J. Am. Chem. Soc.* **83,** 3547.
INGOLD, C. K. (1934). *Chem. Rev.* **15,** 225.
JAHN, H. A. and TELLER, E. (1937). *Proc. R. Soc.* A **161,** 220.
JAMES, H. M. (1935). *J. Chem. Phys.* **3,** 9.
—— and COOLIDGE, A. L. (1933). *J. chem. Phys.* **1,** 825.
—— and JOHNSON, V. A. (1939). *Phys. Rev.* **56,** 119.
JØRGENSEN, C. K. (1959). *Mol. Phys.* **2,** 309.
—— (1962). *Absorption spectra and chemical bonding in complexes.* Pergamon Press, Oxford.

JORGENSEN, W. L. and ALLEN, L. C. (1971). *J. Am. Chem. Soc.* **93,** 567.

JORTNER, J. and RICE, S. A. (1965). In *Modern quantum chemistry* (ed. O. Sinanoglu), Vol. I. Academic Press, New York.

—— —— and WILSON, E. G. (1963). *J. chem. Phys.* **38,** 2302.

JULG, A. (1975). *Topics in Current Chemistry.* Vol. **58,** 1. Springer-Verlag, Berlin.

KARPLUS, M. and PORTER, R. N. (1970). *Atoms and molecules.* Benjamin, New York.

KEALY, T. J. and PAUSON, P. L. (1951). *Nature (Lond.)* **168,** 1039.

KITTEL, C. (1966). *Introduction to solid state physics* (3rd edn) Wiley, New York.

KLESSINGER, M. and MCWEENY, R. (1965). *J. chem. Phys.* **42,** 3343.

KNOX, R. G. (1963). *Theory of excitons.* Academic Press, New York.

KOLLMAN, P. A. and ALLEN, L. C. (1970). *J. Am. Chem. Soc.* **92,** 6101.

KOLOS, W. (1964). *J. chem. Phys.* **41,** 3674.

—— (1968). *Int. J. quantum Chem.* **2,** 471.

—— and WOLNIEWICZ, L. (1968). *J. chem. Phys.* **49,** 404.

—— —— (1969). *J. chem. Phys.* **51,** 1417.

KUHN, H. (1962). *Atomic spectra.* Longmans, London.

LAWS, E. A., STEVENS, R. M., and LIPSCOMB, W. N. (1972). *J. Am. Chem. Soc.* **94,** 4461.

LENNARD-JONES, J. E. (1929). *Trans. Faraday Soc.* **25,** 668.

—— (1949). *Proc. R. Soc.* A **197,** 1.

—— (1952). *J. chem. Phys.* **20,** 1024.

—— and POPLE, J. A. (1951). *Proc. R. Soc.* A **205,** 155.

LIPSCOMB, W. N. (1963). *Boron hydrides.* W. H. Benjamin, New York.

LISTER, D. G. and TYLER, J. K. (1966). *Chem. Commun.* No **6,** 152.

LOHR, L. L. and LIPSCOMB, W. N. (1963). *J. Am. Chem. Soc.* **85,** 240.

LONG, L. H. (1972). *Prog. inorg. Chem.* **15,** 1.

LONGUET-HIGGINS, H. C. (1949). *J. Chim. phys.* **46,** 275.

—— (1950). *J. chem. Phys.* **18,** 265, 275, 283.

—— (1961). *Adv. in Spectrosc.* **2,** 429.

—— and ABRAHAMSON, E. W. (1965). *J. Am. Chem. Soc.* **87,** 2045.

—— and DEWAR, M. J. S. (1952). *Proc. R. Soc.* A **214,** 482.

—— and ORGEL, L. E. (1956). *J. Chem. Soc.* 1969.

—— and ROBERTS, M. DE V. (1955). *Proc. R. Soc.* A **230,** 110.

LÖWDIN, P.-O. (1950). *J. chem. Phys.* **18,** 365.

LOWE, J. P. (1974). *Science* **179,** 527.

LUCKEN, E. A. C. (1963). *Physical methods in heterocyclic chemistry,* Vol. 2, (ed. A. R. Katritzky), p. 89. Academic Press, New York.

MACCOLL, A. (1950). *Trans. Faraday Soc.* **46,** 369.

—— (1954). *J. Chem. Soc.* 352.

MARGENAU, H. and KESTNER, N. R. (1969). *Theory of intermolecular forces.* Pergamon Press, Oxford.

MARGRAVE, J. A. (1961). *J. Am. Chem. Soc.* **83,** 3547.

MASLEN, V. W. and COULSON, C. A. (1957). *J. Chem. Soc.* 4041.

McCONNELL, H. M. and HOLM, C. H. (1957). *J. chem. Phys.* **27,** 314.

McWEENY, R. (1951). *J. chem. Phys.* **19,** 1614. (See **20,** 920 for *errata*).

—— (1952). *Acta Crystallogr.* **5,** 463.

—— (1953). *Acta Crystallogr.* **6,** 631.

—— (1954). *Proc. R. Soc.* A **223,** 63, 306.

—— (1955*a*). *Proc. R. Soc.* A **227,** 288.

—— (1955*b*). *Proc. R. Soc.* A **232,** 114.

—— (1956). *Proc. R. Soc.* A **237,** 355.

—— (1959). *Proc. R. Soc.* A **253,** 242.

McWEENY, R. (1960). *Rev. mod. Phys.* **32**, 335.
—— (1964). In *Molecular orbitals in chemistry, physics, and biology.* (eds P. O. Löwdin and B. Pullman). Academic Press, New York.
—— (1970). *Spins in chemistry*, Appendix 1. Academic Press, New York.
—— (1970). *Jerusalem Symp. on quantum chemistry and biochemistry*, Vol. II, Israel Academy of Sciences and Humanities, Jerusalem.
—— (1972). *Quantum mechanics: principles and formalism.* Pergamon Press, Oxford.
—— (1973). *Quantum mechanics: methods and basic applications.* Pergamon Press, Oxford.
—— MASON, R., and TOWL, A. D. C. (1969). *Discuss. Faraday Soc.* No. **47**, 20.
—— and OHNO, K. (1960). *Proc. R. Soc.* A **255**, 367.
—— and SUTCLIFFE, B. T. (1969). *Methods of molecular quantum mechanics.* Academic Press, London.
MILLER, R. L., LYKOS, P. G., and SCHMEISING, H. N. (1962). *J. Am. Chem. Soc.* **84**, 4623.
MILLER, S. A., TEBBOTH, J. A., and TREMAINE, J. F. (1952). *J. Chem. Soc.* 632.
MILLS, O. S. and ROBINSON, G. (1960). *Proc. Chem. Soc.* 421.
MOFFITT, W. E. (1949). *Proc. R. Soc.* A **196**, 510.
—— (1954*a*). *J. Am. Chem. Soc.* **76**, 3386.
—— (1954*b*). *Rept. Progr. Phys.* **17**, 173.
MORSE, P. M. (1929). *Phys. Rev.* **34**, 57.
MUETTERTIES, E. L. and KNOTH, W. H. (1968). *Polyhedral boranes.* Marcel Dekker, New York.
MULLIKEN, R. S. (1932). *Rev. mod. Phys.* **4**, 1.
—— (1934). *J. chem. Phys.* **2**, 782.
—— (1935). *J. chem. Phys.* **3**, 573.
—— (1939). *J. chem. Phys.* **7**, 339.
—— (1952). *J. phys. Chem.* **56**, 295.
—— (1955). *J. chem. Phys.* **23**, 1833, 1997, 2343.
—— (1959). *Tetrahedron.* **5**, 253.
—— (1972). *Chem. Phys. Lett.* **14**, 137.
MURRELL, J. N. (1974). In *Orbital theories of molecules and solids.* (ed. N. H. March). Clarendon Press, Oxford.
—— and HARGET, A. J. (1972). *Semi-empirical self-consistent-field molecular orbital theory of molecules.* Wiley-Interscience, London.
NASH, M. A., GROSSMAN, S. R., and BRADLEY, D. F. (1968). *Nature, Lond.* 219, 370.
NOBLE, P. N. and KORTZEBORN, R. N. (1970). *J. chem. Phys.* **52**, 5375.
NYGAARD, L., NIELSEN, J. T., KIRCHHEIMER, J., MALTESEN, G., RASTRUP-ANDERSEN, J., and SØRENSEN, G. O. (1969). *J. mol. Struct.* **3**, 491.
NYHOLM, R. S. (1958). *Rec. chem. Prog.* **19**, 45.
ORCHARD, A. F. *et al.* (eds) (1972–4). *Chem. Soc. Spec. Period. Rep. Electronic structure and magnetism of inorganic compounds*, Vol. 1–3, Chemical Society, London.
ORGEL, L. E. (1960). *An introduction to transition-metal chemistry.* Methuen, London.
ORVILLE-THOMAS, W. J. (1957). *Q. Rev. Chem. Soc.* **11**, 162.
PALENIK, G. J. (1969). *Inorg. Chem.* **8**, 2744.
PAOLO, T. DI and SANDORFY, C. (1974). *Nature, Lond.* **252**, 471.
PARKS, J. M. and PARR, R. G. (1958). *J. chem. Phys.* **28**, 335.
PARR, R. G. (1960). *J. Chem. Phys.* **33**, 1184.
—— (1963). *Quantum theory of molecular electronic structure.* W. A. Benjamin, New York.
PASS, G. (1973). *Ions in solution (3): inorganic properties.* Clarendon Press, Oxford.
PAULING, L. (1928). *Chem. Rev.* **5**, 173.

PAULING, L. (1931). *J. Am. Chem. Soc.* **53**, 1367.
—— (1938). *Phys. Rev.* **54**, 899.
—— (1949). *Proc. R. Soc.* A **196**, 343.
—— (1960). *Nature of the chemical bond,* (3rd edn) Cornell University Press, Ithaca, N.Y.
PEACOCK, T. E. and MCWEENY, R. (1959). *Proc. Phys. Soc.* **74**, 385.
PEARSON, R. G. (1959). *Chem. Eng. News.* **37**, 72.
PETERS, C. R. and MILBERG, M. E. (1964). *Acta Crystallogr.* **17**, 229.
PETHRICK, R. A. and WYN-JONES, E. (1969). *Quart. Rev.* **23**, 301.
PEYERIMHOFF, S. D. and BUENKER, R. J. (1970). *Theor. Chim. Acta* **19**, 1.
PIMENTEL, G. C. and MCCLELLAN, A. L. (1960). *The hydrogen bond.* W. H. Freeman, San Francisco.
PITZER, K. S. (1945). *J. Am. Chem. Soc.* **67**, 1126.
PITZER, R. M. and MERRIFIELD, D. P. (1970). *J. chem. Phys.* **52**, 4782.
POLITZER, P. (1976). *J. chem. Phys.* **64**, 4239.
POPLE, J. A. (1951). *Proc. R. Soc.* A **205**, 163.
—— (1953). *Trans. Faraday Soc.* **49**, 1375.
—— (1957). *Q. Rev. Chem. Soc.* **11**, 273.
—— and BEVERIDGE, D. L. (1970). *Approximate molecular orbital theory.* McGraw-Hill, New York.
—— and SANTRY, D. P. (1964). *Mol. Phys.* **7**, 269.
—— and SEGAL, G. A. (1965). *J. chem. Phys.* **43**, S136.
POWER, J. D. (1973). *Phil. Trans. R. Soc.* A **274**, 663.
PRITCHARD, H. O. and SKINNER, H. A. (1955). *Chem. Rev.* **55**, 745.
PUDDEPHATT, R. J. (1972). *The periodic table of the elements.* Clarendon Press, Oxford.
PULLMAN, A. and PULLMAN, B. (1973). In *Wave Mechanics. The first fifty years.* (eds. Price, W. C., Chissick, S., and Ravendale, T.). Butterworth.
QUANE, D. (1970). *J. chem. Ed.* **47**, 396.
RANSIL, B. J. (1959). *J. chem. Phys.* **30**, 1113.
—— (1960). *Rev. mod. Phys.* **32**, 239, 245.
RICHARDS, W. G., WALKER, T. E. H., and HINKLEY, R. K. (1971). *A bibliography of ab initio molecular wavefunctions.* Clarendon Press, Oxford. (*Supplements for 1970–3,* (1974); *for 1974–7* (1978)).
ROOTHAAN, C. C. J. (1951). *Rev. mod. Phys.* **23**, 69.
ROBERTSON, R. E. and MCCONNELL, H. M. (1960). *J. phys. Chem.* **64**, 70.
ROSEN, N. (1931). *Phys. Rev.* **38**, 2099.
RUEDENBERG, K. (1977). *J. chem. Phys.* **66**, 375.
SANDERSON, R. T. (1945). *J. chem. Ed.* **31**, 2.
SANDORFY, C. (1955). *Can. J. Chem.* **33**, 1337.
SCHOMAKER, V. and STEVENSON, D. P. (1941). *J. Am. Chem. Soc.* **63**, 37.
SCHAEFFER, H. F. (1972). *The electronic structure of atoms and molecules.* Addison-Wesley, Reading, Mass.
SCHUSTER, P., ZUNDEL, G., and SANDORFY, C. (1976). *The hydrogen bond.* (3 vols.) North Holland, Amsterdam.
SCROCCO, E. and TOMASI, J. (1973). *Topics in Current Chemistry.* Vol. **42**, 95. Springer-Verlag, Berlin.
SEGAL, G. A. (1976). *Approximate methods for molecular structure calculations.* Plenum Press, New York.
SHANNON, R. D. and PREWITT, C. T. (1969). *Acta Crystallogr.* B **25**, 925.
SHAVITT, I. (1963). In *Methods of computational physics,* Vol. 2. Academic Press, New York.

SIDGWICK, N. V. and POWELL (1940). *Proc. R. Soc.* A **176,** 153.
SIEGBAHN, K. *et al.* (1967). *ESCA: Atomic, molecular, and solid-state structure studied by means of electron spectroscopy.* Almquist and Wicksells, Uppsala.
SLATER, J. C. (1929). *Phys. Rev.* **34,** 1293.
—— (1930). *Phys. Rev.* **36,** 57.
—— (1955). *Phys. Rev.* **98,** 1038.
—— (1960). *Quantum theory of atomic structure,* Vol. 1, Appendix 19. McGraw-Hill, New York.
SOHN, Y. S., HENDRICKSON, D. N., and GRAY, H. B. (1971). *J. Am. Chem. Soc.* **93,** 3603.
STEINER, E. (1976). *The determination and interpretation of molecular wavefunctions.* Cambridge University Press, Cambridge.
STERN, F. (1956). *Phys. Rev.* **104,** 684.
STRAUGHAN, B. P. and WALKER, S. (eds.) (1976). *Spectroscopy,* Vol. 3. Chapman and Hall, London.
STREITWIESER, A. (1961). *Molecular orbital theory for organic chemists.* Wiley, New York.
—— and BRAUMAN, J. (1965). *Supplemental tables of molecular orbital calculations.* Pergamon Press, Oxford.
—— and MULLER-WESTERHOFF, U. (1968). *J. Am. Chem. Soc.* **90,** 7364.
STRYER, L. (1975). *Biochemistry.* Freeman, San Francisco.
SUTTON, L. E. *et al.* (eds.) (1965). *Tables of interatomic distances and configurations in molecules and ions (supplement),* Chem. Soc. Spec. Publ. No. 16.
SWALEN, J. D. and IBERS, J. A. (1962). *Bull. Am. Phys. Soc.* **17,** 43.
TOWNES, C. H. and DAILEY, B. P. (1955). *Discuss. Faraday Soc.* **19,** 14.
————(1949). *J. chem. Phys.* **17,** 782.
TSUCHIDA, R. (1938). *Bull. Chem. Soc. Japan* **13,** 388.
TURNER, D. W., BAKER, C., BAKER, A. D., and BRUNDLE, C. R. (1970). *Molecular photoelectron spectroscopy.* Wiley, New York.
VAN DIJK, F. A. and DYMANUS, A. (1970). *Chem. Phys. Lett.* **5,** 387.
VAN VLECK, J. H. (1935). *J. chem. Phys.* **3,** 807.
VAUGHAN, P. and DONOHUE, J. (1952). *Acta Crystallogr.* **5,** 530.
VISTE, A. and GRAY, H. B. (1964). *Inorg. Chem.* **3,** 1113.
WAHL, A. C. (1964). *J. chem. Phys.* **41,** 2600.
—— (1966). *Science* (N.Y.), **151,** 961.
—— BERTONCINI, P., KAISER, K., and LAND, R. (1970). *Int. J. quantum Chem.* **Symp. 3 (pt. II),** 499.
—— and BLUKIS, U. (1968). *Atoms to molecules.* McGraw-Hill, New York.
—— —— (1968). *J. chem. Ed.* **45,** 787.
—— and DAS, G. (1970). *Adv. quantum Chem.* **5,** 261.
WALDRON, D. R. and BADGER, R. M. (1950). *J. chem. Phys.* **18,** 556.
WALSH, A. D. (1953). *J. chem. Soc.* 2260–2331.
WANG, S. C. (1928). *Phys. Rev.* **31,** 579.
WEINBAUM, S. (1933). *J. chem. Phys.* **1,** 593.
WENTORF, R. H. (1957). *J. chem. Phys.* **26,** 956.
WHARTON, L., BERG, R. A., and KLEMPERER, W. (1963). *J. chem. Phys.* **39,** 2023.
WHELAND, G. W. (1942). *J. Am. Chem. Soc.* **64,** 900.
WHITEHEAD, M. A., BAIRD, N. C., and KAPLANSKY, M. (1965). *Theor. chim. Acta.* **3,** 135.
WILLIAMS, B. G. (ed.) (1977). *Compton scattering: the investigation of electron momentum distribution.* McGraw-Hill, New York.
WILLAMS, J. E., STANG, P. J., and SCHLEYER, P. VON R. (1968). *Ann. Rev. phys. Chem.* **19,** 591.

WILSON, A. and CARROLL, D. F. (1960). *J. Chem. Soc.* 2548.
WILSON, E. B. (1950). *Discuss. Faraday Soc.* **9**, 108.
—— (1959). *Advances in chemical physics.* **2**, 367.
WOLFSBERG, M. and HELMHOLZ, L. (1952). *J. chem. Phys.* **20**, 837.
WOODWARD, R. B. and HOFFMANN, R. (1970). *The conservation of orbital symmetry.* Verlag Chemie, Weinheim/Bergstr.
YOFFE, A. D. (1976). *Chem. Soc. Rev.* **5**, 51.
YOSHIZUMI, H. (1957). *Trans. Faraday Soc.* **53**, 125.
YOSHIMINE, M. and McLEAN, A. D. (1967). *Int. J. quantum Chem.* **1S**, 313.
ZAULI, C. (1960). *J. Chem. Soc.* 2204.
ZIMAN, J. (1963). *Electrons in metals—a short guide to the fermi surface.* Taylor and Francis, London.
ZOER, H., KOSTER, D. A., and WAGNER, A. J. (1969). *Acta Crystallogr.* A **25**, 5107.

Subject Index

IA	IIA	IIIA	IVA	VA	VIA	VIIA	VIII			IB	IIB	IIIB	IVB	VB	VIB	VIIB	
1 H																	2 He
3 Li	4 Be											5 B	6 C	7 N	8 O	9 F	10 Ne
11 Na	12 Mg											13 Al	14 Si	15 P	16 S	17 Cl	18 Ar
19 K	20 Ca	21 Sc	22 Ti	23 V	24 Cr	25 Mn	26 Fe	27 Co	28 Ni	29 Cu	30 Zn	31 Ga	32 Ge	33 As	34 Se	35 Br	36 Kr
37 Rb	38 Sr	39 Y	40 Zr	41 Nb	42 Mo	43 Tc	44 Ru	45 Rh	46 Pd	47 Ag	48 Cd	49 In	50 Sn	51 Sb	52 Te	53 I	54 Xe
55 Cs	56 Ba	57 La	72 Hf	73 Ta	74 W	75 Re	76 Os	77 Ir	78 Pt	79 Au	80 Hg	81 Tl	82 Pb	83 Bi	84 Po	85 At	86 Rn
87 Fr	88 Ra	89 Ac															

IIIA

58 Ce	59 Pr	60 Nd	61 Pm	62 Sm	63 Eu	64 Gd	65 Tb	66 Dy	67 Ho	68 Er	69 Tm	70 Yb	71 Lu
90 Th	91 Pa	92 U	93 Np	94 Pu	95 Am	96 Cm	97 Bk	98 Cf	99 Es	100 Fm	101 Md	102 No	103